From Biological Control to Invasion: the Ladybird *Harmonia axyridis* as a Model Species

Helen E. Roy • Eric Wajnberg
Editors

From Biological Control to Invasion: the Ladybird *Harmonia axyridis* as a Model Species

Foreword by Helen E. Roy and Eric Wajnberg

Previously published in *BioControl*, Volume 53, No. 1, 2008

Springer

Helen E. Roy
Biological Records Centre
Huntingdon, UK

Eric Wajnberg
I.N.R.A., Sophia Antipolis
Cedex, France

Cover illustration: *Harmonia axyridis* egg-laying – Photograph by April Zobel

Library of Congress Control Number: 2008920732

ISBN-13: 978-1-4020-6938-3 e-ISBN-13: 978-1-4020-6939-0

Printed on acid-free paper.

1

springer.com

Contents

From biological control to invasion: the ladybird *Harmonia axyridis* as a model species

Foreword

Helen Roy · Eric Wajnberg

> The one process now going on, that will take millions of years to correct, is the loss of genetic and species diversity. This is the folly our descendants are least likely to forgive us.
> E.O. Wilson (1994)

Biotic homogenisation is considered among the greatest threats to global biodiversity. The rapid increase in introduced exotic species worldwide and the potential of these species to become invasive have now been widely recognised to have ecological and evolutionary consequences (Olden and Poff 2004; Olden et al. 2006). However, many accidentally or intentionally introduced species fail to establish in their new range. Of those alien species that do manage to establish many have negligible effects and some species, often those introduced with agriculture and forestry, are even considered beneficial and desirable (Williamson 1999). The impact of some invaders is unquestionably negative and as such they are designated as invasive alien species (IAS). *Harmonia axyridis* Pallas (Coleoptera: Coccinellidae), "the most invasive ladybird on Earth", is undoubtedly one such species (Roy et al. 2006).

Ladybirds have a long history of use as biological control agents against pest insects (Majerus 1994). Indeed, the Australian vedalia ladybird, *Rodolia cardinalis*, was released in 1888 to control cushiony scale insects, *Icerya purchasi*, which were having devastating impacts on the Californian citrus industry (Majerus 1994). This ladybird, introduced as a classical biological control agent, established and drastically reduced the scale insect population. This marked the advent of modern biological control. Ladybirds are considered "flagships" of biological control and their predatory habits have no doubt contributed to their popularity, particularly with gardeners. Ladybirds are one of the most loved insects worldwide but *H. axyridis* is extremely unpopular outside of its native range for a number

H. Roy (✉)
NERC Centre for Ecology and Hydrology Monks Wood, Huntingdon PE28 2LS, UK
e-mail: hele@ceh.ac.uk

E. Wajnberg
INRA, 400 Route des Chappes, BP 167, 06903 Sophia Antipolis Cedex, France
e-mail: wajnberg@sophia.inra.fr

Biological Control to Invasion: Ladybird Harmonia axyridis. H.E. Roy, E. Wajnberg (eds.),
doi: 10.1007/978-1-4020-6939-0_1, © IOBC 2007

of reasons. First and foremost, *H. axyridis* threatens the diversity of native aphidophagous species through direct competition and as a voracious, polyphagous predator (Majerus et al. 2006). Second, *H. axyridis* is now considered in the USA as a pest species because it consumes soft fruit and is adversely affecting the wine and fruit growing industries (Koch et al. 2004; Kovach 2004). Third, *H. axyridis* occurs at high densities in buildings (thousands being reported in a single household) during overwintering and is thus regarded as a human nuisance (Kovach 2004). Everyone loves ladybirds but, seemingly, not this one!

This book, which is a special issue of BioControl, is a synthesis of current research on *H. axyridis* providing current perspectives and future directions. *Harmonia axyridis* was first used as a classical biological control agent in 1916 (in the USA) but it was not until the 1980s that it became established there and regarded as an IAS (Koch et al. 2003). In Europe, the time between release and establishment was not marked by such a long lag. Through the 1980 and 1990s, *H. axyridis* was released in various European countries and just a decade later reports of establishment throughout Europe were widespread. We hope that lessons learnt from the *H. axyridis* invasion will prove useful in preventing other unwanted insect invaders arriving and becoming permanent residents in new environments.

In this special edition, authors have contributed the latest research on this species from a European perspective, although a contribution by Koch and Galvan (2007) provides an invaluable overview of *H. axyridis* in North America. The first papers within this special edition focus on the distribution of *H. axyridis*. Peter Brown and co-workers comprehensively describe the history of release, spread and distribution of *H. axyridis* Europe-wide. They reflect on the possible origins of *H. axyridis* and provide fascinating insights into the reasons behind the lag between introduction, establishment and major expansion. The following few papers provide more information on specific European countries where high resolution monitoring has enabled detailed analysis on spread, distribution, phenology and phenotypic variation (Adriaens et al. 2007; Brown et al. 2007a, b; Lombaert et al. 2007; Poutsma et al. 2007; Soares et al. 2007). In all these papers it is predicted that the advance of *H. axyridis* across Europe is set to continue, particularly to the north and east.

Risk assessment is an essential component in the development of any biological control strategy (Babendreier 2007). Joop van Lenteren and co-workers provide an overview of the international situation with respect to environmental risk assessment for biological control agents. The authors conclude that *H. axyridis* is an excellent aphid predator but the risks of its release are manifold and it should not have been used as a biological control agent in Northwest Europe. However, this is based on our current understanding and such comprehensive information was not widely available during the early releases. It is imperative that further unnecessary and risky releases of exotic organisms are prevented and van Lenteren et al. (2007) reassuringly report that concerns have been triggered by "the case of *Harmonia axyridis*". We can only hope that it is not too late to mitigate against other historical releases and that in the future, stringent, harmonized, Europe-wide regulation of biological control agents will limit the already rapid, biotic homogenisation.

One of the most significant threats posed by *H. axyridis* is to native biodiversity predominantly through intra-guild predation (when one of two species (or both) competing for the same host or prey also consumes its competitor). Judith Pell and co-workers review the current knowledge on intraguild predation involving *H. axyridis* and consider the implications to ecosystem services. These authors conclude that *H. axyridis* remains in equilibrium with its co-evolved native guild of predators through niche complimentarity, but in its exotic range the native guild of predators are co-evolved with each other but not with *H. axyridis* (Pell et al. 2007). This has serious consequences for predator diversity and ecosystem services and there is a need for rigorous research data to fill the gaps in knowledge. In this special edition there are a number of experimental papers that provide robust data on this subject, including: contributions on intra-guild predation and other feeding interactions; defence strategies that protect *H. axyridis* from becoming intra-guild prey and a fascinating study on phenotypic variation between naturalized populations from around Europe compared to those from biological control rearing facilities (Berkvens et al. 2007a, b; Jansen et al. 2007; Ware and Majerus 2007; Ware et al. 2007).

It is somewhat ironic that *H. axyridis*, being first released as a biocontrol agent, is now widely regarded as an unwanted IAS which itself requires control (Koch et al. 2003; Roy et al. 2006). In this special edition Marc Kenis and coworker comprehensively review the literature on current and potential management strategies against *H. axyridis*. They describe tested protocols, mostly mechanical and conventional chemical techniques, but also consider the possibility of novel strategies involving semiochemicals coupled with natural enemies (Kenis et al. 2007). Experimental research papers included within this edition provide new perspectives on the possibilities for controlling *H. axyridis* (Burgio et al. 2007; Koyama and Majerus 2007; Roy et al. 2007).

In conclusion, there is no doubt that *H. axyridis*, as an IAS, contributes to biotic homogenization and that, in particular, its threat to biodiversity is unacceptable. We hope that this special edition "From biological control to invasion: the ladybird *Harmonia axyridis* as a model species" will ensure this species can be used as a model to prevent, or mitigate against, subsequent releases of high risk organisms, by contributing to our understanding of the mechanisms involved in biological invasions. Biological control is an essential component of sustainable agriculture but the distinction between a successful biological control agent and an invasive species can be narrow. There are many lessons that can be learnt from *H. axyridis* as a model IAS.

Invasive species, and consequent biotic homogenisation, are a global threat to biodiversity and such major drivers of environmental change require global collaboration. This special edition on *H. axyridis* is the result of rich and fruitful collaborations between scientists across Europe and beyond. It has been an enormous pleasure to work with the many scientists, technicians and others who have made this special edition possible. The enthusiasm and passion of all involved has been inspiring and hugely encouraging. The story has just begun and we do not yet know how it will unfold and eventually conclude. However, the contributions in this book demonstrate the commitment to work cooperatively and to increase understanding within this extremely important field of research.

Helen Roy and Eric Wajnberg would like to specifically thank Patrick De Clercq for his contributions as editor on some of the manuscripts. His thorough and thoughtful approach has been very much appreciated.

References

Adriaens T, San Martin y Gomez G, Maes D (2007) Invasion history, habitat preferences and phenology of the invasive ladybird *Harmonia axyridis* in Belgium. BioControl (this issue). doi:10.1007/s10526-007-9137-6

Babendreier D (2007) Pros and cons of biological control. In: Nentwig W (ed) Biological invasions. Springer Publishing, pp 403–418

Berkvens N, Bonte J, Berkvens D, Deforce K, Tirry L, De Clercq P (2007a) Pollen as an alternative food for *Harmonia axyridis*. BioControl (this issue). doi:10.1007/s10526-007-9128-7

Berkvens N, Bonte J, Berkvens D, Tirry L, De Clercq P (2007b) Influence of diet and photoperiod on development and reproduction of European populations of *Harmonia axyridis* (Pallas) (Coleoptera: Coccinellidae). BioControl (this issue). doi:10.1007/s10526-007-9130-0

Brown PMJ, Roy HE, Rothery P, Roy DB, Ware RL, Majerus MEN (2007a) *Harmonia axyridis* in Great Britain: analysis of the spread and distribution of a non-native coccinellid. BioControl (this issue). doi: 10.1007/s10526-007-9124-y

Brown PMJ, Adriaens T, Bathon H, Cuppen J, Goldarazena A, Hägg T, Kenis M, Klausnitzer BEM, Kovář I, Loomans AJM, Majerus MEN, Nedved O, Pedersen J, Rabitsch W, Roy HE, Ternois V, Zakharov IA, Roy DB (2007b) *Harmonia axyridis* in Europe: spread and distribution of a non-native coccinellid. BioControl (this issue). doi:10.1007/s10526-007-9132-y

Burgio G, Lanzoni A, Accinelli G, Maini S (2007) Estimation of mortality by entomophages on exotic *Harmonia axyridis* versus native *Adalia bipunctata* in semi-field conditions in northern Italy. BioControl (this issue). doi:10.1007/s10526-007-9133-x

Jansen JP, Hautier L (2007) Ladybird population dynamics in potato: comparison of native species with an invasive species, *Harmonia axyridis*. BioControl (this issue). doi:10.1007/s10526-007-9134-9

Kenis M, Roy HE, Zindel R, Majerus MEN (2007) Current and potential management strategies against *Harmonia axyridis*. BioControl (this issue). doi:10.1007/s10526-007-9136-7

Koch RL (2003) The multicoloured Asian lady beetle, *Harmonia axyridis*: a review of its biology, uses in biological control and non-target impacts. J Insect Sci 3(32):1–16

Koch RL, Galvan TL (2007) Bad side of a good beetle: the North American experience with *Harmonia axyridis*. BioControl (this issue). doi:10.1007/s10526-007-9121-1

Koch RL, Burkness EC, Wold SJ, Hutchison WD (2004) Phytophagous preferences of the multicolored Asian lady beetle (Coleoptera: Coccinellidae) to autumn ripening fruit. J Econ Entomol 97:539–544

Kovach J (2004) Impact of the multicolored Asian lady beetle as a pest of fruit and people. Am Entomol 50:165–167

Koyama S, Majerus MEN (2007) Interactions between the parasitoid wasp *Dinocampus coccinellae* and two species of coccinellid from Japan and Britain. BioControl (this issue). doi:10.1007/s10526-007-9138-5

Lombaert E, Malausa T, Devred R, Estoup A (2007) Phenotypic variation in invasive and biocontrol populations of the harlequin ladybird, *Harmonia axyridis*. BioControl (this issue). doi:10.1007/s10526-007-9131-z

Majerus MEN (1994) Ladybirds. Harper Collins Publishers, UK

Majerus MEN, Strawson V, Roy HE (2006) The potential impacts of the arrival of the Harlequin ladybird, *Harmonia axyridis* (Pallas) (Coleoptera: Coccinellidae), in Britain. Ecol Ent 31:207–215

Olden JD, Poff NL (2004) Ecological processes driving biotic homogenisation: testing a mechanistic model using fish faunas. Ecology 85:1867–1875

Olden JD, Poff NL, McKinney ML (2006) Forecasting faunal and floral homogenisation associated with human population geography in North America. Biol Conserv 127:261-271

Pell JK, Baverstock J, Roy HE, Ware RL, Majerus MEN (2007) Intraguild predation involving *Harmonia axyridis*: a review of current knowledge and future perspectives. BioControl (this issue). doi: 10.1007/s10526-007-9125-x

Poutsma J, Loomans AJM, Aukema B, Heijerman T (2007) Predicting the potential geographical distribution of the harlequin ladybird, *Harmonia axyridis*, using the CLIMEX model. BioControl (this issue). doi:10.1007/s10526-007-9140-y

Roy HE, Brown P, Majerus MEN (2006) *Harmonia axyridis*: a successful biocontrol agent or an invasive threat? In: Eilenberg J, Hokkanen H (eds) An ecological and societal approach to biological control. Kluwer Academic Publishers, Netherlands

Roy HE, Brown PMJ, Rothery P, Ware RL, Majerus MEN (2007) Interactions between the fungal pathogen *Beauveria bassiana* and three species of Coccinellid: *Harmonia axyridis*, *Coccinella septempunctata* and *Adalia bipunctata*. BioControl (this issue). doi:10.1007/s10526-007-9122-0

Soares AO, Borges I, Borges PAV, Labrie G, Lucas E (2007) *Harmonia axyridis*: what will stop the invader? BioControl (this issue). doi:10.1007/s10526-007-9141-x

van Lenteren JC, Loomans AJM, Babendreier D, Bigler F, (2007) *Harmonia axyridis*: an environmental risk assessment for Northwest Europe. BioControl (this issue). doi:10.1007/s10526-007-9120-2

Ware RL, Majerus MEN (2007) Intraguild predation of immature stages of British and Japanese coccinellids by the invasive ladybird *Harmonia axyridis*. BioControl (this issue). doi:10.1007/s10526-007-9135-8

Ware RL, Ramon-Portugal F, Magro A, Ducamp C, Hemptinne J-L, Majerus MEN (2007) Chemical protection of *Calvia quatuordecimguttata* eggs against intraguild predation by the invasive ladybird *Harmonia axyridis*. BioControl (this issue). doi:10.1007/s10526-007-9129-6

Williamson M (1999) Invasions. Ecography 22:5–12

Wilson EO (1994) Biophilia and the conservation ethic. In: Kellert SR, Wilson EO (eds) The biophilia hypothesis. Island Press, Washington DC

Harmonia axyridis in Europe: spread and distribution of a non-native coccinellid

P. M. J. Brown · T. Adriaens · H. Bathon · J. Cuppen · A. Goldarazena · T. Hägg · M. Kenis · B. E. M. Klausnitzer · I. Kovář · A. J. M. Loomans · M. E. N. Majerus · O. Nedved · J. Pedersen · W. Rabitsch · H. E. Roy · V. Ternois · I. A. Zakharov · D. B. Roy

Abstract Native to Asia, *Harmonia axyridis* (Pallas) (Coleoptera: Coccinellidae) is considered an invasive alien ladybird in Europe and North America, where it was widely introduced as a biological control agent of aphids and coccids. In Europe, *H. axyridis* was sold by various biological control companies from 1995 in France, Belgium and the Netherlands, and was also intentionally released in at least nine other countries. It has spread very rapidly, particularly since 2002, and is now regarded as established in thirteen European countries. The established range extends from Denmark in the north to southern

---

P. M. J. Brown (✉) · H. E. Roy · D. B. Roy
NERC Centre for Ecology and Hydrology – Monks Wood, Huntingdon PE28 2LS, UK
e-mail: pmb@ceh.ac.uk

P. M. J. Brown · H. E. Roy
Department of Life Sciences, Anglia Ruskin University, East Road, Cambridge CB1 1PT, UK

T. Adriaens
Instituut voor Natuur- en Bosonderzoek, Kliniekstraat 25, 1070 Brussel, Belgium

H. Bathon
Federal Biological Research Centre for Agriculture and Forestry, Institute for Biological Control, Heinrichstr. 243, 64287 Darmstadt, Germany

J. Cuppen
Wageningen University, P. O. Box 8080, 6700 DD Wageningen, Netherlands

A. Goldarazena
Neiker-Tecnalia, Instituto Vasco de Investigación y Desarrollo Agrario, Centro Arkaute, Apdo. 46, 01080 Vitoria-Gasteiz Alava, Pais Vasco, Spain

T. Hägg
Kristianstadgatan 15, 21423 Malmo, Sweden

M. Kenis
CABI Europe-Switzerland, 2800 Delemont, Switzerland

B. E. M. Klausnitzer
Institut für Ökologie und Entomologie, Lannerstraße 5, 01219 Dresden, Germany

Biological Control to Invasion: Ladybird Harmonia axyridis. H.E. Roy, E. Wajnberg (eds.),
doi: 10.1007/978-1-4020-6939-0_2, © IOBC 2007

France in the south, and from Czech Republic in the east to Great Britain in the west. In this paper we map the spread and distribution of *H. axyridis* in Europe, and examine the situation on a country-by-country basis. We report first records of the species in five countries; Spain, Sweden, Denmark, Czech Republic and Italy; and first evidence of *H. axyridis* establishment in the latter three countries. Despite releases of *H. axyridis* in Portugal, Spain and Greece, there is little evidence of establishment in southern Europe. It is predicted that the spread and increase within Europe will continue and that *H. axyridis* will become one of the most widely distributed coccinellids in the continent.

Keywords Biological control · Coccinellidae · Halloween beetle · Harlequin ladybird · *Harmonia axyridis* · Introduced species · Invasive species · Multicolored Asian lady beetle · Non-native species

Background to *Harmonia axyridis*

Variously known as the multicolored Asian lady beetle, Halloween beetle and harlequin ladybird, *Harmonia axyridis* (Pallas) (Coleoptera: Coccinellidae), is native to China, Japan, Korea, Mongolia and Siberia (e.g. Dobzhansky 1933; Kuznetsov 1997), although its entire native range, particularly in parts of the former Soviet Union, has not been clearly recorded. Although usually stated in the literature to be semi-arboreal (e.g. Hodek 1973), it occupies many habitats, and in parts of both its native and introduced ranges has been recorded in meadows, heathlands and riparian zones (Adriaens et al. 2007), reedbeds (Ware et al. 2005; Brown et al. 2007) and crop systems (Colunga-Garcia and Gage 1998).

The species has a long history of use as a classical biological control agent of aphids and coccids in North America, where it was first introduced in 1916 (Gordon 1985). It has been widely used for pest control in crops as diverse as pecans (Tedders and Schaefer 1994) and

I. Kovář
The Natural History Museum, Kunratice 1, 148 00 Praha 4, Czech Republic

A. J. M. Loomans
Department of Entomology, National Reference Laboratory, Plant Protection Service, P. O. Box 9102, 6700 HC Wageningen, Netherlands

M. E. N. Majerus
Department of Genetics, University of Cambridge, Downing Street, Cambridge CB2 3EH, UK

O. Nedved
Faculty of Science, University of South Bohemia, Branišovská 31, 370 05 Ceske Budejovice, Czech Republic

J. Pedersen
Zoological Museum, University of Copenhagen, Universitetsparken 15, 2100 Kobenhavn, Denmark

W. Rabitsch
Federal Environment Agency, Spittelauer Lände 5, 1090 Wien, Austria

V. Ternois
Observatoire permanent pour le suivi de la Coccinelle asiatique en France, CPIE du Pays de Soulaines, Domaine de Saint-Victor, 10200 Soulaines-Dhuys, France

I. A. Zakharov
Vavilov Institute of General Genetics RAS, Moscow 119991, Russia

red pines (McClure 1987). As a biological control agent *H. axyridis* has incidentally succeeded in controlling pest aphid species on other crops, including apples (Brown and Miller 1998) and citrus fruits (Michaud 2002). Despite releases in fourteen US states between 1964 and 1982 (Gordon 1985), *H. axyridis* was not reported as established in the country until 1988 (Chapin and Brou 1991). There is uncertainty over the source of the established US population, which may originate from a single intentional release or accidental introduction (Krafsur et al. 1997). Day et al. (1994) suggested that the source could have been accidental seaport introductions.

Harmonia axyridis has recently been recorded in South America (de Almeida and da Silva 2002) and South Africa (N. Mgocheki, personal communication; J. Hatting, personal communication), and has been introduced in Egypt (Ferran et al. 2000) where it has become established around Cairo (S. Elnagdy, personal communication).

In Europe early introductions of *H. axyridis* occurred in the east, including to Ukraine from 1964 (for control of aphids on fruit trees) (Katsoyannos et al. 1997) and Belarus from 1968 (Sidlyarevich and Voronin 1973). In western Europe, *H. axyridis* was first used as a biological control agent in 1982 in France and first marketed in 1995 (Coutanceau 2006). Various companies subsequently made the species commercially available (Adriaens et al. 2003).

In a risk assessment of 31 exotic natural enemies of pest species used in biological control in Europe, *H. axyridis* had the second highest environmental risk index. This was based on its wide host range (i.e. multiple prey species), ability to establish and disperse, and direct and indirect effects on non-target species (van Lenteren et al. 2003). van Lenteren et al. (2007) concluded that there are no easy ways to mitigate or reduce the risk of *H. axyridis* and that it should not have been released in northwest Europe.

The aim of this paper is to present a thorough examination of the history, spread and distribution of *H. axyridis* in Europe. The data are an amalgamation of records of the occurrence of *H. axyridis* in Europe, mapped at 50 km resolution. Records were verified by the authors, derived from published papers, and/or made by experienced naturalists.

Countries in which *H. axyridis* was introduced as a biological control agent

Belgium

Harmonia axyridis was used as a biological control agent in Belgium from 1997 (Adriaens et al. 2003). A large-scale ladybird field survey (Coccinula—Belgian ladybird working group, http://www.inbo.be/content/page.asp?pid=EN_FAU_INS_LAD_start) was launched in 1999 in the Walloon region, and 2001 in the Flemish region (Adriaens et al. 2003). By 2007 Coccinula had in excess of 500 volunteer surveyors (Adriaens et al. 2007), hence coccinellids are very well recorded in Belgium. *Harmonia axyridis* was first found in the wild in Belgium in autumn 2001, in both Ghent and Brussels (Adriaens et al. 2007) and the invasion probably originated from populations in the north of the country. By 2003, *H. axyridis* was found in large overwintering aggregations comprising 500+ individuals (Adriaens et al. 2007) and was invading semi-natural ecosystems (Adriaens et al. 2003). *Harmonia axyridis* was recorded most commonly in gardens and parks, road verges, forests and woodland fringes, but it also occurred in other habitats such as heathlands, meadows and wetlands (Adriaens et al. 2007). By 2006, *H. axyridis* was recorded in all regions of Belgium (Fig. 1). The Coccinula database (data from 1999 to 2007) shows that *H. axyridis* was the most abundant coccinellid in Flanders, in terms of the total number of individuals recorded (excluding overwintering groups, which would have made *H. axyridis* abundance even higher). *Harmonia axyridis* was recorded in 76% (426 of 560) of sampled 5 km^2 in Flanders, making it the second highest occurring coccinellid after *Coccinella septempunctata* L. (Coleoptera: Coccinellidae). In Belgium as a whole, *H. axyridis* was recorded in 49% (557 of 1139) of sampled 5 km^2, making its occurrence fifth highest after *C. septempunctata, Propylea quatuordecimpunctata* (L.) (Coleoptera: Coccinellidae), *Adalia bipunctata* (L.) (Coleoptera: Coccinellidae) and *Thea vigintiduopunctata* L. (Coleoptera: Coccinellidae). A full account of the spread of *H. axyridis* in Belgium is provided by Adriaens et al. (2007).

Czech Republic

The flightless strain of *H. axyridis* was introduced into hop gardens in northwestern Czech Republic in 2003, but the species did not apparently establish at that time. The easterly spread of *H. axyridis* from Germany is assumed to be the main cause of the species arriving

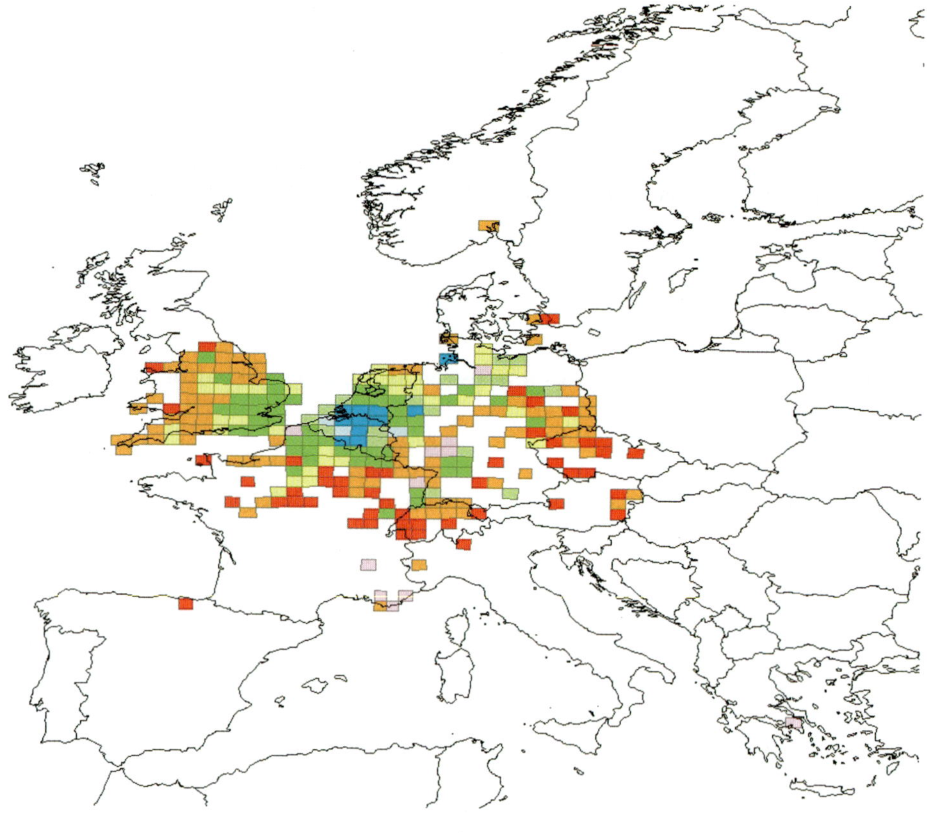

Fig. 1 Recorded occurrence of *H. axyridis* in 50 km^2 in Europe. (Year of first known record: red = 2007; orange = 2006; yellow = 2005; dark green = 2004; pale green = 2003; dark blue = 2002; pale blue = 2001; pink = pre 2001)

in the Czech Republic, where it was first recorded in the wild in 2006 in Prague and nearby rural areas (O. Nedved and I. Kovář, unpublished data). In 2007, intentional monitoring was started (advertised at http://zoo.bf.jcu.cz/kz/harmonia.htm), resulting in records from all parts of Bohemia and northern Moravia, the eastern-most recorded locality of *H. axyridis* in Europe (Fig. 1). Records of ovipositing females and mature larvae confirmed that the species had established. Most *H. axyridis* were found on aphid infested woody plants: cherry *Prunus avium* L. (Rosaceae), mock-orange *Philadelphus coronarius* L. (Hydrangeaceae), steeple-bush *Spiraea douglasii* Hook (Rosaceae), willow *Salix alba* L. (Salicaceae), crab apple *Malus sylvestris* (L.) Mill. (Rosaceae) and pear *Pyrus communis* L. (Rosaceae). Based on the findings of mature larvae and ovipositing females, five prey species; *Aphis philadelphi* Börner (Hemiptera: Aphididae), *Aphis spiraecola* Patch (Hemiptera: Aphididae), *Dysaphis plantaginea* (Passerini) (Hemiptera: Aphididae), *Tuberolachnus salignus* (Gmelin) (Hemiptera: Aphididae) and *Cacopsylla pyri* (L.) (Hempitera: Psyllidae); may be considered as suitable (essential *sensu* Hodek 1973) prey for *H. axyridis*.

France

Harmonia axyridis was first introduced to France from China by the Institut National de la Recherche Agronomique (INRA) (http://www.inra.fr) in 1982 (Iperti and Bertand 2001), but was in quarantine until 1990 (Kabiri 2006). Between 1990 and 1997 field experiments and releases were carried out, especially in southeastern France, but also in the north of the country and near Paris (Coutanceau 2006). Mass production of *H. axyridis* in France began in 1992 and the first open-air commercial field releases were in 1995 (Kabiri 2006). The species is regarded by Coutanceau (2006) to have become acclimatized in France by 1991. The earliest known record in the wild is of two individuals, both of the colour form *spectabilis* (Fig. 2), found at Saint-Aubin (Lot-et-Garonne) in southwestern France in November 1991 (Coutanceau 2006). *Harmonia axyridis* was recorded at five overwintering sites in southern France between 1992 and 1993 (Coutanceau 2006) and at Pas-de-Calais in northern France, where it was released in hops in 1994 (Trouvé 1995). There are very few known records of *H. axyridis* in France from 1995 to 2003, but from 2004 it was widely established (Coutanceau 2006) and by 2007 was found across much of northern France (Fig. 1). It is unknown whether the sparseness of records prior to 2004 is a result of lack of recording effort, or genuine scarcity of *H. axyridis* in France at that time. Certainly, since 2004 the recording effort in France has increased, principally due to the launch of a dedicated survey (Observatoire *H. axyridis*–http://perso.orange.fr/vinc.ternois/cote_nature/Harmonia_axyridis). This is now a well organised survey with in excess of 120 contributors in 15 French regions, each having a regional co-ordinator.

Germany

Although officially never sold in Germany, *H. axyridis* was released in Frankfurt between 1997 and 1998 to control aphids on roses (H. Bathon, unpublished data). It was first recorded in the wild in 1999 in Hamburg (Tolasch 2002) and Frankfurt-Niederrad (H. Bathon, unpublished data). By 2000, *H. axyridis* was common in the Rhein-Main region and there was major expansion of the species in Germany by 2002 (Klausnitzer 2002). By 2006, *H. axyridis* occurred in all regions of western Germany and was common

Fig. 2 *Harmonia axyridis* (**a**) f. *succinea* © Gavin Hatt, John Innes Centre Entomology, (**b**) f. *spectabilis* © Ken Dolbear, (**c**) f. *conspicua* © Gavin Hatt, John Innes Centre Entomology

in many cities (F. Köhler, personal communication). The species is well recorded in central eastern Germany from 2006, but there are few records from the northeast or southeast of the country (Fig. 1). This may be due to lower recording effort rather than genuine absence of the species.

Greece

French stock of *H. axyridis* was introduced into four citrus-growing areas of Greece in 1994 (Katsoyannos et al. 1997). From 1995 to 1999 over 100,000 adult *H. axyridis* were released in central and southern Greece and on several islands, to control aphids on various crops (including citrus, vegetables, beans and maize) (Kontodimas et al. in press). Further releases took place from 1997 to 2002, mainly in urban areas (Attica and Peloponessos regions, Corfu, Rhodes and Crete) (Kontodimas et al. in press). Despite these major releases and four overlapping generations of *H. axyridis* being reported in Greece (Katsoyannos et al. 1997), there is little evidence of establishment of the species, although small numbers (<50) of overwintered adults were found in Attica in spring 1998 and 1999 (Kontodimas et al. in press) (Fig. 1).

Italy

Harmonia axyridis was used as a biological control agent in greenhouses in northern Italy in the 1990s. The first known sighting of *H. axyridis* naturalised in Italy, was in Turin in October 2006, when approximately 30 specimens, including pupae, were found on trees of an *Acer* sp. (Aceraceae) (I. Zakharov, unpublished data) (Fig. 1). It is unknown whether these ladybirds originated from escaped specimens within Italy, or are immigrants from southern France or Switzerland. In July 2007, a few adult *H. axyridis* were found at two sites in the Piedmont region (M. Kenis, unpublished data), presenting further evidence of establishment in northeastern Italy.

Netherlands

Harmonia axyridis was marketed as a biological control insect in the Netherlands from approximately 1996 (Cuppen et al. 2004) until 2003. The first record of *H. axyridis* naturalised in the Netherlands was in October 2002, when a pupa was found on an ivy *Hedera helix* L. (Araliaceae) leaf in Nijmegen. A larva of *H. axyridis* was confirmed in Rotterdam the following month. Initial distribution was reported primarily in the southern half of the country. Numbers of *H. axyridis* rose dramatically from 2004 (Loomans 2004) and survey effort was increased from that year, with records requested via three Dutch websites (http://www.stippen.nl, http://www.knnv.nl and http://www.nev.nl). The species appears to have spread from south to north, and by 2006 all mainland 50 km^2 in the Netherlands had records of *H. axyridis* (Fig. 1). Over 2000 records of the species were received between 2002 and 2007.

Portugal

Between 1984 and 1985, *H. axyridis* was used to control aphids on citrus crops in the Algarve province of Portugal, and on the Portuguese administered islands of the Azores (Katsoyannos et al. 1997; Soares et al. 2007). There is no evidence of subsequent establishment. However, recent work by Soares and Serpa (2007) concluded that if re-introduced to the Azores, *H. axyridis* would present a risk to the native species *Coccinella undecimpunctata* L. (Coleoptera: Coccinellidae).

Spain

In 1995 *H. axyridis* was introduced to control aphids on greenhouses in Almeria, southeastern Spain (SIFA 2004). It established in greenhouses (Jacas et al. 2006), but there is no evidence of subsequent establishment in the wild. In 2003 and 2004 two specimens of *H. axyridis* were found in a garden on Tenerife, Canary Islands and it was found again on the island in 2006 (S. Eizaguirre, personal communication). There are no official reports of introductions of *H. axyridis* for agricultural purposes from the Canary government, and the species is not believed to be present in large numbers (Machado 2005). In 2007, two adult *H. axyridis* were collected from flowers of a *Tilia platyphyllos* Scop. (Tiliaceae) tree in a park in Loiu, Bilbao, Basque country of northern Spain (A. Goldarazena, unpublished data) (Fig. 1). This is the first record of naturalised

H. axyridis in the Iberian peninsula. There is no evidence that *H. axyridis* was released for biological control purposes in the Basque country (R. Amenabar, personal communication). The specimens are likely to have either originated from France, or from imported goods entering the nearby Bilbao airport.

Switzerland

Harmonia axyridis was available commercially in Switzerland for a short period in the late 1990s, and was used in a small field trial on roses in Grossdietwil in 1996 (Andermatt 1996). However, *H. axyridis* did not apparently establish as a result, and applications for the commercial release of the species were refused by the Swiss Pesticide Regulatory Authority in 1997 (van Lenteren et al. 2007). No adults were found in the wild in Switzerland until 2004, when a single specimen was found at an exhibition of Asian plants in Basle (Klausnitzer 2004). *Harmonia axyridis* was not found during surveys in 2005, but in 2006 it was recorded in ten cantons in northern Switzerland (Eschen et al. 2007), and larvae of the species were found for the first time in Switzerland. The higher abundance of the beetle near the northern border suggests that it invaded the country from southern Germany and/or eastern France. In a large survey for ladybirds in northwestern Switzerland, *H. axyridis* was found to be the seventh most abundant species on trees and shrubs, and the first cases of aggregations on buildings were notified in autumn 2006 in Basle, where the species was the most abundant coccinellid at some sites (Eschen et al. 2007). The monitoring continued in 2007 and a dedicated website was launched (http://www.cabi-bioscience.ch/harmonia). By late 2007, the ladybird was found in most Swiss cantons (Fig. 1).

Countries in which *H. axyridis* has been found in the wild without evidence of deliberate introduction

Austria

In Europe and North America, the attraction of *H. axyridis* to buildings for use as overwintering sites (Koch 2003) has increased the number of sightings of the species. The first published record in Austria (October 2006; Wiener Neustadt, eastern Austria) was one such sighting, where several adult specimens of *H. axyridis* were found on a house wall searching for an overwintering site (Rabitsch and Schuh 2006). An earlier record subsequently came to light, in which *H. axyridis* larvae were found on plants on a balcony in nearby Vienna in July 2006, and there are several later records of adults in Vienna in 2006 and 2007 (W. Rabitsch, unpublished data). *Harmonia axyridis* is clearly established in Austria, and independent observations in the west (Götzis, October 2006; Rankweil, April 2007; Vorarlberg, August 2007), centre (Salzburg, May 2007), and south (Herberstein, July 2007) of the country (W. Rabitsch, unpublished data), support the hypothesis of spread from Switzerland and/or Germany (Fig. 1). There is no known evidence that *H. axyridis* was used for biological control purposes in Austria, making it less likely that the spread resulted from (illegally) imported specimens escaped from greenhouses or gardens. The scattered documentation of *H. axyridis* in Austria reflects the largely unnoticed expansion in the country. No large aggregations have yet been reported, with the maximum number from a single site (Vienna) being approximately 20 adult *H. axyridis*.

Denmark

The first known record of *H. axyridis* in Denmark was in Copenhagen in July 2006 (J. Pedersen, unpublished data). Then and on five subsequent occasions in August 2006, October 2006 and June 2007, a total of 14 adults of *H. axyridis* were found in a light trap. Similarly, a single adult was found in a light trap at Mandemarke, in each of August 2006 and July 2007 (J. Pedersen, unpublished data) (Fig. 1). All specimens were f. *succinea* (Fig. 2). Denmark is the most northerly country in Europe where *H. axyridis* is thought to have established. It is not known to have been introduced there and is assumed to have spread north from Germany.

Great Britain and the Channel Islands

The first record of *H. axyridis* in Great Britain, in September 2004, initiated considerable media and public interest, and led to the launch of the web-based Harlequin Ladybird Survey (http://www.harlequin-survey.org) (Majerus et al. 2006; Roy et al. 2006). Thanks to thousands of contributors across Britain, the species has been recorded and mapped at high resolution (Roy et al. 2005; Brown et al. 2007) and its rapid spread north and west from the southeast is clear (Fig. 1). By mid 2007, *H. axyridis* was recorded in 88% of English and 38% of Welsh vice counties. It was also found on the island of Jersey, Channel Islands. We predict that it will continue to spread north and reach Scotland by the end of 2008. Three colour forms have been recorded in Britain; f. *succinea*, f. *spectabilis* and f. *conspicua* (Fig. 2). A detailed account and analysis of the spread of *H. axyridis* in Great Britain is provided by Brown et al. (2007).

Liechtenstein

Harmonia axyridis was found for the first time in the tiny principality of Liechtenstein in August 2007 (A. Loomans, unpublished data). The whole principality is contained within a single 50 km^2 in which there were earlier records for both Austria and Switzerland.

Luxembourg

Considering that the border areas of France, Belgium and Germany that surround Luxembourg were all known to have populations of *H. axyridis* (Fig. 1), it was inevitable that the species would arrive in this small country. It was present by September 2004, when first recorded on *Acer pseudoplatanus* L. (Aceraceae) in the south of Luxembourg (Schneider and Loomans 2006). Further sightings of *H. axyridis* were made in October, November and December 2004 in urban locations (Schneider and Loomans 2006) and large reproducing populations were recorded in August 2005 in the north (Clervaux) and south (Luxembourg City) of the country (M. Majerus, unpublished data).

Norway

Whilst there is no evidence that *H. axyridis* is established in Norway, it has been recorded in Oslo, having been found on horticultural plants imported from the Netherlands in April 2006 (Staverloekk et al. in press) (Fig. 1).

Sweden

Harmonia axyridis was first recorded in Sweden in Malmö in April 2007 (T. Hägg, unpublished data), when a single adult was found in a house (Fig. 1). Subsequently, a dead adult was found at the same location, and it is thought that the species had been overwintering in the house, suggesting that it arrived in late 2006, possibly from Germany or Denmark. There are no other known records of *H. axyridis* in Sweden.

Establishment and spread

Harmonia axyridis has spread in Europe at a very fast rate (Table 1; Fig. 3). It is not possible to fully explain the mechanisms of spread leading to the current distribution, because of the spatial and temporal aspects of the deliberate releases of *H. axyridis*. This involved multiple introduction sites in at least twelve European countries over a period of approximately forty years (Table 1).

Time lags may occur throughout the invasion process, including the arrival, establishment and impacts of the invading species (Crooks 2005). Our data indicates a variable time lag between initial establishment and major expansion, at least in the countries where *H. axyridis* was deliberately introduced. In France this time lag was approximately

Table 1 Summary of *H. axyridis* occurrence in Europe

Country	Year of first record in the wild	Deliberately introduced? (Earliest year of introduction)	Evidence of establishment?	No. 50 km^2 with *H. axyridis* by August 2007
Ukraine	Unknown	Yes (1964)	Unknown	Unknown
Belarus	Unknown	Yes (1968)	Unknown	Unknown
Portugal	N/A	Yes (1984)	No	0
France	1991	Yes (1982)	Yes	63
Greece	1998	Yes (1994)	Limited	1
Germany	1999	Yes (1997)	Yes	75
Belgium	2001	Yes (1997)	Yes	21
Netherlands	2002	Yes (1996)	Yes	26
Spain, including Canary Islands	2003	Yes (1995)	No	2
Switzerland	2004	Yes (1996)	Yes	15
Luxembourg	2004	No	Yes	2
England and Channel Islands	2004	No	Yes	60
Italy	2006	Yes (1990s)	Yes	1
Czech Republic	2006	Yes (2003)	Yes	11
Austria	2006	No	Yes	7
Denmark	2006	No	Yes	2
Wales	2006	No	Yes	5
Norway	2006	No	No	1
Liechtenstein	2007	No	Yes	1
Sweden	2007	No	No	1

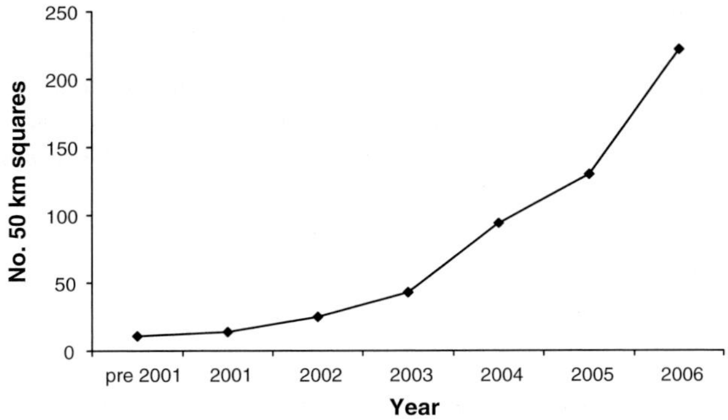

Fig. 3 Cumulative occurrence of *H. axyridis* in 50 km^2 in Europe

13 years (i.e. 1991–2004), in the Netherlands approximately 6 years (i.e. 1996–2002), and in Belgium approximately 4 years (i.e. 1997–2001). In countries where the species has not known to have been introduced, there seems to be very little time lag between the first record of establishment and major expansion (e.g. less than one year in England). The period between establishment and rapid spread in the USA is difficult to assess because of multiple intentional releases over many years (Gordon 1985), uncertainty over the date of establishment, and ambiguity over whether an intentional release or an accidental introduction was the source of establishment (Day et al. 1994; Koch 2003).

The reasons for the time lag between establishment of *H. axyridis* and major expansion in Europe are as yet unknown. However, it is possible that it is related to the genetic make-up of releases, and to differences between environmental conditions in the locations where released stocks originated and where they were released. The genetic make-up of released stocks would be a function of selection acting to locally adapt populations in their native range before collection, and random genetic drift and selection acting on laboratory cultures before release. It is likely that some cultures passed through genetic bottlenecks while in culture. Moreover, there are reports of rapid changes in some obvious genetically controlled traits in laboratory cultures. For example, the phenotypic frequencies of melanic forms (f. *conspicua* and f. *spectabilis*) of a laboratory population increased from 0.5 to >0.99 in 50 generations (Berkvens et al. 2007). As many laboratory cultures are maintained on diets that *H. axyridis* would not encounter in the wild (i.e. *Ephestia* eggs), it is inevitable that released beetles will have been exposed to abnormal selection pressures in captivity. In consequence, it is highly unlikely that released individuals will be precisely adapted to the conditions that they face in their introduced range. In many cases, releases therefore do not lead to establishment, as recorded for *H. axyridis* in the USA (Gordon 1985) and Greece (Kontodimas et al. in press). However, if we assume that at least some individuals do survive and reproduce, the expectation would be that these would take time to start increasing in number significantly, for two reasons. First, because of their as yet imperfect adaptation to local conditions, and second because at low population size they are likely to suffer the effects of inbreeding depression, which is considerable in many aphidophagous coccinellids (Hodek 1973; Majerus 1994, 2003). There will thus be a period after establishment during which the average fitness of members of the population increase as the population is purged of deleterious recessives (Haldane 1927), and slowly becomes better locally adapted to the new conditions, through selection acting on the genotypic variation generated by recombination and mutation (Fisher 1930). It is only once the population has become locally adapted that it will start to increase in number.

However, the rate of increase may then be rapid for two reasons. First, the population may have been purged of deleterious recessive alleles, as already mentioned. Second, while in culture the released beetles may have been inadvertently selected for rapid reproduction, high fecundity and broad dietary range; biological control companies reared large numbers as efficiently as possible for commercial reasons. Once numbers in a released population increased sufficiently to approach carrying capacity, increased rate of dispersal would be inevitable as the beetles sought new resources. The rate of this process may vary greatly depending on factors such as the size of the original samples collected from the native range, the number of different populations that such samples were collected from, the length of time and number of generations that cultures were kept in captivity for before release, the number of individuals that founded each culture generation, and the number of beetles finally released. If releases involved stocks with quite different origins, then the speed of local adaptation would be faster than if all released beetles were from a single stock. This is because matings between individuals from different locally adapted populations would produce a greater range of genetic variants among their offspring for selection to act upon.

The initial need for adaptation before increase in number and spread may explain the lack of a time lag between establishment and major expansion in countries where *H. axyridis* was not deliberately introduced; individuals of *H. axyridis* arriving in these countries from parts of the introduced range had already been through the adaptation phase. So after a period of adaptation of one or several populations in France, Netherlands, Belgium and/or Germany, a rapid spread of *H. axyridis* occurred in Europe.

Genetic variation and the origins of *H. axyridis* in Europe

Three main colour forms of adult *H. axyridis* have been found in Europe: f. *succinea*, f. *spectabilis* and f. *conspicua* (Fig. 2). The typical form *axyridis*, which predominates in central Asia (Dobzhansky 1933; I. Zakharov and M. Majerus, personal observation), has not yet been recorded in Europe. The *succinea* complex has been divided into many subforms (e.g. *siccoma*—0 spots; *frigida*—6 spots; *novemdecimsignata*—19 spots) (Dobzhansky 1933), but is regarded as a single form here. Thus f. *succinea* has elytra with a ground colour of yellow, orange, or red, and 0–21 black spots, which may or may not be fused. Forma *spectabilis* has black elytra with four yellow, orange, or red spots or other shaped markings, which sometimes contain a central black spot. Forma *conspicua* is as f. *spectabilis*, but with only the anterior pair of spots.

There is strong evidence of a link between climate and colour form in coccinellids (e.g. Majerus 1994, 1998). Whilst the basis of difference in distinct colour form (e.g. f. *succinea* cf. f. *spectabilis*) is genetic, difference in colouration within a form is at least partly environmental. For example, in many species of coccinellid, inverse correlations between temperature and extent of melanic patterning have been reported (e.g. Dobzhansky 1933; Abbas et al. 1988; Majerus 1998). In the case of *H. axyridis*, the amount of melanism in f. *succinea* was found to increase at high elevations (Nalepa et al. 1996).

European populations of *H. axyridis* generally include a mix of the three colour forms, but with f. *succinea* predominating (Fig. 4). Interestingly, in England the percentage of melanic specimens declined markedly from the year of arrival of *H. axyridis*, 2004 (45%, $n = 344$) (Majerus and Roy 2005) to the second year, 2005 (20%, $n = 6180$) (M. Majerus, unpublished data). The broad consistency in the colour form frequency data from Europe (Fig. 4) provides some evidence of the genetic similarity of populations in different

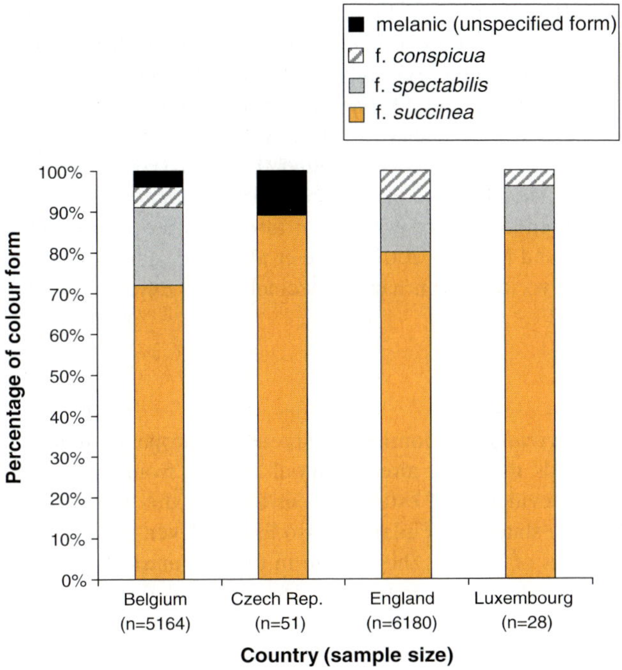

Fig. 4 Percentages of *H. axyridis* colour forms in various countries (Belgium data from 2004–2006 (Adriaens et al. 2007), Czech Republic data from 2006–2007 (O. Nedved, unpublished data), England data from 2005 (M. Majerus, unpublished data), Luxembourg data from 2004 (Schneider and Loomans 2006))

European countries. Further, we suggest that this possible genetic similarity is the result of spread of the species from very limited points of origin in Europe. Genetic analyses of *H. axyridis* samples from different countries are needed in order to confirm this.

Traits that give *H. axyridis* a competitive advantage over native coccinellids

In Europe, *H. axyridis* has been shown to be multi-voltine, with evidence of four generations per year in Greece (Katsoyannos et al. 1997) and two generations per year in Great Britain (Brown et al. 2007). Production of multiple generations would help to explain the rapid spread of *H. axyridis* in Europe. In northern Europe, many native coccinellids are uni-voltine (Majerus 1994; Klausnitzer and Klausnitzer 1997), thus *H. axyridis* may have a significant advantage over them in terms of potential population growth. Dispersal of *H. axyridis* in northern Europe appears to occur primarily in autumn, when the species moves to aggregate at overwintering sites. This pattern is similar to that observed in North America (Koch 2003). Large overwintering aggregations have been observed in several European countries including Belgium, the Netherlands, France, Germany and England.

Recent European research has provided further evidence of the adaptability and resilience of *H. axyridis*. In laboratory experiments, 35–48% of *H. axyridis* larvae fed a pollen-only diet successfully reached adulthood, and although fitness was reduced, approximately

40% of these adult females produced viable eggs (Berkvens et al. 2007). Roy et al. (2007) found that only the highest dose (10^9 spores ml^{-1}) of the fungal pathogen *Beauveria bassiana* (Balsamo) Vuillemin (Ascomycota: Hypocreales) killed *H. axyridis*, whereas two other species included in the study (*A. bipunctata* and *C. septempunctata*) experienced high mortality (70% and 80% respectively) at a medium dose (10^7 spores ml^{-1}) of the pathogen. Koyama and Majerus (2007) found that *H. axyridis* had lower susceptibility than *C. septempunctata* to the parasitoid wasp *Dinocampus coccinellae* (Schrank) (Hymenoptera: Braconidae). Recent work on intraguild predation (Pell et al. 2007; Roy et al. in press; Ware and Majerus 2007; Ware et al. in press) indicates that *H. axyridis* will have a serious negative impact on native coccinellids in Europe.

Conclusion

We predict that *H. axyridis* will continue to spread in Europe, particularly northwards and eastwards. Northwards there are already initial records from Norway and Sweden, and whilst there is little evidence of establishment to date, this is likely in the near future. Spread into the Baltic states and Finland is also likely. Given that *H. axyridis* survives as a native species in parts of Siberia, the climate in these countries should not be a barrier to establishment. Eastwards it is entirely possible (indeed likely) that the species is already present, but undetected, in countries such as Poland, Slovakia, Hungary and Slovenia. Ireland is the only remaining country left to be invaded from the westward spread, and we predict that *H. axyridis* will reach there, from Great Britain, in the relatively near future, and once established will spread across the whole country. The spread southwards is less certain, although as the species is adapted to Mediterranean and sub-tropical climates in parts of its native range (southern Japan and China), it may gradually adapt to the warmer climes of southern Europe, the Iberian peninsular and north Africa, and eventually establish and spread there. Indeed there are already initial reports of establishment in Egypt (S. Elnagdy, personal communication). Over a very short time period *H. axyridis* is likely to become one of the most widely distributed coccinellids in Europe. It remains to be seen what effect this will have on native coccinellids, but mounting evidence suggests a seriously detrimental impact.

Acknowledgements We extend our thanks to all of the following: the thousands of volunteer recorders and naturalists across Europe who have provided records; others who have provided records and support, including; Manfred Bernhard, Gernot Hoch, Erwin Holzer, Andreas Kapp (Austria); Burkhard Hinnersmann, Frank Köhler, Manfred Uhlig (Germany); Theodoor Heijermans, Kees van der Krieke, Paul van Wielink (Netherlands); Arnstein Staverloekk (Norway); David Calvo (Spain); Natural Environment Research Council, Paul Mabbott, Laura-Jane Michie, Remy Ware, Ian Wright (UK); Gavin Hatt and Ken Dolbear for permission to use their images. We are grateful for support from the DAISIE (Delivering Alien Invasive Species Inventories for Europe) project funded by the European Community's Sixth Framework Programme (contract no. GOCE-2004–511202).

References

Abbas I, Nakamura K, Katakura H, Sasaji H (1988) Geographical variation of elytral spot patterns in the phytophagous ladybird, *Epilachna vigintioctopunctata* (Coleoptera: Coccinellidae) in the province of Sumatra Barat, Indonesia. Res Population Ecol 30:43–56

Adriaens T, Branquart E, Maes D (2003) The multicoloured Asian ladybird *Harmonia axyridis* Pallas (Coleoptera: Coccinellidae), a threat for native aphid predators in Belgium? Belgian J Zool 133: 195–196

Adriaens T, San Martin y Gomez G, Maes D (2007) Invasion history, habitat preferences and phenology of the invasive ladybird *Harmonia axyridis* in Belgium. BioControl (this issue). doi:10.1007/s10526-007-9137-6

Andermatt M (1996) Marienkäferlarven als blattlaus-bekämpfungsmittel? Andermatt Biocontrol 3:10–11

Berkvens N, Bonte J, Berkvens D, Deforce K, Tirry L, De Clercq P (2007) Pollen as an alternative food for *Harmonia axyridis*. BioControl (this issue). doi:10.1007/s10526-007-9128-7

Brown MW, Miller SS (1998) Coccinellidae (Coleoptera) in apple orchards of eastern West Virginia and the impact of invasion by *Harmonia axyridis*. Entomol News 109:143–151

Brown PMJ, Roy HE, Rothery P, Roy DB, Ware RL, Majerus MEN (2007) *Harmonia axyridis* in Great Britain: analysis of the spread and distribution of a non-native coccinellid. BioControl (this issue). doi: 10.1007/s10526-007-9124-y

Chapin JB, Brou VA (1991) *Harmonia axyridis* (Pallas), the third species of the genus to be found in the United States (Coleoptera, Coccinellidae). Proc Entomol Soc Wash 93:630–635

Colunga-Garcia M, Gage SH (1998) Arrival, establishment, and habitat use of the multicolored Asian lady beetle (Coleoptera: Coccinellidae) in a Michigan landscape. Environ Entomol 27:1574–1580

Coutanceau J-P (2006) *Harmonia axyridis* (Pallas, 1773): une coccinelle asiatique introduite, acclimatée et en extension en France. Bull Soc Entomol France 111:395–401

Crooks JA (2005) Lag times and exotic species: The ecology and management of biological invasions in slow-motion. Ecoscience 12:316–329

Cuppen J, Heijerman T, van Wielink P, Loomans A (2004) Het lieveheersbeestje *Harmonia axyridis* in Nederland: een aanwinst voor onze fauna of een ongewenste indringer (Coleoptera: Coccinellidae)? Nederlandse Faunistische Mededelingen 20:1–12

Day WH, Prokrym DR, Ellis DR, Chianese RJ (1994) The known distribution of the predator *Propylea quattuordecimpunctata* (Coleoptera, Coccinellidae) in the United States, and thoughts on the origin of this species and 5 other exotic lady beetles in eastern North America. Entomol News 105:244–256

de Almeida LM, da Silva VB (2002) First record of *Harmonia axyridis* (Pallas) (Coleoptera, Coccinellidae): a lady beetle native to the Palaearctic region. Rev Bras Zool 19:941–944

Dobzhansky T (1933) Geographical variation in ladybeetles. Am Nat 67:97–126

Eschen R, Babendreier D, Nauer S, Bigler F, Kenis M (2007) Surveys for ladybirds (Coleoptera: Coccinellidae) in Switzerland and confirmation of the presence of the invasive alien ladybird species *Harmonia axyridis* (Pallas). Mitteilungen der Schweizerischen Entomologischen Gesellschaft 80:7–14

Ferran A, El-Arnaouty S, Beyssat-Arnaouty V, Galal H (2000) Introduction and release of the coccinellid *Harmonia axyridis* Pallas for controlling *Aphis craccivora* Koch on faba beans in Egypt. Egypt J Biol Pest Control 10:129–136

Fisher RA (1930) The genetical theory of natural selection. Oxford University Press, Oxford

Gordon RD (1985) The Coleoptera (Coccinellidae) of America north of Mexico. J New York Entomol Soc 93:1–912

Haldane JBS (1927) A mathematical theory of natural and artificial selection, part V: selection and mutation. Proc Cambridge Philosophical Soc 28:838–844

Hodek I (1973) Life history and biological properties. In: Hodek I (ed) Biology of Coccinellidae. Dr W Junk, The Hague

Iperti G, Bertand E (2001) Hibernation of *Harmonia axyridis* (Coleoptera: Coccinellidae) in south-eastern France. Acta Soc Zool Bohem 65:207–210

Jacas J-A, Urbaneja A, Viñuela E (2006) History and future of introduction of exotic arthropod biological control agents in Spain: a dilemma? BioControl 51:1–30

Kabiri F (2006) *Harmonia axyridis*: a problem or a distraction? In: Regulation of biological control agents (REBECA), Wageningen, Netherlands, 4–6 April 2006. http://www.rebeca-net.de/downloads/Risk%20assessment%20macrobials%20Kabiri.pdf

Katsoyannos P, Kontodimas DC, Stathas GJ, Tsartsalis CT (1997) Establishment of *Harmonia axyridis* on citrus and some data on its phenology in Greece. Phytoparasitica 25:183–191

Klausnitzer B (2002) *Harmonia axyridis* (Pallas, 1773) in Deutschland (Col.,Coccinellidae). Entomologische Nachrichten und Berichte 46:177–183

Klausnitzer B (2004) *Harmonia axyridis* (Pallas, 1773) in Basel-Stadt (Coleoptera, Coccinellidae). Entomologische Gesellschaft Basel 54:115–122

Klausnitzer B, Klausnitzer H (1997) Marienkäfer (Coccinellidae). 4. überarbeitete auflage. Die Neue Brehm-Bücherei, Magdeburg

Koch RL (2003) The multicolored Asian lady beetle, *Harmonia axyridis*: a review of its biology, uses in biological control, and non-target impacts. J Ins Sci 3:1–16

Kontodimas DC, Stathas GJ, Martinou AF Status of the aphidophagous predator *Harmonia axyridis* (Pallas) (Coleoptera: Coccinellidae) in Greece. Eur J Entomol (in press)

Koyama S, Majerus MEN (2007) Interactions between the parasitoid wasp *Dinocampus coccinellae* and two species of coccinellid from Japan and Britain. BioControl (this issue). doi:10.1007/s10526-007-9138-5

Krafsur ES, Kring TJ, Miller JC, Nariboli P, Obrycki JJ, Ruberson JR, Schaefer PW (1997) Gene flow in the exotic colonizing ladybeetle *Harmonia axyridis* in North America. Biol Control 8:207–214

Kuznetsov VN (1997) Lady beetles of Russian far east. The Sandhill Crane Press, Gainesville, Florida

Loomans A (2004) Veelkleurig Aziatisch lieveheersbeestje *Harmonia axyridis* bezig aan een opmars in Nederland. Coccinula 10:62–63

Machado A (2005) El sarantontón asiático *Harmonia axyridis* (Pallas, 1773) presente en Canarias (Coleoptera: Coccinellidae). Vieraea 34:71–72

Majerus MEN (1994) Ladybirds. No. 81, New Naturalist series. HarperCollins, London

Majerus MEN (1998) Melanism: evolution in action. Oxford University Press, Oxford

Majerus MEN (2003) Sex wars: genes, bacteria, and sex ratios. Princeton University Press, Princeton, New Jersey

Majerus MEN, Mabbott P, Rowland F, Roy HE (2006) The harlequin ladybird, *Harmonia axyridis*, arrives in Britain. Entomol Mon Magaz 142:87–92

Majerus MEN, Roy HE (2005) Scientific opportunities presented by the arrival of the harlequin ladybird, *Harmonia axyridis*, in Britain. Antenna 29:196–208

McClure MS (1987) Potential of the Asian predator, *Harmonia axyridis* Pallas (Coleoptera, Coccinellidae), to control *Matsucoccus resinosae* Bean and Godwin (Homoptera, Margarodidae) in the United States. Environ Entomol 16:224–230

Michaud JP (2002) Invasion of the Florida citrus ecosystem by *Harmonia axyridis* (Coleoptera: Coccinellidae) and asymmetric competition with a native species, *Cycloneda sanguinea*. Environ Entomol 31:827–835

Nalepa CA, Kidd KA, Ahlstrom KR (1996) Biology of *Harmonia axyridis* (Coleoptera: Coccinellidae) in winter aggregations. Ann Entomol Soc Am 89:681–685

Pell JK, Baverstock J, Roy HE, Ware RL, Majerus MEN (2007) Intraguild predation involving *Harmonia axyridis*: a review of current knowledge and future perspectives. BioControl (this issue). doi: 10.1007/s10526-007-9125-x

Rabitsch W, Schuh R (2006) First record of the multicoloured Asian ladybird *Harmonia axyridis* (Pallas, 1773) in Austria. Beiträge zur Entomofaunistik 7:161–164

Roy HE, Baverstock J, Ware RL, Clark SJ, Majerus MEN, Baverstock KE, Pell JK Intraguild predation of the aphid pathogenic fungus *Pandora neoaphidis* by the invasive coccinellid *Harmonia axyridis*. Ecol Entomol (in press)

Roy HE, Brown P, Majerus MEN (2006) *Harmonia axyridis*: a successful biocontrol agent or an invasive threat? In: Eilenberg J, Hokkanen H (eds) An ecological and societal approach to biological control. Springer, Dordrecht

Roy HE, Brown PMJ, Rothery P, Ware RL, Majerus MEN (2007) Interactions between the fungal pathogen *Beauveria bassiana* and three species of coccinellid: *Harmonia axyridis*, *Coccinella septempunctata* and *Adalia bipunctata*. BioControl (this issue). doi:10.1007/s10526-007-9122-0

Roy HE, Rowland F, Brown PMJ, Majerus MEN (2005) Ecology of the harlequin ladybird: a new invasive species. Br Wildl 16:403–407

Schneider N, Loomans AJM (2006) Sur la présence au Luxembourg de la coccinelle arlequin *Harmonia axyridis* (Pallas, 1773) (Insecta, Coleoptera, Coccinellidae). Bulletin de la Société des Naturalistes Luxembourgeois 106:71–74

Sidlyarevich VI, Voronin KE (1973) Trials on using *Leis axyridis* under glass. Zashchita Rastenii 6:24

SIFA (Servicio de información fitosanitaria de Almería) (2004) Organismos de control biológico

Soares AO, Borges I, Borges PAV, Labrie G, Lucas É (2007) *Harmonia axyridis*; What will stop the invader? BioControl (this issue). doi:10.1007/s10526-007-9141-x

Soares AO, Serpa A (2007) Interference competition between ladybird beetle adults (Coleoptera: Coccinellidae): effects on growth and reproductive capacity. Popul Ecol 49:37–43

Staverloekk A, Saethre MG, Haagvar E A review of the biology of the invasive harlequin ladybird *Harmonia axyridis* (Pallas, 1773) (Coleoptera, Coccinellidae). Norwegian J Entomol (in press)

Tedders WL, Schaefer PW (1994) Release and establishment of *Harmonia axyridis* (Coleoptera, Coccinellidae) in the southeastern United States. Entomol News 105:228–243

Tolasch T (2002) *Harmonia axyridis* (Pallas, 1773) breitet sich in Hamburg aus – ausgangspunkt für en besiedlung mitteleuropas? Entomologische Nachrichten und Berichte 46:185–188

Trouvé C (1995) Essais de lutte biologique contre le puceron du houblon (*Phorodon humuli*) dans le nord de la France: résultats 1993 et 1994. In: 47th international symposium on crop protection, Ghent, 9 May 1995. Abstract book, p 109 van Lenteren JC, Babendreier D, Bigler F, Burgio G, Hokkanen H, Kuske S, Loomans A, Menzler-Hokkanen I, van Rijn P, Thomas M, Tommassini M, Zeng Q-Q (2003) Environmental risk assessment of exotic natural enemies used in inundative biological control. BioControl 48:3–38

van Lenteren JC, Loomans AJM, Babendreier D, Bigler F (2007) *Harmonia axyridis*: an environmental risk assessment for Northwest Europe. BioControl (this issue). doi:10.1007/s10526-007-9120-2

Ware RL, Evans N, O'Farrell K, Majerus MEN, Malpas L, Michie LJ Intraguild predation by the invasive ladybird *Harmonia axyridis*: 1: British and Japanese coccinellid eggs. Neobiota (in press)

Ware RL, Majerus MEN (2007) Intraguild predation of immature stages of British and Japanese coccinellids by the invasive ladybird *Harmonia axyridis*. BioControl (this issue). doi:10.1007/s10526-007-9135-8

Ware RL, Majerus MEN, Roy HE, Symington F (2005) The harlequin ladybird arrives in Britain: a threat to our native species? Bull Amateur Entomol Soc 64:175–186

Bad side of a good beetle: the North American experience with *Harmonia axyridis*

Robert L. Koch · Tederson L. Galvan

Abstract The multicolored Asian lady beetle, *Harmonia axyridis* (Pallas) (Coleoptera: Coccinellidae) is a generalist predator of aphids and other soft-bodied insects and has been utilized in biological control programs around the world. Over the last two decades, this species has spread throughout much of the continental USA and southern Canada. Despite the benefits it offers as a biological control agent, *H. axyridis* is perhaps most well known for its adverse impacts. In this paper we provide a review of the North American experience with *H. axyridis*, focusing on these adverse impacts, which can be classified into three general categories: impacts on non-target arthropods, impacts on fruit production and impacts as a household invader. The impacts of *H. axyridis* on non-target arthropods and, to lesser extent, the impacts as a household invader possibly could have been anticipated, due to its generalist feeding preferences and overwintering behavior in Asia. However, it is unlikely that the impacts on fruit production could have been anticipated. Therefore, even in retrospect, it is difficult to predict the potential impacts that an introduced natural enemy might have in its adventive range.

Keywords Biological control · Coccinellidae · Coleoptera · Invasive species

Introduction

"Hordes of maddening ladybugs headed this way" (Gerhardt 2000). "They're everywhere! They're everywhere!" (Fortune 2000). "Steroid Asian lady beetles move into town" (Koehler 2002). These are titles of newspaper articles referring to the invasion and impacts of the multicolored Asian lady beetle, *Harmonia axyridis* (Pallas) (Coleoptera:

R. L. Koch (✉)
Plant Protection Division, Minnesota Department of Agriculture, 625 Robert Street North, Saint Paul, MN 55155, USA
e-mail: robert.koch@state.mn.us

T. L. Galvan
Department of Entomology, University of Minnesota, 1980 Folwell Avenue, Saint Paul, MN 55108, USA

Biological Control to Invasion: Ladybird Harmonia axyridis. H.E. Roy, E. Wajnberg (eds.),
doi: 10.1007/978-1-4020-6939-0_3,

Coccinellidae), in North America. This semi-arboreal coccinellid is a generalist predator of aphids and other soft-bodied arthropods (Hodek and Honěk 1996; Koch 2003). In its native Asian range, *H. axyridis* is a voracious predator of various pest species (Yasumatsu and Watanabe 1964; Hodek and Honěk 1996). Because of this biological control potential, *H. axyridis* has been utilized as a biological control agent in various locations around the world. Intentional releases for biological control purposes (e.g., Gordon 1985) coupled with potential unintentional introductions via international trade (Day et al. 1994; Roy et al. 2005) contributed to the expansion of its range to include North America (Chapin and Brou 1991; Coderre et al. 1995; Dreistadt et al. 1995), South America (Almeida and Silva 2002; Saini 2004), Europe and Africa (Brown et al. 2007).

Koch et al. (2006c) provide a detailed review of the invasion history of *H. axyridis* in North America. Releases in North America for biological control date back to 1916, with more intensive releases in 1970s and 1980s (Gordon 1985; Tedders and Schaefer 1994; LaMana and Miller 1996). Established populations in North America were first detected in Louisiana in the southeastern USA in 1988 (Chapin and Brou 1991). Despite the well documented releases of *H. axyridis* in North America, it has been argued that these early populations resulted from unintentional introductions via international commerce rather than the documented intentional releases for biological control (Day et al. 1994; Day and Tatman 2006). Currently, it remains unclear whether the North American populations stemmed from intentional or unintentional introductions or a combination of both (e.g., Krafsur et al. 1997). Regardless of the source of these initial populations, the range of *H. axyridis* expanded rapidly to cover much of the continental USA and southern Canada (Koch et al. 2006c). In addition, populations have been detected in Mexico (Koch et al. 2006c). The rate of range expansion was estimated to be 442 km per year (McCorquodale 1998) and suggested to be greater in areas with more arboreal habitat (Hesler et al. 2004) or with biomes more similar to those of its native range (Koch et al. 2006c). However, such estimates may be confounded by various factors, such as potential undocumented releases and varying levels of sampling intensity (Koch et al. 2006c).

Beneficial impacts of *H. axyridis* as a biological control agent have been observed in various systems and locations around the world (Koch 2003; Pervez and Omkar 2006). In North American pecans, *H. axyridis* is a successful biological control agent for the pecan aphid complex (*Monellia caryella* (Fitch), *Moneliopsis pecanis* Bissel and *Melanocallis caryaefoliae* (Davis)), which was the target for some of the early releases of this predator (Tedders and Schaefer 1994; LaRock and Ellington 1996). This predator now occurs in and may be contributing to pest suppression in various agricultural and natural systems in North America (Colunga-Garcia and Gage 1998; Koch 2003; Nault and Kennedy 2003; É. Lucas, personal communication). For example, this predator appears to be an important natural enemy of *Aphis glycines* Matsumura, an invasive soybean pest in North America (Fox et al. 2004; Rutledge et al. 2004; Fox et al. 2005; Costamanga and Landis 2006; Mignault et al. 2006). *Harmonia axyridis* has also been documented preying on *A. glycines* on its overwintering host, *Rhamnus cathartica* L., in autumn (Hesler et al. 2004).

However, like other exotic natural enemies that have had some negative impacts after establishment (Howarth 1991; Simberloff and Stiling 1996; Louda et al. 2003), *H. axyridis* is showing adverse impacts in its adventive range (Koch 2003). In this paper we provide a review of the North American experience with *H. axyridis*, focusing on its adverse impacts. These impacts can be classified into three general categories: impacts on non-target arthropods, impacts on fruit production and impacts as a household invader. Could such impacts have been anticipated?

Impacts on non-target arthropods

Intraguild impacts

Harmonia axyridis has become a prominent member of the generalist predatory guild in many of the agricultural (Brown and Miller 1998; Colunga-Garcia and Gage 1998; Michaud 2002; Nault and Kennedy 2003; Alyokhin and Sewell 2004; Koch et al. 2006a; Hesler et al. 2004; Musser et al. 2004; Park and Obrycki 2004; É. Lucas, personal communication) and natural habitats (Wallace and Hain 2000; Sebolt and Landis 2004) it has invaded in North America. The establishment of *H. axyridis* alone, or sometimes in combination with other exotic coccinellids (e.g., *Coccinella septempunctata* L. and *Propylea quatordecimpunctata* (L.)), has been associated with a numerical and/or proportional decline of some native coccinellid species in apple orchards (Brown and Miller 1998), citrus groves (Michaud 2002), potato fields (Alyokhin and Sewell 2004), and across an agricultural landscape (Colunga-Garcia and Gage 1998). In contrast, the replacement of the previously established *C. septempunctata* with *H. axyridis* as the most abundant coccinellid in apple orchards may have actually increased the abundance of some native coccinellids (Brown 2003). Harmon et al. (2007) recently highlighted the dramatic decline of *Adalia bipunctata* (L.) over a broad geographic range after the invasion of *C. septempunctata* and *H. axyridis*. However, it has been pointed out that such correlations from observational studies do not demonstrate causality (Day and Tatman 2006; Harmon et al. 2007). Despite dramatic system and species-specific cases such as *A. bipunctata*, analyses of multiple long-term data sets were not able to show a significant general adverse effect of exotic coccinellids on populations of native coccinellids in the USA and Canada (Harmon et al. 2007). Potential impacts on populations of guild members other than coccinellids have received little attention. Brown (2003) found no decrease in the abundance of *Aphidoletes aphidomyza* Rondani or various Chrysopidae after the establishment of *H. axyridis*.

The success of *H. axyridis* in North America has been attributed to various factors (e.g., Michaud 2002; Snyder et al. 2004; Yasuda et al. 2004; Shapiro-Ilan and Cottrell 2005; Labrie et al. 2006). Among the factors potentially contributing to the rapid rise to dominance of *H. axyridis* over native species, intraguild predation has been most intensively studied in North America. Intraguild predation involves predation among individuals of the same trophic level (Rosenheim et al. 1995). As mentioned by Yasuda et al. (2004), the intraguild predator gains direct nutritional benefits and indirect benefits through reducing competition (Dixon 2000). In such interactions, *H. axyridis* generally dominates (Koch 2003). Native predators (i.e., coccinellids and chrysopids) generally exhibit lower rates of predation on *H. axyridis* eggs compared to the predation rates of *H. axyridis* on the eggs of native species (Phoofolo and Obrycki 1998; Cottrell 2004, 2005; Flowers et al. 2005). *Harmonia axyridis* larvae are relatively larger and more aggressive than larvae of native coccinellids (Cottrell and Yeargan 1998; Michaud 2002; Yasuda et al. 2004). When paired with heterospecific coccinellid larvae of the same size or smaller than itself, *H. axyridis* larvae are generally intraguild predators (Cottrell and Yeargan 1998; Yasuda et al. 2004). However, when paired with a heterospecific larva considerably larger than itself, *H. axyridis* larvae may be intraguild predators (Michaud 2002) or prey (Cottrell and Yeargan 1998). Snyder et al. (2004) found no relationship between size advantage and intraguild predation rates. They suggest that the success of *H. axyridis* as an intraguild predator of native coccinellids may be due to greater abilities to capture intraguild prey and avoid and escape intraguild predators (Snyder et al. 2004). *Harmonia axyridis* has been documented feeding on various stages of North American predatory species, such as *A. bipunctata* (Sato et al. 2005), *C. septempunctata* (Snyder et al. 2004; Yasuda et al. 2004), *Coccinella transversoguttata* Falderman (Snyder et al. 2004; Yasuda et al. 2004), *Coleomegilla maculata* DeGeer (Cottrell and Yeargan 1998; Cottrell 2004, 2005), *Cycloneda sanguinea* (L.) (Michaud 2002), *Hippodamia convergens* Guérin-Méneville (Snyder et al. 2004; Yasuda et al. 2004), *Laricobius nigrinus* Fender (Flowers et al. 2005), *Olla v-nigrum* Mulsant (Cottrell 2004, 2005), *Sasajiscymnus tsugae* Sasaji and McClure (Flowers et al. 2005), *Chrysoperla carnea* Stephens (Phoofolo and Obrycki 1998; Gardiner and Landis 2007), and *A. aphidomyza* (Gardiner and Landis 2007). Not only is this predator known to feed on these various prey, it can complete development feeding on some of them (Cottrell and Yeargan 1998; Michaud 2002; Cottrell 2004).

The presence of an intraguild predator can disrupt pest suppression in some systems (Rosenheim et al. 1995). However, the presence of *H. axyridis* as an intraguild predator in a system does not necessarily result in reduced pest suppression (Brown and Miller 1998; Lucas et al. 2002; Gardiner and Landis 2007). For example, as mentioned above, *H. axyridis* is an important predator of *A. glycines* on soybean and an intraguild predator of *A. aphidomyza* and *C. carnea*. The high predation rate of *H. axyridis* on the pest appears to compensate for the resulting reduction in the abundance of other natural enemies (Gardiner and Landis 2007).

Interguild impacts

Despite the paucity of literature on the topic, non-target prey of *H. axyridis* are not limited to other predators. It was suggested that *H. axyridis* may impact native, non-pest aphid species (Boettner et al. 2000). In prairie remnants, this predator was indeed found preying on native aphids (e.g., *Aphis monardae* Oestlund on *Monarda fistulosa* L. and *Aphis asclepiadis* (Fitch) on *Asclepias syriaca* L.) (K. Wyckhuys and R. Koch, unpublished data). In addition, Sebolt and Landis (2004) identified *H. axyridis* as a predator of the chrysomelid, *Galerucella calmariensis* L., which is a weed biological control agent used in North American wetlands. They concluded that the predator complex including *H. axyridis* might reduce populations of the weed biological control agent, but would be unlikely to prevent its establishment (Sebolt and Landis 2004). Furthermore, Koch et al. (2003) documented *H. axyridis* preying on the eggs and larvae of a non-pest lepidopteran, the monarch butterfly, *Danaus plexippus* L. A quantitative risk assessment, combining the likelihoods of exposure (i.e., co-occurrence) and effect (i.e., predation), showed the potential for *H. axyridis* to adversely affect the production of *D. plexippus* in agricultural systems (Koch et al. 2006b). Continued development of such risk assessment methodologies will improve our understanding of the non-target impacts of exotic species such as *H. axyridis*.

It comes as little surprise to see an exotic natural enemy with such a broad host range having adverse impacts on non-target species (intraguild and interguild). In a biological control context, generalist feeding preferences can enable a natural enemy to persist on alternate prey during times when target prey are unavailable, thus potentially increasing the likelihood of biological control success. However, it is this same attribute that increases the likelihood of impacts on non-target species. A heightened awareness for the potential of generalist natural enemies to impact non-target species is driving the biological control community to develop rigorous methods for identification and evaluation of highly specific natural enemies, which have lower associated risk of impacting non-target species.

Impacts as a household invader

Mass aggregations of coccinellids at overwintering sites are well documented (e.g., Hagen 1962; Sakurai et al. 1993; Majerus 1994; Nalepa et al. 1996; Kuznetsov 1997). *Harmonia axyridis* has shown hypsotactic aggregation behavior in Asia and North America, accumulating in cracks and crevices of mountain tops, rocks, houses, forest huts, barns, and other human-made structures (Obata 1986; Sakurai et al. 1993; Kidd et al. 1995; Nalepa et al. 1996; Kuznetsov 1997; Huelsman and Kovach 2004).

In the USA, the infamous autumn flight of *H. axyridis* to the aggregation sites has been studied by Nalepa et al. (2000, 2005). These authors suggested a five-step aggregation behavior for *H. axyridis*: (1) exodus from feeding locations, (2) arrival on buildings or mountaintops, (3) decision to stay or leave, (4) if staying, insects search for cracks or crevices, and (5) then settle in these places (Nalepa et al. 2000, 2005). In addition, *H. axyridis* also shows a preference for places with high color contrasts, and for flying during the day when temperatures are above 21°C (Nalepa et al. 2005). Observations from Ohio suggest that the flight begins in autumn on the first day with temperatures greater than 18°C just after a low temperature near 0°C (Huelsman et al. 2002). Beyond the annoyance caused by the shear numbers of *H. axyridis* on and inside homes, infestations cause problems in houses by staining carpets, curtains, furniture, and the walls; additionally, in the food industry, and health and research institutions *H. axyridis* can be a contaminant pest (Riddick et al. 2000; Huelsman and Kovach 2004; Nalepa et al. 2004). In addition, *H. axyridis* will occasionally bite humans (Kuznetsov 1997; Kovach 2004). Furthermore, seasonal allergic reactions to the presence of *H. axyridis* in homes have been described in several case studies (Yarbrough et al. 1999; Ray and Pence 2004; Goetz 2007). Allergy types caused by infestations of this beetle are mainly allergic rhinoconjuntivitis, and less often asthma, urticaria (Goetz 2007) or angioedema (Davis et al. 2006).

Control of *H. axyridis* with insecticides in buildings and houses may be undesirable because of potential negative effects of pesticide exposure to humans and pets. In addition, this coccinellid is beneficial as a predator during the spring and summer. Therefore, methods based on volatile chemicals to repel *H. axyridis* from houses and buildings and to attract them to traps (i.e., "push-pull" strategy) have been examined (Riddick et al. 2000). Among the potential repellents tested, camphor and menthol have shown good results for repelling *H. axyridis* (Riddick et al. 2000). In addition to these two compounds, DEET (*N,N*-diethyl-3-methylbenzamide) has shown repellency to *H. axyridis* (Riddick et al. 2004). However, further work is needed to develop formulations suitable for exterior use on homes (Riddick et al. 2004). In Europe, the presence of some type of aggregation pheromone was suggested for guiding *A. bipunctata* to the same overwintering locations every year (Majerus 1997). Nalepa et al. (2000) suggest that *H. axyridis* does not rely on volatile chemical cues to locate overwintering locations. However, further work like that of Brown et al. (2006) is needed to explore the role of volatiles emitted by adult *H. axyridis*. In their study, they used gas chromatography–mass spectrometry to identify volatiles (e.g., a sesquiterpene, (-)-β-caryophyllene) emitted by live adults (Brown et al. 2006).

The utilization of houses and other buildings is not unique to *H. axyridis*. *Adalia bipunctata* is known to overwinter in houses (Majerus and Kearns 1989). Furthermore, *C. septempunctata* became a nuisance pest during population explosions in 1952 and 1976 in England (Majerus and Kearns 1989). However, the magnitude and/or frequency of these problems are greater for *H. axyridis* than what has been reported for other species. It should not come as a surprise that *H. axyridis* is becoming a household nuisance pest in its adventive range, because this pest exhibits a similar behavior in parts of its native range

(Obata 1986; Kuznetsov 1997). Kuznetsov (1997) described this behavior: "Sometimes huge numbers of beetles accumulated in houses. They penetrate everywhere, settle on people and annoy them with slight pinching."

Impacts as a pest of fruit production

Fruit feeding and contamination

Harmonia axyridis has been reported feeding on fruits, such as grapes, apples, peaches, plums, pears, pumpkins and raspberries, in North America (Hesler et al. 2004; Koch et al. 2004; Kovach 2004). The feeding behavior of *H. axyridis* switches from carnivory to phytophagy in a manner that is not completely understood. Whether *H. axyridis* is causing primary injury to the fruits (i.e., breaking the skin of fruit) or is feeding on fruits injured by other agents remains unclear for some fruit types. In Minnesota, *H. axyridis* adults were not able to cause primary injury to grapes, apples or pumpkins, but they could inflict primary injury to raspberries (Koch et al. 2004; T. Galvan, unpublished data). In contrast, Kovach (2004) reported that *H. axyridis* could cause primary injury to apples, peaches and grapes in Ohio. Whether or not this beetle is breaking the skin of the fruits themselves, they display a strong preference for previously damaged versus undamaged fruits (Koch et al. 2004; Kovach 2004). Consuming food of plant origin (i.e., pollen, nectar, and extrafloral nectaries) is not unusual in predatory Coccinellidae (Hemptinne and Desprets 1986; Hodek and Honěk 1996; Ricci and Ponti 2002). Such feeding increases the chances of survival when prey are scarce, raises reserve levels for overwintering, and may improve coccinellid fitness even when prey are available (Smith 1960; Hagen 1962; Hemptinne and Desprets 1986; Harmon et al. 2000). Frugivory (i.e., fruit feeding) in predatory coccinellids appears to be rare, but has been documented for *C. septempunctata* and *A. bipunctata* (Hodek and Honěk 1996). We are unaware of predatory Coccinellidae causing agricultural problems of economic importance, except for the recent case of *H. axyridis* on wine grapes in North America.

Adults of *H. axyridis* move to vineyards 2–3 weeks before harvest, which begins in early September in Minnesota and Wisconsin, when prey populations in summer crops decline (Galvan et al. 2006c). During this period the grape berries are nearing maturity and may be injured by other fruit feeders, including paper wasps, yellow-jackets, other wasps, and birds or physiological splitting, which is common in some varieties close to harvest (Galvan et al. 2006b). Splitting is caused by a sudden increase in absorption and/or adsorption of water, atmospheric humidity, or temperature (Opara et al. 1997). Varieties with a tight cluster structure tend to have high rates of splitting (T. Galvan, unpublished data). This previous damage offers accessible feeding sites for *H. axyridis*. Field studies showed a strong correlation between injured wine grape berries and *H. axyridis* infestations in grape clusters (Galvan et al. 2006b). Depending on the proportion of clusters with damage, *H. axyridis* populations can easily build to infest 65% of all clusters for some varieties (Galvan et al. 2006b).

We hypothesize that *H. axyridis* feeds on grapes and other fruits in autumn to build reserves for overwintering. A couple months prior to overwintering, coccinellids start to prepare for winter by a mechanism called pre-diapause, where beetles accumulate fat and glycogen reserves, which is triggered by a decline in food resources (Hagen 1962; Hodek and Čerkasov 1963; Hodek 1986). The combination of these environmental and physiological changes, which has been called the "adaptation syndrome" (de Wilde 1970), in the months preceding winter contributes to coccinellid survival during diapause (Hagen 1962; Hodek and Čerkasov 1963). Koch et al. (2004) confirmed a highly significant preference for sugar water versus water alone, which suggested that the beetles may be seeking more than just a source of moisture prior to overwintering. Since wine grapes have a high level of sugar content (20–25%) in the weeks prior to harvest (Plocher and Parke 2001), vineyards are an excellent location to increase sugar reserves. The importance of sugars to overwintering survival was suggested by studies in which beetles fed sugar water had higher survival rates than those given water alone (T. Galvan, unpublished data).

The primary problem with *H. axyridis* on wine grapes is attributed to it acting as a contaminant. Adults tend to aggregate on clusters with damaged grapes. If *H. axyridis* are disturbed or crushed during harvest and processing of the grapes, they release a yellow fluid (reflex bleeding), which creates an unpleasant odor and taste in the resulting wine (Pickering et al. 2004; Galvan et al. 2007a). The reflex bleeding fluid is released from the tibio-femoral joints of adults (Al Abassi et al. 1998; Dixon 2000; Laurent et al. 2001). The fluid contains alkaloids used for defense and 3-alkyl-2-methoxypyrazines that could be used as an aggregation pheromone or in Müllerian mimicry due to their strong smell (Moore et al. 1990; Al Abassi et al. 1998; Dixon 2000; Cudjoe et al. 2005). One 3-alkyl-2-methoxypyrazine, the 2-isopropyl-3-methoxypyrazine (IPMP), has been suggested to be one of the key compounds responsible for the taint produced by *H. axyridis* in wines (Pickering et al. 2005). In addition to IPMP, two other 3-alkyl-2-methoxypyrazines, the 2-sec-butyl-3-methoxypyrazine (SBMP) and 2-isobutyl-3-methoxypyrazine (IBMP) could also be affecting the 2-sec-butyl-3 wine flavor since they are found in *H. axyridis* (Cudjoe et al. 2005; Cai et al. 2007). These compounds are also well known for their contribution to vegetative, herbaceous, green bell pepper and earthy character of wines such as Cabernet Sauvignon and Sauvignon blanc (Allen et al. 1994; Sala et al. 2002).

Integrated pest management in vineyards

An integrated pest management program for *H. axyridis* on wine grapes should be based on the timing of pest infestation (phenology), use of procedures to estimate levels of pest infestation (sampling methods), knowledge of the relationship between levels of infestation and resulting taint (sensory thresholds), and control methods. For phenology, *H. axyridis* adults begin to move to grape clusters between 2 and 3 weeks prior to harvest (Galvan et al. 2006b). This is the key management window. Even though *H. axyridis* populations increase or decrease throughout the growing season, grape growers should not worry about this pest until 2 or 3 weeks before harvest when the proportion of injured berries starts to increase, which then provides an opportunity for *H. axyridis* feeding on the grapes (Galvan et al. 2006b). *Harmonia axyridis* is found in vineyards as early as June when they are feeding on aphids or other small soft-bodied insects. Grape growers can follow *H. axyridis* population fluctuation in vineyards using yellow sticky traps, which can be used as an early warning tool as well.

Enumerative (e.g., number of beetles per cluster) and binomial (e.g., proportion of clusters infested) sampling plans were developed for *H. axyridis* in wine grapes (Galvan et al. 2007a). For enumerative plans, the average sample size (with a sample unit of one cluster) required to reach desired precision levels ranged from 180 to 546 clusters. By contrast, binomial plans resulted in a much lower required sample size of about 25 clusters depending on the threshold used (Galvan et al. 2007a). Because of this increased efficiency, binomial sampling should be used for practical pest management purposes regarding this pest.

Sensory-based action thresholds were developed for *H. axyridis* in Frontenac wine grapes using a tasting panel and wine from artificially infested grapes (Galvan et al. 2007b). The estimated sensory threshold at which 10% of the population of wine consumers was able to detect the characteristic taint of *H. axyridis* is 1.9 beetles per kg of grapes, or 0.27 beetles per grape cluster of Frontenac (Galvan et al. 2007b). This sensory threshold is similar to the 0.2 *H. axyridis* adults per cluster suggested by Pickering et al. (2006a) for the white grape variety Riesling. Sensory thresholds can be interpreted as action thresholds for *H. axyridis* in wine grapes (i.e., number of *H. axyridis* per kg of grapes), which when combined with existing control methods and sampling plans, form the basis of integrated pest management for this insect in wine grapes.

Although remediation of tainted wine by adding oak chips, activated charcoal, and deodorized oak has decreased *H. axyridis*-related taint or IPMP concentration, it has not completely removed the taint from contaminated wine (Pickering et al. 2006b). Therefore, the use of control measures such as insecticides to manage *H. axyridis* before it can become a wine contaminant is essential for reducing the economic impact of this pest on the wine industry. In field and laboratory studies, carbaryl, bifenthrin, zeta-cypermethrin, thiamethoxam, and imidacloprid showed either toxic and/or repellent effects on *H. axyridis* (Galvan et al. 2006a). However, of these insecticides, only carbaryl, thiamethoxam and imidacloprid are currently labeled in the USA for use on wine grapes within 7 days of harvest, which is the period in which *H. axyridis* typically reaches high densities (Galvan et al. 2006b).

We are unaware of reports of *H. axyridis* feeding on fruits in its native Asian range. Therefore, the pest status that *H. axyridis* has attained in fruit production, particularly wine grapes, is clearly an unanticipated adverse consequence of the establishment of this beetle in North America. Even though other coccinellids have been documented feeding on fruits, none have attained pest status in fruit production. This unique pest situation deserves attention to improve our understanding of the biology and to further develop integrated pest management programs for this system. In addition, the impact of *H. axyridis* on other fruits such as raspberries deserves further attention.

Conclusions

Despite the benefits offered as a biological control agent, *H. axyridis* is becoming known as a case of biological control gone awry. In North America, *H. axyridis* is having impacts on non-target arthropods, impacts on fruit production and impacts as a household invader. Because of these, Babendreier (2007) used the phrase "*Harmonia* cases" to refer to situations when the release of biological control agents results in adverse effects. Further work is needed to ensure the safety of future releases of biological control agents. For the case of *H. axyridis* in North America, the impacts on non-target arthropods and, to lesser extent, the impacts as a household invader possibly could have been anticipated, due to its generalist feeding preferences and overwintering behavior in Asia, respectively. However, it is unlikely that the impacts on fruit production could have been anticipated. Therefore, even in retrospect, it is difficult to predict the potential impacts that an introduced natural enemy might have in its adventive range.

Acknowledgments We thank Helen Roy (CEH Monks Wood) for inviting us to contribute a paper to this special issue. We are grateful to Éric Lucas (Université du Québec à Montréal) for providing information about his recent research, and Geir Friisoe (Minnesota Department of Agriculture), Sandra Nogueira Koch (University of Minnesota) and three anonymous reviewers for providing reviews of this paper.

References

Al Abassi S, Birkett MA, Petterson J, Pickett JA, Woodcock CM (1998) Ladybird beetle odour identified and found to be responsible for attraction between adults. Cell Mol Life Sci 54:876–879

Allen MS, Lacey MJ, Boyd S (1994) Determination of methoxypyrazines in red wine by stable isotope dilution gas chromatography–mass spectrometry. J Agric Food Chem 42:1734–1738

de Almeida LM, da Silva VB (2002) First record of *Harmonia axyridis* (Pallas) (Coleoptera, Coccinellidae): a lady beetle native to the Palaearctic region. Rev Bras Zool 19:941–944

Alyokhin A, Sewell G (2004) Changes in a lady beetle community following the establishment of three alien species. Biol Invasions 6:463–471

Babendreier D (2007) Pros and cons of biological control. In: Nentig W (ed) Biological invasions. Springer-Verlag, Berlin, pp 403–418

Boettner GH, Elkinton JS, Boettner CJ (2000) Effects of a biological control introduction on three nontarget native species of saturniid moths. Conserv Biol 14:1798–1806

Brown MW (2003) Intraguild responses of aphid predators on apple to the invasion of an exotic species, *Harmonia axyridis*. BioControl 48:141–153

Brown MW, Miller SS (1998) Coccinellidae (Coleoptera) in apple orchards of eastern West Virginia and the impact of invasion by *Harmonia axyridis*. Entomol News 109:136–142

Brown AE, Riddick EW, Aldrich JR, Holmes WE (2006) Identification of (-)-β-caryophyllene as a gender-specific terpene produced by the multicolored Asian lady beetle. J Chem Ecol 32:2489–2499

Brown PMJ, Adriaens T, Bathon H, Cuppen J, Goldarazena A, Hägg T, Kenis M, Klausnitzer BEM, Kovar I, Loomans AJM, Majerus MEN, Nedved O, Pederson J, Rabitsch W, Roy HE, Ternois V, Zakharov IA, Roy DB (2007) *Harmonia axyridis* in Europe: spread and distribution of a non-native coccinellid. BioControl (this issue). doi:10.1007/s10526-007-9132-y

Cai L, Koziel JA, Matthew EO (2007) Determination of characteristic odorants from *Harmonia axyridis* beetles using in vivo solid-phase and multidimensional gas chromatography–mass spectrometry–olfactometry. J Chromatogr A 1147:66–78

Chapin JB, Brou VA (1991) *Harmonia axyridis* (Pallas), the third species of the genus to be found in the United States (Coleoptera: Coccinellidae). Proc Entomol Soc Wash 93:630–635

Coderre D, Lucas É, Gagné I (1995) The occurrence of *Harmonia axyridis* (Pallas) (Coleoptera: Coccinellidae) in Canada. Can Entomol 127:609–611

Colunga-Garcia M, Gage SH (1998) Arrival, establishment, and habitat use of the multicolored Asian lady beetle (Coleoptera: Coccinellidae) in a Michigan landscape. Environ Entomol 27:1574–1580

Costamagna A, Landis DA (2006) Predators exert top-down control of soybean aphid across a gradient of agricultural management systems. Ecol Appl 16:1619–1628

Cottrell TE (2004) Suitability of exotic and native ladybeetle eggs (Coleoptera: Coccinellidae) for development of lady beetle larvae. Biol Control 31:362–371

Cottrell TE (2005) Predation and cannibalism of lady beetle eggs by adult lady beetles. Biol Control 34: 159–164

Cottrell TE, Yeargan KV (1998) Intraguild predation between an introduced lady beetle, *Harmonia axyridis* (Coleoptera: Coccinellidae), and a native lady beetle, *Coleomegilla maculata* (Coleoptera: Coccinellidae). J Kans Entomol Soc 71:159–163

Cudjoe E, Wiederkehr TB, Brindle ID (2005) Headspace gas chromatography–mass spectrometry: a fast approach to the identification and determination of 2-alkyl-3-methoxypyrazine pheromones in ladybugs. Analyst 130:152–155

Davis RS, Vandewalker ML, Hutcheson PS, Slavin RG (2006) Facial angioedema in children due to ladybug (*Harmonia axyridis*) contact: 2 case reports. Ann Allergy Asthma Immunol 97:440–442

Day WH, Tatman KM (2006) Changes in abundance of native and adventive Coccinellidae (Coleoptera) in alfalfa fields, in northern New Jersey (1993–2004) and Delaware (1999–2004), U.S.A Entomol News 117:491–502

Day WH, Prokrym DR, Ellis DR, Chianese RJ (1994) The known distribution of the predator *Propylea quatuordecimpunctata* (Coleoptera: Coccinellidae) in the United States, and thoughts on the origin of this species and five other exotic lady beetles in eastern North America. Entomol News 105:224–256

de Wilde J (1970) Hormones and insect diapause. Mem Soc Endocrinol 18:487–514
Dixon AFG (2000) Insect predator-prey dynamics: ladybird beetles and biological control. Cambridge University Press, Cambridge, 257 pp
Dreistadt SH, Hagen KS, Bezark LG (1995) *Harmonia axyridis* (Pallas) (Coleoptera: Coccinellidae), first western United States record for this Asiatic lady beetle. Pan-Pac Entomol 71:135–136
Flowers RW, Salom SM, Kok LT (2005) Competitive interactions among two specialist predators and a generalist predator of hemlock woolly adelgid, *Adelges tsugae* (Homoptera: Adelgidae), in the laboratory. Environ Entomol 34:664–675
Fortune B (2000) They're everywhere! They're everywhere! Lexington Herald. 27 October 2000
Fox TB, Landis DA, Cardoso FF, Difonzo CD (2004) Predators suppress *Aphis glycines* Matsumura population growth in soybean. Environ Entomol 33:608–618
Fox TB, Landis DA, Cardoso FF, Difonzo CD (2005) Impact of predation on establishment of the soybean aphid, *Aphis glycines* in soybean, *Glycine max*. BioControl 50:545–563
Galvan TL, Burkness EC, Hutchison WD (2006a) Efficacy of selected insecticides for management of the multicolored Asian lady beetle on wine grapes near harvest. Plant Health Prog. doi:10.1094/PHP-2006-1003-01-RS
Galvan TL, Burkness EC, Hutchison WD (2006b) Influence of berry injury on infestations of the multicolored Asian lady beetle in wine grapes. Plant Health Prog. doi:10.1094/PHP-2006-0607-01-BR
Galvan TL, Burkness EC, Hutchison WD (2006c) Wine grapes in the Midwest: reducing the risk of the multicolored Asian lady beetle. Public. 08232. University of Minnesota Extension Service, Saint Paul, MN, USA
Galvan TL, Burkness EC, Hutchison WD (2007a) Enumerative and binomial sequential sampling plans for the multicolored Asian lady beetle (Coleoptera: Coccinellidae) in wine grapes. J Econ Entomol 100:1000–1010
Galvan TL, Burkness EC, Vickers Z, Stenberg P, Mansfield AK, Hutchison WD (2007b) Sensory-based action threshold for the multicolored Asian lady beetle-related taint in wine grapes. Am J Enol Vitic (in press)
Gardiner MM, Landis DA (2007) Impact of intraguild predation by adult *Harmonia axyridis* (Coleoptera: Coccinellidae) on *Aphis glycines* (Hemiptera: Aphididae) biological control in cage studies. Biol Control 40:386–395
Gerhardt G (2000) Hordes of maddening ladybugs headed this way. Rocky Mountain News. 30 November 2000
Goetz DW (2007) *Harmonia axyridis* ladybug hypersensitivity in clinical allergy practice. Allergy Asthma Proc 28:50–57
Gordon RD (1985) The Coccinellidae (Coleoptera) of America north of Mexico. J New York Entomol Soc 93:1–912
Hagen KS (1962) Biology and ecology of predaceous Coccinellidae. Ann Rev Entomol 7:289–3226
Harmon JP, Ives AR, Losey JE, Olson AC, Rauwald KS (2000) *Coleomegilla maculata* (Coleoptera: Coccinellidae) predation on pea aphids promoted by proximity to dandelions. Oecologia 125:543–548
Harmon JP, Stephens E, Losey J (2007) The decline of native ladybirds (Coleoptera: Coccinellidae) in the United States and Canada. J Insect Conserv 11:85–94
Hemptinne JL, Desprets A (1986) Pollen as a spring food for *Adalia bipunctata*. In: Hodek I (ed) Ecology of Aphidophaga. Dr W Junk, Dordrecht, pp 29–35
Hesler LS, Kieckhefer RK, Catangui MA (2004) Surveys and field observation of *Harmonia axyridis* and other Coccinellidae (Coleoptera) in eastern and central South Dakota. Trans Am Entomol Soc 130: 113–130
Hodek I (1986) Life cycle strategies, diapause and migration in aphidophagous Coccinellidae. In: Hodek I (ed) Ecology of Aphidophaga. Dr. W. Junk, Dordrecht, pp 155–166
Hodek I, Čerkasov J (1963) Imaginal dormancy in *Semiadalia undecimnotata* Schneid. (Coccinellidae, Col.) II. Changes in water, fat and glycogen content. Acta Soc Zool Bohem 27:298–318
Hodek I, Honěk A (1996) Ecology of Coccinellidae. Kluwer Academic Publishers, Dordrecht, 464 pp
Howarth FG (1991) Environmental impacts of classical biological control. Ann Rev Entomol 36:485–509
Huelsman M, Kovach J (2004) Behavior and treatment of the multicolored Asian lady beetle (*Harmonia axyridis*) in the urban environment. Am Entomol 50:163–164
Huelsman MF, Kovach J, Jasinski J, Young C, Eisley B (2002) Multicolored Asian lady beetle (*Harmonia axyridis*) as a nuisance pest in households in Ohio. In: Jones SC, Zhai J, Robinson WH (eds) Proceedings of 4th international conference on urban pests, pp 243–250
Kidd K, Nalepa CA, Day ER, Waldvogel MG (1995) Distribution of *Harmonia axyridis* (Pallas) (Coleoptera: Coccinellidae) in North Carolina and Virginia. Proc Entomol Soc Wash 97:729–731

Koch RL (2003) The multicolored Asian lady beetle, *Harmonia axyridis*: a review of its biology, uses in biological control and non-target impacts. J Insect Sci 3:1–16
Koch RL, Hutchison WD, Venette RC, Heimpel GE (2003) Susceptibility of immature monarch butterfly, *Danaus plexippus* (Lepidoptera: Nymphalidae: Danainae), to predation by *Harmonia axyridis* (Coleoptera: Coccinellidae). Biol Control 28:265–270
Koch RL, Burkness EC, Burkness SJ, Hutchison WD (2004) Phytophagous preferences of the multicolored Asian lady beetle (Coleoptera: Coccinellidae) to autumn ripening fruit. J Econ Entomol 97:539–544
Koch RL, Burkness EC, Hutchison WD (2006a) Spatial distribution and fixed-precision sampling plans for the ladybird *Harmonia axyridis* in sweet corn. BioControl 51:741–751
Koch RL, Venette RC, Hutchison WD (2006b) Predicted impact of an invasive generalist predator on monarch butterfly (Lepidoptera: Nymphalidae) populations: a quantitative risk assessment. Biol Invasions 8:1179–1193
Koch RL, Venette RC, Hutchison WD (2006c) Invasions by *Harmonia axyridis* (Pallas) (Coleoptera: Coccinellidae) in the western hemisphere: implications for South America. Neotrop Entomol 35: 421–434
Koehler D (2002) Steroid Asian lady beetles move into town. Grand Forks Herald. 13 September 2002
Kovach J (2004) Impact of multicolored Asian lady beetles as a pest of fruit and people. Am Entomol 50:159–161
Krafsur ES, Kring TJ, Miller JC, Nariboli P, Obrycki JJ, Ruberson JR, Schaefer PW (1997) Gene flow in the exotic colonizing ladybeetle *Harmonia axyridis* in North America. Biol Control 8:207–214
Kuznetsov VN (1997) Lady beetles of the Russian Far East. The Sandhill Crane Press, Inc., Gainesville, 248 pp
Labrie G, Lucas É, Coderre D (2006) Can developmental and behavioral characteristics of the multicolored Asian lady beetle *Harmonia axyridis* explain its invasive success? Biol Invasions 8:743–754
LaMana ML, Miller JC (1996) Field observations on *Harmonia axyridis* Pallas (Coleoptera: Coccinellidae) in Oregon. Biol Control 6:232–237
LaRock DR, Ellington JJ (1996) An integrated pest management approach, emphasizing biological control, for pecan aphids. Southwest Entomol 21:153–167
Laurent P, Lebrun B, Braekman J, Daloze D, Pasteels JM (2001) Biosynthetic studies on adaline and adalinine, two alkaloids from ladybird beetles (Coleoptera: Coccinellidae). Tetrahedron 57:3403–3412
Louda SM, Pemberton RW, Johnson MT, Follett PA (2003) Nontarget effects—the Achilles' heel of biological control? Retrospective analysis to reduce risk associated with biocontrol introductions. Ann Rev Entomol 48:369–396
Lucas É, Gagné I, Coderre D (2002) Impact of the arrival of *Harmonia axyridis* on adults of *Coccinella septempunctata* and *Coleomegilla maculata* (Coleoptera: Coccinellidae). Eur J Entomol 99:457–463
Majerus MEN (1994) Ladybirds in winter. In: Majerus M (ed) Ladybirds. New naturalist 81. Harper Collins, London, 320 pp
Majerus MEN (1997) How is *Adalia bipunctata* (Linn.) (Coleoptera: Coccinellidae) attracted to overwintering sites? Entomologist 116:212–217
Majerus M, Kearns P (1989) Ladybirds. Richmond Publishing Co. Ltd., Slough, 103 pp
McCorquodale DB (1998) Adventive lady beetles (Coleoptera: Coccinellidae) in eastern Nova Scotia, Canada. Entomol News 109:15–20
Michaud JP (2002) Invasion of the Florida Citrus ecosystem by *Harmonia axyridis* (Coleoptera: Coccinellidae) and asymmetric competition with a native species, *Cycloneda sanguinea*. Environ Entomol 31:827–835
Mignault M-P, Roy M, Brodeur J (2006) Soybean aphid predators in Quebec and the suitability of *Aphis glycines* as prey for three Coccinellidae. BioControl 51:89–106
Moore BP, Brown WV, Rothschild M (1990) Methylalkylpyrazines in aposematic insects, their hostplants and mimics. Chemoecology 1:43–51
Musser FR, Nyrop JP, Shelton AM (2004) Survey of predators and sampling method comparison in sweet corn. J Econ Entomol 97:136–144
Nalepa CA, Kidd KA, Ahlstrom KR (1996) Biology of *Harmonia axyridis* (Coleoptera: Coccinellidae) in winter aggregations. Ann Entomol Soc Am 89:681–685
Nalepa CA, Kidd KA, Hopkins DI (2000) The multicolored Asian ladybeetle (Coleoptera: Coccinellidae): orientation to aggregation sites. J Entomol Sci 35:150–157
Nalepa CA, Kennedy GG, Brownie C (2004) Orientation of multicolored Asian lady beetles to buildings. Am Entomol 50:165–166
Nalepa CA, Kennedy GG, Brownie C (2005) Role of visual contrast in the alighting behavior of *Harmonia axyridis* (Coleoptera: Coccinellidae) at overwintering sites. Environ Entomol 34:425–431

Nault BA, Kennedy GG (2003) Establishment of multicolored Asian lady beetle in eastern North Carolina: seasonal abundance and crop exploitation within an agricultural landscape. BioControl 48:363–378

Obata S (1986) Determination of hibernation site in the ladybird beetle, *Harmonia axyridis* Pallas (Coleoptera: Coccinellidae). Kontyu 54:218–223

Opara LU, Studman CJ, Banks NH (1997) Fruit skin splitting and cracking. Hort Rev 19:217–262

Park YL, Obrycki JJ (2004) Spatio-temporal distribution of corn leaf aphids (Homoptera: Aphididae) and lady beetles (Coleoptera: Coccinellidae) in Iowa cornfields. Biol Control 31:210–217

Pervez A, Omkar (2006) Ecology and biological control application of multicoloured Asian ladybird, *Harmonia axyridis*: a review. Biocontrol Sci Technol 16:111–128

Phoofolo MW, Obrycki JJ (1998) Potential for intraguild predation and competition among predatory Coccinellidae and Chrysopidae. Entomol Exp Appl 89:47–55

Pickering GJ, Lin J, Riesen R, Reynolds A, Brindle I, Soleas G (2004) Influence of *Harmonia axyridis* on the sensory properties of white and red wine. Am J Enol Vitic 55:153–159

Pickering GJ, Lin J, Reynolds A, Soleas G, Riesen R, Brindle I (2005) The influence of *Harmonia axyridis* on wine composition and aging. J Food Sci 70:128–135

Pickering GJ, Lin J, Ker K (2006a) Origin and remediation of Asian lady beetle (*Harmonia axyridis*) taint in wine. In: Dris R (ed) Crops: growth, quality and biotechnology. III. Quality management of food crops for processing technology. WFL Publisher, Helsinki, pp 785–794

Pickering GJ, Lin J, Reynolds A, Soleas G, Riesen R (2006b) The evaluation of remedial treatments for the wine affected by *Harmonia axyridis*. Inter J Food Sci Tech 41:77–86

Plocher TA, Parke RJ (2001) Northern winework: growing grapes and making wine in cold climates. Northern Winework, Inc., Hugo

Ray JN, Pence HL (2004) Ladybug hypersensitivity: report of a case and review of the literature. Allergy Asthma Proc 25:133–136

Ricci C, Ponti L (2002) Seasonal food of *Ceratomegilla notata* (Coleoptera: Coccinellidae) in mountain environments of Northern Italian Alps. Eur J Entomol 102:527–530

Riddick EW, Aldrich JR, de Milo A, Davis JC (2000) Potential for modifying the behavior of the multicolored Asian lady beetle (Coleoptera: Coccinellidae) with plant-derived natural products. Ann Entomol Soc Am 93:1314–1321

Riddick EW, Aldrich JR, Davis JC (2004) DEET repels *Harmonia axyridis* (Pallas) (Coleoptera: Coccinellidae) adults in laboratory bioassays. J Entomol Sci 39:373–386

Rosenheim JA, Kaya HK, Ehler LE, Marois JJ, Jaffee BA (1995) Intraguild predation among biological control agents: theory and evidence. Biol Control 5:303–335

Roy H, Rowland F, Brown P, Ware R, Majerus M (2005) Ecology of the harlequin ladybird—a new invasive species. Brit Wildlife 16:403–407

Rutledge CE, O'Neil RJ, Fox TB, Landis DA (2004) Soybean aphid predators and their use in integrated pest management. Ann Entomol Soc Am 97:240–248

Saini ED (2004) Presencia de *Harmonia axyridis* (Pallas) (Coleoptera: Coccinellidae) en la provincia de Buenos Aires. Aspectos biológicos y morfológicos. RIA 33:151–160

Sakurai H, Kumada Y, Takeda S (1993) Seasonal prevalence and hibernating-diapause behavior in the lady beetle, *Harmonia axyridis*. Res Bull Fac Agric Gifu Univ 58:51–55

Sala C, Mestres M, Martí MP, Busto O, Guasch J (2002) Headspace solid-phase microextraction analysis of 3-alkyl-2-methoxypyrazines in wines. J Chromatogr A 953:1–6

Sato S, Yasuda H, Evans EW (2005) Dropping behaviour of larvae of aphidophagous ladybirds and its effects on incidence of intraguild predation: interactions between the intraguild prey, *Adalia bipunctata* (L.) and *Coccinella septempunctata* (L.), and the intraguild predator, *Harmonia axyridis* Pallas. Ecol Entomol 30:220–224

Sebolt DC, Landis DA (2004) Arthropod predators of *Galerucella calmariensis* L. (Coleoptera: Chrysomelidae): an assessment of biotic interference. Environ Entomol 33:356–361

Shapiro-Ilan DI, Cottrell TE (2005) Susceptibility of lady beetles (Coleoptera: Coccinellidae) to entomopathogenic nematodes. J Invert Path 89:150–156

Simberloff D, Stiling P (1996) How risky is biological control. Ecology 77:1965–1974

Smith BC (1960) A technique for rearing Coccinellid beetles on dry foods, and influence of various pollens on the development of *Coleomegilla maculate lengi* Timb. (Coleoptera: Coccinellidae). Can J Zool 38:1047–1049

Snyder WE, Clevenger GM, Eigenbrode SD (2004) Intraguild predation and successful invasion by introduced ladybird beetles. Oecologia 140:559–565

Tedders WL, Schaefer PW (1994) Release and establishment of *Harmonia axyridis* (Coleoptera: Coccinellidae) in the southeastern United States. Entomol News 105:228–243

Wallace MS, Hain FP (2000) Field surveys and evaluation of native and established predators of the hemlock wooly adelgid (Homoptera: Adelgidae) in the Southeastern United States. Environ Entomol 29:638–644

Yarbrough JA, Armstrong JL, Blumberg MZ, Phillips AE, McGahee E, Dolen WK (1999) Allergic rhinoconjunctivitis caused by *Harmonia axyridis* (Asian lady beetle, Japanese lady beetle, or lady bug). J Allergy Clin Immunol 104:704–709

Yasuda H, Evans EW, Kajita Y, Urakawa K, Takizawa T (2004) Asymmetrical larval interactions between introduced and indigenous ladybirds in North America. Oecologia 141:722–731

Yasumatsu K, Watanabe C (1964) A tentative catalogue of insect natural enemies of injurious insects in Japan—part 1. Parasite–predator host catalogue. Entomological Laboratory, Faculty of Agriculture Kyushu University, Fukuoka

Harmonia axyridis: an environmental risk assessment for Northwest Europe

Joop C. van Lenteren · Antoon J. M. Loomans · Dirk Babendreier · Franz Bigler

Abstract In this paper, we summarize the international situation with respect to environmental risk assessment for biological control agents. Next, we apply a recently designed, comprehensive risk evaluation method consisting of a stepwise procedure to evaluate the environmental risks of *Harmonia axyridis* in Northwest Europe. This resulted in the very clear conclusion that *H. axyridis* is a potentially risky species for Northwest Europe, because it is able to establish, it has a very wide host range including species from other insect orders and even beyond the class of Insecta, it may feed on plant materials, it can cover large distances (>50 km per year), it does move into non-target areas, it may attack many non-target species including beneficial insects and insects of conservation concern, its activities have resulted in the reduction of populations of native predators in North America, it is known as a nuisance in North America and recently also in Northwest Europe, and it may develop as a pest of fruit in North America. Considering the *H. axyridis* case, current knowledge would lead to the conclusion that, although the predator is capable to effectively control several pest species, its risks are manifold and it should, thus, not have been released in Northwest Europe. At the time of the first releases in Nortwest Europe in 1995, the available scientific literature made clear that *H. axyridis* is a large sized polyphagous predator and has a great reproductive capacity in comparison with other ladybird beetles, and that there was a need to study non-target effects because of its

J. C. van Lenteren (✉)
Laboratory of Entomology, Wageningen University, PO Box 8031, Wageningen 6700 EH, The Netherlands
e-mail: Joop.vanLenteren@wur.nl

A. J. M. Loomans
Section Entomology, Plant Protection Service, PO Box 9102, Wageningen 6700 HC, The Netherlands
e-mail: a.j.m.loomans@minlnv.nl

D. Babendreier · F. Bigler
Agroscope Reckenholz-Tänikon Research Station ART, Reckenholzstrasse 191, Zurich 8046, Switzerland
e-mail: dirk.babendreier@art.admin.ch

F. Bigler
e-mail: franz.bigler@art.admin.ch

Biological Control to Invasion: Ladybird Harmonia axyridis. H.E. Roy, E. Wajnberg (eds.),
doi: 10.1007/978-1-4020-6939-0_4,

polyphagous behaviour. In retrospect, this information should have been sufficient to reject import and release of this species, but it was apparently ignored by those who considered release of this predator in Northwest Europe. The case of *Harmonia* releases in Northwest Europe underlines that there is an urgent need for harmonized, world-wide regulation of biological control agents, including an information system on risky natural enemy species.

Keywords *Harmonia axyridis* · Environmental risk assessment · Host range · Dispersal · Establishment · Non-target effects · Quick scan risk evaluation · Comprehensive risk evaluation

Introduction

In the past 100 years many exotic natural enemies have been imported, mass-reared and released as biological control agents for pest control (Albajes et al. 1999; van Lenteren 2000, 2003; Lynch et al. 2000; USDA 2001; Mason and Huber 2002; Copping 2004). Although the majority of these releases have not resulted in unwanted side effects, some serious cases of non-target hazards by exotic biological control agents against insects and weeds have been recently reported (e.g. Boettner et al. 2000; Follett and Duan 2000; Wajnberg et al. 2000; Louda et al. 2003; van Lenteren et al. 2006a). Due to the current popularity of biological control, new Invertebrate Biological Control Agents (IBCAs) will become available. To reduce the chance of releasing exotic natural enemies that might pose a risk for the environment, guidelines are being developed to assist in environmental risk assessment.

Various organizations have developed standards, including guidelines for the export, import, shipment, evaluation and release of biological control agents (e.g. EPPO 2002; IPPC 2005). Environmental effects of biological control agents form a central element of these guidelines and a growing number of countries already apply risk assessment procedures prior to the import and release of a new natural enemy. Earlier, we collected, studied and summarized procedures to assess natural enemies currently used by about 25 countries and codes of conduct or guidelines produced by various organizations (van Lenteren and Loomans 2006). Within an EU funded project (van Lenteren et al. 2003) an OECD working group (Anonymous 2004) and an IOBC Commission (Bigler et al. 2005), guidelines have been developed to harmonize information requirements for import and release of invertebrate biological control agents. Based on all this information, we designed a new comprehensive method. Subsequently, we also developed a quick scan to be used for natural enemies that are already in use (van Lenteren and Loomans 2006). In this way, we hope to provide biological control experts and risk assessors with the tools for a proper and uniform evaluation of the information provided in the application. In this paper, we summarize the development of risk assessment procedures for natural enemies, we then describe a stepwise risk assessment procedure, and we will apply a quick scan and a comprehensive method to evaluate the environmental risks of *Harmonia axyridis* in Northwest Europe.

Environmental risk assessment of natural enemies

Risk assessment procedures for biological control agents are usually characterized by questions on four issues:

1. Characterization and identification of biological control agent
2. Health risks
3. Environmental risks
4. Efficacy

The kind of information needed to evaluate these issues are addressed in Anonymous (2004), van Lenteren et al. (2003) and Bigler et al. (2005), and information on the methods to be used to assess non-target effects are addressed in Babendreier et al. (2005) and Bigler et al. (2006). In this paper we will concentrate on the third issue, but also shortly address the other issues. Assessment of risks related to releases of natural enemies demands integration of many aspects of their biology, as well as information on ecological interactions. A comprehensive risk assessment comprises the following steps:

1. Identification and evaluation of potential risk of releasing a natural enemy,
2. A plan to minimize risk and mitigate unwanted effects of biological control agents (e.g. Moeed et al. 2006), and
3. A risk/benefit analysis of the proposed release of the natural enemy, together with risk/benefit analyses of current and alternative pest management methods (e.g. Bigler and Kölliker-Ott 2006).

The last step is essential, because the risk/benefit posed by the release of an exotic natural enemy might particularly be considered acceptable in comparison with the risks posed by other control methods. For definitions of terms used in this paper, we refer to Anonymous (2003) and Bigler et al. (2006).

Risk identification and calculation of risk index

Normally, for a risk assessment, one will identify and evaluate the potential negative effects, and determine the probabilities that these will materialize (e.g. Moeed et al. 2006; Bigler et al. 2006). The negative impacts of a biological control agent can be defined as any negative effect, which can be named and measured, such as direct and indirect negative effects on non-target organisms and negative effects on the environment. The risk of negative effects of the release of a biological control agent is the product of the likelihood (L) of impact and the magnitude (M) of impact. The likelihood and magnitude of five groups (ecological determinants) of risks are usually considered: establishment, dispersal, host range, direct effects, and indirect non-target effects. Next, qualitative scales for likelihood and magnitude need to be described (Table 1), after which one may quantify the scales for likelihood and magnitude (Tables 15.2 and 15.3 in van Lenteren and Loomans 2006). In an early version of an environmental risk assessment, a numerical value was added to each descriptor of likelihood and magnitude to be able to quantify risk (see van Lenteren et al. 2003). The overall risk index for each natural enemy was obtained by first multiplying the values obtained for likelihood and magnitude, followed by summing-up the resulting values obtained for establishment, dispersal, host range, direct and indirect effects. Based on an evaluation of 31 cases of natural enemy introductions into Europe, the following risk categories were proposed (van Lenteren et al. 2003):

1. Low risk category: for organisms falling in this category, a proposal of no objection against release of the agent can usually be issued;

Table 1 Qualitative scales for likelihood, magnitude and level of risk of adverse effects (after Hickson et al. 2000; van Lenteren and Loomans 2006)

Likelihood	Description
Very unlikely	Not impossible but only occurring in exceptional circumstances
Unlikely	Could occur but is not expected to occur under normal conditions
Possible	Equally likely or unlikely
Likely	Will probably occur at some time
Very likely	Is expected to occur

Magnitude	Description
Minimal	Insignificant (repairable or reversible) environmental impact
Minor	Reversible environmental impact
Moderate	Slight effect on native species
Major	Irreversible environmental effects but no species loss, remedial action available
Massive	Extensive irreversible environmental effects

Level of risk of adverse effect

Likelihood	Magnitude				
	Minimal	Minor	Moderate	Major	Massive
Very unlikely	Insignificant	Insignificant	Low	Medium	Medium
Unlikely	Insignificant	Low	Low	Medium	High
Possible	Low	Low	Medium	Medium	High
Likely	Low	Low	Medium	High	High
Very likely	Medium	Medium	High	High	High

2. Intermediate risk category: for organisms falling in this category, the advise will be issued to come up with specific additional information before a conclusion concerning release can be drawn;
3. High risk category: for organisms falling in this category, generally a proposal to not to release the agent will be issued.

Low risk indices were found for many parasitoids, several predatory mites, and one predatory insect. Intermediate risk indices were found for all guilds of natural enemies: parasitoids, predatory insects, predatory mites, parasitic nematodes and entomopathogenic fungi. Entomopathogens (*Beauveria, Metarhizium* and *Steinernema*) all score intermediate because of their broad host range, but their very limited dispersal capacities strongly reduce risk. The highest risk indices were found for predatory insects (*Harmonia axyridis* Pallas, *Hippodamia convergens* Guérin-Méneville, *Podisus maculiventris* (Say), *Orius insidiosus* (Say)) and parasitoids (*Encarsia pergandiella* (Howard), *Trichogramma brassicae* Bezdenko and *Cales noacki* Howard). This was not a surprise as they would all be classified by biological control experts in the high-risk category based on what is known of their biology.

Because this was the first quantitative risk assessment developed for natural enemies, it was foreseen that the quantification system might have to be adapted based on growing experience. The main problems encountered with this risk assessment were the following:

1. Information for the likelihood and magnitude of all five areas of assessment needed to be available before an evaluation could be made. This makes the assessment in a number of cases unnecessarily costly.
2. The assessment did not identify candidate natural enemies that appear to be clearly unacceptable for import and release based on data for one group of risks early in the process. This should be improved to prevent unnecessary data collection.
3. The numerical values calculated with this assessment did not allow a very clear separation between risk categories. This may result in interpretation and decision making that can easily be manipulated.
4. The overall risk index was obtained by adding five different categories which are in fact not completely independent from each other and should not be rated equally.
5. The overall score of a certain species for a certain ecoregion might lead to establishing an absolute value and unnecessary strict administrative need for measures.

Therefore, we designed a new environmental risk assessment, which is now a stepwise procedure and includes weight factors to solve the problems mentioned above (van Lenteren et al. 2006a; van Lenteren and Loomans 2006).

Risk management

The next step of a risk assessment process is to discuss risk management, including risk mitigation and risk reduction. If an exotic biological control agent is expected to cause significant adverse effects on non-target organisms a permit for releases will not be issued. In some cases, risks may be minimized by imposing restrictions concerning for example the types of crops on which the use of the organism is or is not allowed (e.g. treatment of flowering plants with a myco-insecticide), by requesting specific application techniques (e.g. soil incorporation only for insect pathogenic nematodes), or by specifying the eco-regions were the organism is allowed for use (e.g. use of tropical natural enemies in greenhouses in temperate climates).

Risk/benefit analysis

The last step in making a justified environmental risk analysis for a new biological control agent is to conduct a risk/benefit analysis which should include a comparative performance of pest management methods. The environmental benefits of use of the proposed biological control agent should be compared to environmental effects of currently used and other alternative control methods. Then, the environmental risk analysis is used in the overall risk/benefit assessment where the data concerning characterization, health risks, environmental risks and efficacy of all the control methods for a specific pest will be compared (for details see van Lenteren et al. 2003, 2006a; Bigler and Kölliker-Ott 2006).

Stepwise risk assessment procedure

Recently, as a follow up to the first quantitative risk assessment, an environmental risk assessment method was developed consisting of a stepwise procedure which can be used for all types of invertebrate biological control agents in augmentative and classical biological control, for relevant species or biotypes (e.g. in the case of biotypes that diapause or not, or biotypes with and without wings), whether they are native, established exotics or not yet established exotics (Table 2, summarized in Fig. 1; van Lenteren and Loomans 2006). Native species are included in the evaluation procedure as well, because in cases where natural enemies are released in very large numbers for immediate control of the target pest, like innundative biological control, direct dispersal (overflow, drift) from the release area into the surrounding environment is of main concern for direct non-target effects, irrespective whether the natural enemy species is exotic or not. Contrary to the first quantitative risk assessment described in the previous section, here the decision to advise release or not is taken after each relevant step in the process, thus preventing unnecessary research and resulting in early elimination of clearly risky natural enemies. Definitions for specific terms used in the evaluation process are given in Table 3.

At *step 1*, exotic and native natural enemies are distinguished. For native natural enemies only one more step (6) in the procedure needs to be followed. Dispersal (step 5) of native agents may be an important issue to be considered in order to address step 6 accordingly. For example, direct and indirect effects of a polyphagous biological control agent may be limited because of very limited dispersal. However, because experimental procedures to establish the dispersal potential of natural enemies might be quite lengthy, this is not included here as a standard procedure for native natural enemies. For exotic

Table 2 Schedule for an environmental risk assessment of an invertebrate biological control agent

1.	Origin—native	GO TO 6
	Origin—exotic, either absent OR present in target area	GO TO 2
2.	Augmentative Biological Control (ABC) programme—establishment not intended	GO TO 3
	Classical Biological Control (CBC) programme—establishment intended	GO TO 4
3.	Establishment unlikely (likelihood L = 1–2) no weight factor included	GO TO 6
	Establishment possible to very likely (L = 3–5), apply magnitude (M) as a weight factor	
	if risk threshold not crossed (ERI = less than 12)	GO TO 4
	if risk threshold crossed (ERI = 12 or more) (upon request of applicant, GO TO 4)	NO release
4.	If monophagous OR if oligophagous/polyphagous AND only related AND no valued non-targets attacked	Release
	If oligophagous/polyphagous AND related and unrelated non-targets attacked AND/OR valued non-targets attacked (upon request of applicant, GO TO 5)	No release
5.	Dispersal local (L = 1–2)	GO TO 6
	Dispersal outside target area (L = 3 or more) AND extensive (M 2 or more) apply magnitude (M) as a weight factor	
	if risk threshold is not crossed (ERI = 5 or less)	GO TO 6
	if risk threshold is crossed (ERI = 6 or more)	NO release
6.	Direct and indirect effects inside dispersal area of natural enemy unlikely (L = 1–2) AND at most transient and limited (M = 1–2)	Release
	Direct and indirect effects inside 'dispersal area' likely (L = 3–5) OR permanent (M = 3–5)	NO release

The determinants of the Environmental Risk Index (ERI = Likelihood (L) × Magnitude (M)) should be calculated per step as indicated by van Lenteren et al. (2003) and where appropriate with weight factors as given in Fig. 2 (after van Lenteren and Loomans 2006)

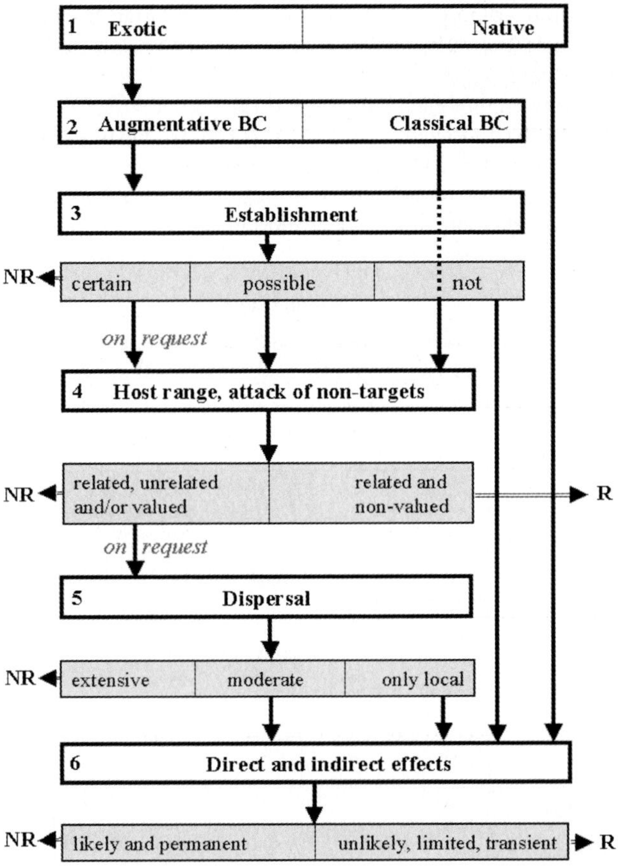

Fig. 1 Simplified scheme of an environmental risk assessment of an invertebrate biological control agent. R, NR: release, no release is recommended respectively (after van Lenteren and Loomans 2006)

Table 3 Definitions of terms used in environmental risk assessment

Term	Definition
Exotic	Non-indigenous to the country of release
Local	Restricted to the vicinity (<100 m) of the target area (establishment, dispersal)
Transient	Restricted to only the season of release (establishment, direct and indirect effects)
Permanent	Effect expected to occur during many seasons/years
Monophagous	No non-target species attacked (likelihood = 1)
Oligophagous	1–10 non-target species attacked (likelihood = 2 or 3)
Polyphagous	>10 species attacked (likelihood = 4 or 5)
Related	Within same genus

natural enemies, whether already present or absent in the target area, more steps need to be followed.

At *step 2*, natural enemies that are aimed for augmentative biological control (ABC) programmes where establishment of the organism in the area of release is not intended, are separated from natural enemies aimed for classical biological control (CBC) where establishment is the aim. For ABC natural enemies one then needs to demonstrate that they cannot establish in step 3.

If the natural enemy cannot establish (*step 3*, Likelihood = 1–2), one more step of the procedure (6) needs to be followed. However, if it can establish, the Environmental Risk Index (ERI = Likelihood (L) × Magnitude (M)) should be calculated for establishment (Fig. 2a). If a risk threshold is crossed (L = 3–5 AND M = 3–5, Fig. 2a), the natural enemy should not be released, and is thus eliminated early in the evaluation process.

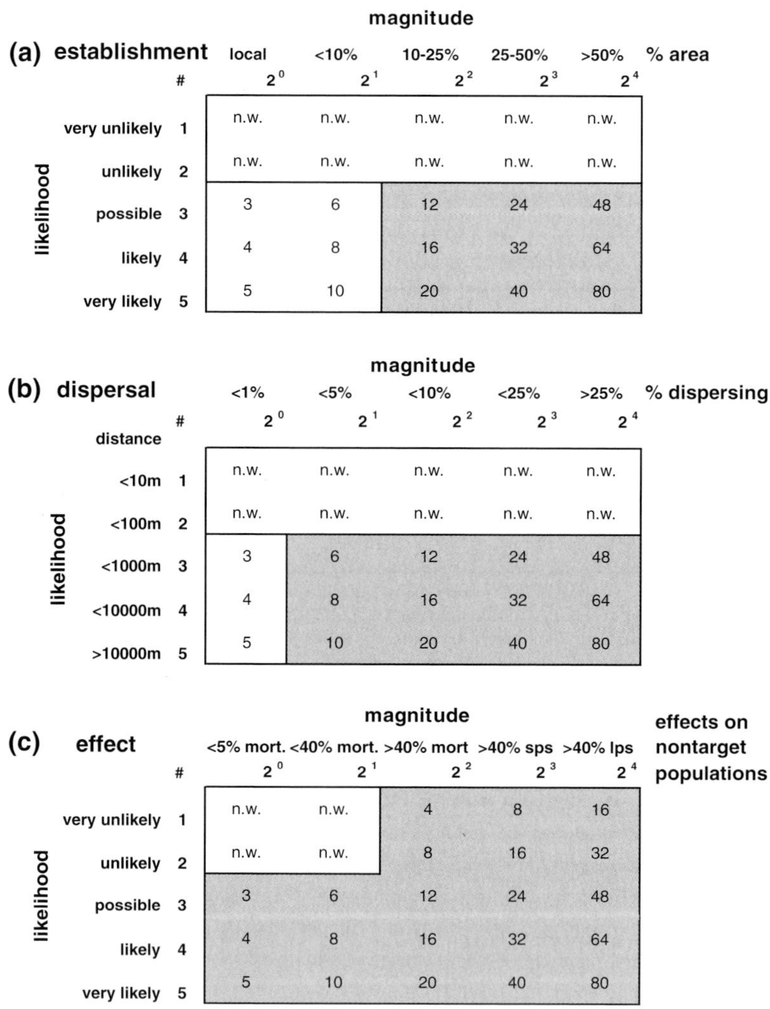

Fig. 2 Ecological Risk Index matrix to determine the level of risk of adverse effects of an IBCA for three ecological determinants: establishment (top), dispersal (middle) and direct and indirect effects (bottom). Ecological Risk Indices calculated as Likelihood (L) (vertical) × Magnitude (M) (horizontal) with their respective calculation factors: 1–5 for likelihood, 2^x as a weight factor for magnitude; n.w. = no weight factor included, mort. = mortality, sps = short term population suppression, lps = long term population suppression (see van Lenteren and Loomans 2006 for descriptions of determinants). White = below threshold, grey = above threshold (after van Lenteren and Loomans 2006)

However, if the applicant desires, he can provide data from studies on host range (step 4), dispersal (step 5) and direct/indirect non-target effects (step 6) and ask for the decision to be reconsidered. If the risk threshold is not crossed, the same procedure needs to be followed as for CBC natural enemies in step 4.

At *step 4*, the host range issue (see van Lenteren et al. 2006b) is addressed. If the ABC or CBC agent is either monophagous, or oligophagous/polyphagous and attacks only related AND no valued non-targets, i.e. species not of conservation concern, it should be considered for release. On the other hand, if the agent is oligophagous/polyphagous and does attack related and unrelated non-targets AND/OR valued non-targets, the agent should not be considered for release. However, if the applicant desires, he can provide data from studies on dispersal (step 5) and direct/indirect non-target effects and ask for the decision to be reconsidered. In that case, continue with step 5. On request, dispersal can be considered relevant for risk assessment of augmentative releases (see Mills et al. 2006).

At *step 5*, questions about dispersal of ABC and CBC (where appropriate and on request) agents are addressed. If dispersal is local and mainly in the area of release (L = 1 or 2, see Fig. 15.2b in van Lenteren and Loomans 2006), the procedure can be continued at step 6. But if dispersal is outside the target area (L = 3 or more) AND is extensive (M 2 or more) and thus the environmental risk index (ERI) crosses the value of 6 (Fig. 2b), the agent should not be released. If the ERI is 5 or less, the procedure can be continued at step 6.

At *step 6*, issues related to direct and indirect non-target effects are addressed as releases of exotic agents may negatively affect the abundance of native non-target species or other natural enemies that exploit the same resource (see Messing et al. 2006). If direct and indirect effects inside the 'dispersal area' are unlikely (L = 1–2) AND at most transient and limited (M = 1–2), the agent can be released. However, if direct and indirect effects inside the 'dispersal area' are likely (L = 3–5) OR permanent (M = 3–5), the agent should not be released (Fig. 2c).

To calculate risk levels for establishment, dispersal and direct/indirect non-target effects, the criteria are applied as given in van Lenteren et al. (2003), but weight factors are added, and the resulting values can be obtained from Fig. 2. If the ERI is below the risk threshold, the value will be in a white box (=continue procedure/release recommended). When the ERI is above the threshold, the value will be in a grey box (=discontinue procedure/no release recommended). Although threshold values as indicated in Fig. 2 are currently still largely based on expert judgement, these values need justification and fine-tuning. Here, accuracy and stringency are likely to increase as more data become available through experimental research. The final part of this new risk assessment, i.e. the risk management and the risk/benefit analysis, is the same as described in the previous section.

The stepwise risk assessment procedure has successfully been applied to the 150 species of natural enemies that are currently commercially available in Northwest Europe (producers information on the web; producers price lists; Loomans 2004). This includes the 92 natural enemies in the EPPO list of commercially available agents (EPPO 2002). The following conclusions could be drawn after this exercise (van Lenteren and Loomans 2006):

1. All native species that were evaluated are considered safe for release.
2. Exotic species intended for use in augmentative biological control that are likely to establish and cross the risk threshold are detected very early in the evaluation process, and will be excluded from release without the need to study host range, dispersal and direct/indirect non-target effects.

3. Exotic species that are monophagous, or oligophagous/polyphagous with attack of only related and no attack of valued non-targets are also detected early in the evaluation without the need to study dispersal and direct/indirect non-target effects; they can be released
4. Exotic species that are oligophagous/polyphagous and attack related and unrelated non-targets and/or valued non-targets will be excluded from release without the need to study dispersal and direct/indirect non-target effects.

Some exotic IBCAs that are not on the EPPO list, but are actually released commercially in Europe (e.g. *H. axyridis*, *H. convergens* and *O. insidiosus*), had a high ecological risk index in our previous assessment (see van Lenteren et al. 2003), indicating a high potential risk. When we evaluate these exotic IBCAs for release using the stepwise assessment procedure, they are considered unsuitable for release at step 3 or 4. On the other hand a species such as *T. brassicae*, also with a high risk index in our previous assessment (see van Lenteren et al. 2003) is not eliminated early in the stepwise procedure and can be released (establishment possible, polyphagous, but dispersal is local and direct and indirect effects within dispersal area unlikely) (see Babendreier et al. 2003; Kuske et al. 2003; Mills et al. 2006). The early elimination of obviously risky species, and the acceptance of other species, that scored – erroneously – a high index in the previous assessment by van Lenteren et al. (2003), clearly show improvements of the stepwise assessment proposed in van Lenteren and Loomans (2006).

Risk identification and risk indices for *Harmonia axyridis*

In this section, we present risk identification and the calculation of risk indices for *H. axyridis* prepared by two quantitative assessment procedures summarized above, and we will also attempt to prepare a risk assessment based on data available by the mid 1990s when *Harmonia* was not yet considered a problem.

Risk identification and risk index for *H. axyridis* based on the van Lenteren et al. (2003) approach

We will use the qualitative scales for likelihood and magnitude presented in Table 1 as a basis. This table was used by van Lenteren et al. (2003) to develop lists of descriptors as a first step towards quantification of risk; these lists are summarized in Table 4. The next step was to give a numerical value to each criterion. For likelihood, very unlikely was given a 1, unlikely a 2, etc.; for magnitude, minimal received a 1, minor a 2, etc. The overall risk index for each natural enemy is obtained by multiplying the figures for likelihood and magnitude, and then by adding the resulting figures obtained for dispersal, establishment, host specificity, direct and indirect effects. The data for *H. axyridis* are summarized in Table 5. van Lenteren et al. (2003) estimated the risks for Italy, but these estimates are also valid for Northwest Europe. Of a possible maximum of 125, the risk index for *H. axyridis* scored 101, and was the second highest value determined for 31 cases presented in the van Lenteren et al. (2003) paper.

Table 4 Descriptions of likelihood and magnitude for establishment, dispersal, host range, direct and indirect effects (after van Lenteren et al. 2003)

Likelihood	Establishment[a] in non-target habitat	Dispersal[b] potential	Host range[c]	Direct and Indirect effects
Very unlikely	Very unlikely	<10 m	0 species	Very unlikely
Unlikely	Unlikely	<100 m	1–3 species	Unlikely
Possible	Possible	<1,000 m	4–10 species	Possible
Likely	Likely	<10,000 m	11–30 species	Likely
Very likely	Very likely	>10,000 m	>30 species	Very likely
Magnitude	**Establishment[d] in non-target habitat**	**Dispersal[e] potential**	**Host range[f]**	**Direct[g] and Indirect[h] effects**
Minimal	Local (transient in time and space)	<1%	Species	<5% mortality
Minor	<10%	<5%	Genus	<40% mortality
Moderate	10–25%	<10%	Family	>40% mortality and/or > 10% short term population suppression
Major	25–50%	<25%	Order	>40% short term population suppression, or > 10% permanent population suppression
Massive	>50%	>25%	Phylum	>40% long term population suppression or local extinction

[a] The propensity to overcome adverse conditions and availability of refuges

[b] Distance moved per release

[c] The propensity to realise its ecological host range in the release area

[d] Percentage of potential non target habitat where biological control agent may establish

[e] Percentage of released biological control agent dispersing from target release area

[f] Taxon range that biological control agent attacks

[g] Direct effect: mortality, population suppression or local extinction of directly affected non-target organisms

[h] Indirect effect: mortality, population suppression or local extinction of one or more species of non-target species that are indirectly influenced by the released biological control agent

Risk identification, risk index and risk assessment for *H. axyridis* based on the stepwise approach

We will follow the schedule for an environmental risk assessment of an invertebrate biological control agent presented in Table 2 and summarized in Fig. 1. *Harmonia axyridis* is an exotic natural enemy (question at step 1), thus we go to step 2. In Northwest Europe, the use of *H. axyridis* was proposed for augmentative releases without the goal to have the biological control agent established, so we go to step 3. The species can establish, which

Table 5 Calculation of risk index for *Harmonia axyridis* made in 2003 with the van Lenteren et al. (2003) approach

	Establishment	Dispersal	Host range	Direct effects	Indirect effects	Risk indix (sum L × M's)	References
Likelihood (L)	5	4	5	5	5		Burgio et al. 2002
Magnitude (M)	4	4	5	4	4		Tedders and Schaefer 1994
L × M	20	16	25	20	20	101	

means that we have to estimate the Environmental Risk Index (ERI = likelihood × magnitude) for establishment. Based on literature data summarized in Koch et al. (2006) and our own field experience we estimate the likelihood of establishment as "very likely" (the best proof is, of course, that *H. axyridis* has already established in a dozen Northwest European countries (Brown et al. 2007a)). The estimate for magnitude of establishment is that *H. axyridis* will establish in 25–50% of the potential non-target habitats. This estimate is based on our own field experience since 2003 indicating that *H. axyridis* might have spread to and established since in up to 50% of potential non-target habitats. When applying these two estimates to the section on establishment in Fig. 2 we come to an ERI of 40, which would mean that the risk threshold is crossed (the value is in the grey marked section of the figure) and that it should be advised not to release this natural enemy. However, if the producer of natural enemies desires, he can provide data from studies on the host range of the organism (step 4). Let us suppose that he did provide these data. We collected host range data from the literature (e.g. Koch et al. 2006; Loomans, unpublished) and found that *H. axyridis* may feed on many aphid species, as well as on numerous other insect prey (e.g. Hemiptera, Psyllidae, Coccoidea, Chrysomelidae, Curculionidae, Coccinellidae and Lepidoptera), spider mites (Tetranychidae), dead insects and also on plant material (e.g. damaged fruit, pollen and nectar). It seems safe to conclude that the organism is highly polyphagous, attacks related and unrelated non-target species and attacks valued non-target species. Thus, the conclusion would once more be that the species should not be released. However, the producer of the biological control agent might be willing to provide data on dispersal (step 5) and direct/indirect non-target effects (step 6), and ask that the decision to not release be reconsidered . At step 5 questions about dispersal of the species are addressed. Direct and indirect non-target effects might be limited if the species does not leave the area of release. Dispersal data for *H. axyridis* show, however, that the species may cover large distances (up to 442 km per year in North America, McCorquodale 1998; 50–100 km in Northwest Europe, Loomans 2007; Brown et al. 2007b) and does move into non-target areas, including nature reserves. This results in an estimate for likelihood of dispersal of more than 10 km per release and for magnitude of dispersal of more than 25% of the released biological control agent from target release area. The ERI of 80 for dispersal of this species crosses the threshold (Fig. 2) and for the third time the conclusion would be that the species should not be released. But let us continue with the procedure and also try to answer the questions at step 6. The literature provides a number of cases of negative direct and indirect effects in the dispersal area of the species for *H. axyridis*. The species attacks many non-target organisms including beneficial insects and insects of conservation concern (Ware and Majerus 2007), has resulted in the reduction of populations of native predators in North America, is known as a nuisance in North America and recently also in Northwest Europe, and is a pest of fruit production in North America (e.g. Koch et al. 2006 and references therein). The estimate for likelihood of effects on non-target populations is "very likely". The most difficult aspect of this whole procedure is to make an estimate for magnitude of non-target effects. Based on all current knowledge, we estimate that the magnitude is between less than 40% mortality of one or more non-target organisms and more than 40% long term population suppression of one or more non-target organisms. Even the lowest estimate results in an ERI of 10 and, thus, the risk threshold is crossed for the fourth time. It is obvious from the information that we have now, that application of this stepwise approach would have led to the very clear conclusion that *H. axyridis* is a potentially risky species for Northwest Europe.

The next step in the risk assessment procedure is to discuss risk management, including risk mitigation and risk reduction. Based on the biology of *H. axyridis*, it can be concluded that there are no easy ways to mitigate or reduce risk (Kenis et al. 2007). It has been suggested to release flightless strains of this predator in order to reduce risk of dispersal into non-target ecosystems (Ferran et al. 1998). Although the flightless strain could result in a significant reduction in dispersal and spread, it does not necessarily reduce its non-target impact. However, the potential consequences of such releases are not yet fully evaluated. Moreover, there are other, native coccinellid species that have a similar capacity for control of aphids.

The last step in making a justified environmental risk analysis for a new biological control agent is to conduct a risk/benefit analysis which should include a comparative performance of pest management methods. In the *H. axyridis* case, current knowledge would lead to the conclusion that, although the predator is capable to effectively control several pest species (a strong benefit; e.g. Landis et al. 2004), its risks are manifold (reduction in population size of native ladybird beetles, attack of many of non-target species, frugivorous behaviour, large aggregations are nuisance to humans, allergic reactions in and biting of humans; e.g. Koch et al. 2006), and it should, thus, not have been released in Nortwest Europe.

Environmental risk assessment for *H. axyridis* based on pre-1995 data

Harmonia axyridis is of Asian origin, is a predator of aphids and other soft-bodied arthropods, is frequently associated with trees in natural and agricultural settings when prey is available, but also occurs in herbaceous habitats (Koch et al. 2006). The predator has been used in biological control programmes since 1916 in the USA, when the first intentional releases were made in California, with later and more frequent releases in the USA and Canada during the 1970s and 1980s (Gordon 1985). The ladybird beetle has also been introduced intentionally in Europe, Africa, Central and South America (see Koch et al. 2006 and references therein). Established populations were first detected in North America in 1988 (Chapin and Brou 1991).

INRA (France) imported *H. axyridis* in 1982 and kept the species in quarantine until 1992. The first intentional, experimental releases were made from 1990–1997 in Southeast France followed by commercial releases in France in 1994, and in 1995 in Northwest Europe. Mass production of *H. axyridis* was started in 1992 by the French company Biotop (Kabiri 2006). The Belgian company Biobest and the Dutch company Koppert Biosystems started to sell *H. axyridis* (produced by Biotop) in 1995 and 1996, respectively (Kabiri 2006). Interestingly, this predator, while released in Italy and southern France, does not seem to have established and created problems there in spite of predictions on its establishment potential in that region in a later period (Iperti and Bertrand 2001; Bazzocchi et al. 2004). In Greece (Katsoyannos et al. 1997) the establishment potential was considered low in citrus orchards, but no observations where made at that time outside the release area. The first European record of a feral *Harmonia* population originates from 1999, in the town Frankfurt-Niederrad (Germany), where *H. axyridis* releases were made nearby in previous years for aphid control in roses (H. Bathon, personal communication, July 2007), and subsequently many records were made across West European countries (Brown et al. 2007a).

The earliest paper on potential negative side effects of *H. axyridis* dates from 1995 and is from North America (Coderre et al. 1995). Were the biological data about this predator at that time such that one could have concluded it was a highly risky species? To answer this question, we have searched the literature for information about the biology of *H. axyridis* and negative side effects. In the most recent review of the predator by Koch et al. (2006) we found quite a number of papers on *H. axyridis* published before 1995, but most of these concerned taxonomy, distribution patterns and use in biological control. An internet search using Google Scholar with the keywords *Harmonia axyridis* in the title of the paper and for the period before 1995 revealed more than 120 papers. When we combined the species name with risk(s), or nontarget we did not find any paper. As a control, we used risk(s) or nontarget in combination with biological control, and we always found several papers. A check of the more than 120 papers found with *H. axyridis* in the title and published before 1995 showed six papers that might contain information about potential risks. This literature search, together with the pre-1995 papers listed in Koch et al. (2006) and contact with some of the authors of papers resulted in the following information.

1. In a number of papers it is mentioned that *H. axyridis* is a large sized polyphagous predator and has a great reproductive capacity in comparison with other ladybird beetles
2. In some papers, not only the polyphagous habit is mentioned, but also prey species are listed indicating a wide prey range (Vasil'ev 1963; Hodek 1973; Iablokoff-Khnzorian 1982; Schanderl et al. 1985; McClure 1987)
3. In one paper, the need to explicitly study non-target effects because of the polyphagous habit of *H. axyridis* is mentioned (Coderre et al. 1995).

Based on this, we may conclude that the potential risk (climate matching and polyphagy, including attack of benficial insects) of *H. axyridis* was clear before the first releases were made in Northwest Europe. In retrospect, this information should have been sufficient to reject import and release of this species, but it was apparently ignored by those who considered release of this predator in Northwest Europe.

Interestingly, in 1996, a first application for commercial release of *H. axyridis* in Switzerland was submitted to the Swiss Pesticide Regulatory Authority, which is responsible for licensing pesticides and biological control agents. Based on information in the literature on origin, potential for establishment, host range and potential of spread, a license for commercial use was refused in January 1997. Other applications for commercial release in Switzerland submitted in later years were not approved for the same reasons. Despite the fact that the licence for commercialising *H. axyridis* was rejected by the Swiss authority, the insect had been commercialised and released in Switzerland before and during the submission process (e.g. Anonymous 1996). Thus, rejecting an application for release is not always sufficient to prevent release. In addition, *H. axyridis* has crossed the Swiss border and was first found there in 2006 (Eschen et al. 2007). *H. axyridis* has now spread over many regions in Switzerland and has established. When the EU funded project to develop the quantitative risk assessment method (see Section Environmental risk assessment of natural enemies) was initiated, we discussed risks of releasing exotic natural enemies with two specialists (Dr. B. Aukema and Ir. H. Stigter) of the Dutch Plant Protection Service. These specialists warned us for releases of *H. axyridis* and *H. convergens* based on their wide host range (Loomans, personal communication). These two cases show (1) the importance and urgent need of harmonized regulation of biological control agents in Europe, (2) the need of a generally accessible system which provides information on natural enemies that are considered safe or not safe for release in certain ecoregions of Europe, and (3) the requirement of a group of experts which can advise European bodies (e.g. EC, EPPO, EFSA) and national authorities about the risks of import and release of exotic natural enemies.

Discussion

In this paper environmental risk evaluation methods were summarized and applied to *H. axyridis*. Proposed methods of risk assessment gradually have shifted, coming from a descriptive, more qualitative framework, largely based on expert judgment in general (e.g. Hickson et al. 2000), via an overall qualitative and quantitative method (van Lenteren et al. 2003) to a stepwise evaluation procedure, using quantitative information when needed and where possible (Bigler et al. 2006; van Lenteren and Loomans 2006). This not only allows better insight into relevant ecological factors, but also constitutes a more objective approach for evaluating the risks of biological control agents. Methods to determine establishment, dispersal, host range, direct and indirect effects on non-target organisms are discussed in Babendreier et al. (2005) and Bigler et al. (2006). When we apply the most recent, stepwise risk assessment procedure to *H. axyridis*, we have to conclude that, based on current knowledge, (1) this predator is a potentially high risk species for Northwest Europe, (2) there are no easy and reliable ways to mitigate or reduce risk of releases of this predator, and (3) a risk/benefit analysis which includes a comparative performance of pest management methods would result in the advice not to release this predator. However, the predator has already been released, is established and is spreading rapidly (Brown et al. 2007a, b). The simple fact that regulation concerning import and release of exotic natural enemies does not exist in some countries and is not well organized in other countries has resulted in this problematic situation (Bigler et al. 2005). As a result, the topic of implementation of a registration procedure for natural enemies is currently hotly debated by the biological control industry, scientists and regulators (Blum et al. 2003; GreatRex 2003; Hokkanen 2003; van Lenteren et al. 2003, 2006a; Anonymous 2004; Bigler et al. 2005, 2006).

The biological control industry foresees lengthy, cumbersome procedures leading to high costs, and, thus, in some cases, the impossibility to marketing a potentially useful natural enemy because of too high costs. Such costs will strongly depend on the biological and ecological characteristics of a natural enemy. When dealing with a natural enemy that has a very narrow host range, testing and the preparation of a dossier can be limited to about six person months. However, preparation of a dossier for an exotic polyphagous natural enemy that is able to establish, such as *H. axyridis*, could take up to several years, particularly if experiments on dispersal and direct/indirect ecological effects are needed. We estimate that a comprehensive dossier can be appraised in up to six person weeks by governmental agencies. Based on the experience with classical biological control agents reviewed by peers, evaluations, however, take at least 6 months to complete (Sheppard et al. 2003).

Regulators within ministries of environment and agriculture want to prevent unnecessary and risky releases of exotic organisms, and their concerns have been triggered by the *Harmonia* case. Current activities in the field of regulation will hopefully result in a light and harmonized registration procedure that is not prohibitive for the biological control industry and will result in the pre-selection of safe natural enemies (see e.g. Bigler et al. 2005). A proposed quick scan method for organisms already in use (van Lenteren and Loomans 2006) should be considered as a kick-start from a situation with no regulations for the use of biological control agents, to one where import and release are regulated to ensure safe use. This quick scan method applied for Northwest Europe resulted in continuation of release of a large number of exotic species. Use of such a quick scan method results in the continuation of ongoing successful and safe biological control programmes, without the risk of returning to chemical control programmes. We estimate that preparation of a dossier for a quick scan will take two person weeks, and appraisal one to six person days per biological control agent. The end result of such a quick scan method applied in various countries may result in lists of species that can be used in certain, specified regions (ecoregions) of the world. These species will be exempted from a comprehensive environmental risk analysis. The availability of regularly updated 'white lists' might stimulate the application of biological control worldwide.

The case of *Harmonia* releases in Northwest Europe underlines once more that there is an urgent need for harmonized, Europe-wide (indeed global) regulation of biological control agents, including an information system on risky natural enemy species.

Acknowledgements Horst Bathon (Germany), Giovanni Burgio (Italy), and Eric Lucas (Canada) are thanked for providing important references and information on *H. axyridis*.

References

Albajes R, Gullino ML, van Lenteren JC, Elad Y (eds) (1999) Integrated pest and disease management in greenhouse crops. Kluwer Academic Publishers, Dordrecht

Anonymous (1996) Marienkäferlarven als Blattlaus-Bekämpfungsmittel. Andermatt Biocontrol J 3:10–11

Anonymous (2003) Glossary of terms. The second report on harmonisation of risk assessment procedures, Appendix 2 Scientific Steering Committee, European Commission, Brussels

Anonymous (2004) Guidance for information requirements for regulation of invertebrates as biological control agents. OECD series on pesticides 21. http://www.oecd.org/dataoecd/6/20/28725175.pdf Cited 4 August 2007

Babendreier D, Kuske S, Bigler F (2003) Parasitism of non-target butterflies by *Trichogramma brassicae* Bezdenko (Hymenoptera: Trichogrammatidae) under field cage and field conditions. Biol Control 26:139–145

Babendreier D, Bigler F, Kuhlmann U (2005) Methods used to assess non-target effects of invertebrate biological control agents of arthropod pests. BioControl 50:821–870

Bazzocchi GG, Lanzoni A, Accinelli G, Burgio G (2004) Overwintering, phenology and fecundity of *Harmonia axyridis* in comparison with native coccinellid species in Italy. BioControl 49:245–260

Bigler F, Kölliker-Ott UM (2006) Balancing environmental risks and benefits: a basic approach. In: Bigler F, Babendreier D, Kuhlmann U (eds) Environmental impact of invertebrates for biological control of arthropods: methods and risk assessment. CAB Int Wallingford, pp 273–286

Bigler F, Bale JS, Cock MJW, Dreyer H, GreatRex R, Kuhlmann U, Loomans AJM, van Lenteren JC (2005) Guidelines on information requirements for import and release of invertebrate biological control agents (IBCAs) in European countries. Biocontrol News Inf 26(4):115N–123N

Bigler F, Babendreier D, Kuhlmann U (eds) (2006) Environmental impact of invertebrates for biological control of arthropods: methods and risk assessment. CAB Int Wallingford

Blum B, Ehlers R, Haukeland-Salinas S, Hokkanen H, Jung K, Kuhlmann U, Menzler-Hokkanen I, Ravensberg W, Strasser H, Warrior P, Wilson M (2003) Letter to the editor. Biological control: safety and regulatory policy. BioControl 48:477–484

Boettner GH, Elkinton JS, Boettner CJ (2000) Effects of a biological control introduction on three nontarget native species of Saturniid moths. Conserv Biol 14:1798–1806

Brown PMJ, Adriaens T, Bathon H, Cuppen J, Goldarazena A, Hägg T, Kenis M, Klausnitzer BEM, Kovar I, Loomans AJ, Majerus MEN, Nedved O, Pedersen J, Rabitsch W, Roy HE, Ternois V, Zakharov IA, Roy DB (2007a) *Harmonia axyridis* in Europe: spread and distribution of a non-native coccinellid. BioControl (this issue). doi:10.1007/s10526-007-9132-y

Brown PMJ, Roy HE, Rothery P, Roy DB, Ware RL, Majerus MEN (2007b) *Harmonia axyridis* in Great Britain: analysis of the spread and distribution of a non native coccinellid. BioControl (this issue). doi: 10.1007/s10526-007-9124-y

Burgio G, Santi F, Maini S (2002) On intraguild predation and cannibalism in *Harmonia axyridis* Pallas and *Adalia bipunctata* L. (Coleoptera, Coccinellidae). Biol Control 24:110–116

Chapin JB, Brou VA (1991) *Harmonia axyridis* (Pallas), the third species of the genus to be found in the United States (Coleoptera: Coccinellidae). Proc Entomol Soc Washington 93:630–635

Coderre D, Lucas E, Gagné I (1995) The occurrence of *Harmonia axyridis* (Pallas) (Coleoptera: Coccinellidae) in Canada. Can Entom 127:609–611

Copping LG (2004) The manual of biocontrol agents. BCPC Publications Sales, 7 Omni Business Centre, Omega Park, Alton, Hampshire

EPPO (2002) List of biological control agents widely used in the EPPO region. PM6/3(2). Bulletin OEPP/EPPO Bulletin 32(3):447–461, http://www.eppo.org/standards/biocontrol/bio_list.htm Cited 4 August 2007

Eschen R, Babendreier D, Naue S, Bigler F, Kenis M (2007) Surveys for ladybirds (Coleoptera: Coccinellidae) in Switzerland and confirmation of the presence of the invasive alien ladybird species *Harmonia axyridis* (Pallas). Mitteilungen der Schweizerischen Entomologischen Gesellschaft 80:7–14

Ferran A, Giuge L, Tourniaire R, Gambier J, Fournier D (1998) An artificial non-flying mutation to improve the efficiency of the ladybird *Harmonia axyridis* in biological control of aphids. BioControl 43:53–64

Follett PA, Duan JJ (2000) Nontarget effects of biological control. Kluwer Academic Publishers, Dordrecht

Gordon RD (1985) The Coccinellidae (Coleoptera) of America North of Mexico. J New York Entomol Soc 93:1–912

GreatRex R (2003) Comments on the OECD proposal for 'guidance for registration requirements for invertebrates as biological control agents' http://www.ibma.ch/pdf/comments_on_the_oecd_proposal.pdf Cited 4 August 2007

Hickson R, Moeed A, Hannah D (2000) HSNO, ERMA and risk management. New Zealand Sci Rev 57:72–77

Hodek I (1973) Biology of Coccinellidae. Academia Publishing House, Prague

Hokkanen HMT (2003) Demonstrating the safety of biocontrol. BioControl 48:1

Iablokoff-Khnzorian SM (1982) Les coccinelles; Coléoptères—Coccinellidae. Société Nouvelle des Éditions Boubée, Paris

Iperti G, Bertand E (2001) Hibernation of *Harmonia axyridis* (Coleoptera: Coccinellidae) in South-Eastern France. Acta Societas Zoologica Bohemica 65:207–210

IPPC (2005) Revision of ISPM No. 3. Guidelines for the export, shipment, import and release of biological control agents and beneficial organisms. International Plant Protection Convention. Draft for country consultation 2004, https://www.ippc.int/IPP/En/default.jsp Cited 4 August 2007

Kabiri F (2006) Information about production of *Harmonia axyridis* in Europe. http://www.rebeca-net.de/downloads/Risk%20assessment%20macrobials%20Kabiri.pdf Cited 4 August 2007

Katsoyannos P, Kontodimas DC, Stathas GJ, Tsartsalis CT (1997) Establishment of *Harmonia axyridis* on citrus and some data on its phenology in Greece. Phytoparasitica 25:183–191

Kenis M, Roy HE, Zindel R, Majerus MEN (2007) Current and potential management strategies against *Harmonia axyridis*. BioControl (this issue). doi:10.1007/S10526-007-9136-7

Koch RL, Venette RC, Hutchinson W (2006) Invasions by *Harmonia axyridis* (Pallas)(Coleoptera: Coccinellidae) in the western hemisphere: implications for South America. Neotrop Entomol 35:421–434

Kuske S, Widmer F, Edwards PJ, Turlings TCJ, Babendreier D, Bigler F (2003) Dispersal and persistence of mass released *Trichogramma brassicae* (Hymenoptera: Trichogrammatidae) in non-target habitats. Biol Control 27:181–193

Landis DA, Fox TB, Costamagna AC (2004) Impact of multicolored Asian ladybeetle as a biological control agent. Am Nat 50:153–154

Loomans AJM (2004) Biologische bestrijders en de Flora- en Faunawet: criteria voor risicoinschatting en toelating biologische bestrijders in Nederland. Gewasbescherming 35(1):33–37

Louda SM, Pemberton RW, Johnson MT, Follett PA (2003) Nontarget effects: the Achilles heel of biological control? Retrospective analyses to reduce risk associated with biocontrol introductions. Annu Rev of Entomol 48:365–396

Lynch LD, Hokkanen HMT, Babendreier D, Bigler F, Burgio G, Gao Z-H, Kuske S, Loomans A, Menzler-Hokkanen I, Thomas MB, Tommasini G, Waage J, van Lenteren JC, Zeng Q-Q (2000) Indirect effects in the biological control of arthropods with arthropods. In: Wajnberg E, Scott JC, Quimby PC (eds) Evaluating indirect ecological effects of biological control. CAB Int Wallingford, pp 99–125

Mason PG, Huber JT (2002) Biological control programmes in Canada, 1981–2000. CAB Int Wallingford

McClure MS (1987) Potential of the Asian predator, *Harmonia axyridis* (Coleoptera: Coccinellidae), to control *Matsuccoccus matsumurae* (Homoptera: Margarodidae) in the United States. Environ Entomol 16:224–230

McCorquodale DB (1998) Adventive lady beetles (Coleoptera: Coccinellidae) in eastern Nova Scotia, Canada. Entomol News 109(11):15–20

Messing R, Roitberg B, Brodeur J (2006) Measuring and predicting indirect impacts of biological control: competition, displacement and secondary interactions. In: Bigler F, Babendreier D, Kuhlmann U (eds) Environmental impact of invertebrates for biological control of arthropods: methods and risk assessment. CAB Int Wallingford, pp 64–77

Mills NJ, Babendreier D, Loomans AJM (2006) Methods for monitoring the dispersal of natural enemies from point source releases associated with augmentative biological control. In: Bigler F, Babendreier D, Kuhlmann U (eds) Environmental impact of invertebrates for biological control of arthropods: methods and risk assessment. CAB Int Wallingford, pp 114–131

Moeed A, Hickson R, Barratt BIP (2006) Principles of environmental risk assessment with emphasis on the New Zealand perspective. In: Bigler F, Babendreier D, Kuhlmann U (eds) Environmental impact of invertebrates for biological control of arthropods: methods and risk assessment. CAB Int Wallingford, pp 241–253

Schanderl H, Ferran A, Larroque MM (1985) Les besoins trophiques et thermique des larves de la coccinelle *Harmonia axyridis* Pallas. Agronomie 5:417–422

Sheppard A., Hill R, DeClerck-Floate RA, McClay A, Olckers T, Quimby PC, Zimmermann HG (2003) A global review of risk-benefit-cost analysis for the introduction of classical biological control against weeds: a crisis in the making? Biocontrol News Inf 24:91N–108N

Tedders WL, Schaefer PW (1994) Release and establishment of *Harmonia axyridis* (Coleoptera: Coccinellidae) in the southeastern United States. Entomol News 105:228–243

USDA (2001) The ROBO database. Available at http://www.ars-grin.gov/nigrp/robo.html Cited 4 August 2007

van Lenteren JC (2000) Measures of success in biological control of arthropods by augmentation of natural enemies. In: Gurr G, Wratten S (eds) Measures of success in biological control. Kluwer Academic Publishers, Dordrecht, pp 77–103

van Lenteren JC (ed) (2003) Quality control and production of biological control agents: theory and testing procedures. CAB Int Wallingford

van Lenteren JC, Loomans AJM (2006) Environmental risk assessment: methods for comprehensive evaluation and quick scan. In: Bigler F, Babendreier D, Kuhlmann U (eds) Environmental impact of invertebrates for biological control of arthropods: methods and risk assessment. CAB Int Wallingford, pp 254–272

van Lenteren JC, Babendreier D, Bigler F, Burgio G, Hokkanen HMT, Kuske S, Loomans AJM, Menzler-Hokkanen I, van Rijn PCJ, Thomas MB, Tommasini MG, Zeng Q-Q (2003) Environmental risk assessment of exotic natural enemies used in inundative biological control. BioControl 48:3–38

van Lenteren JC, Bale J, Bigler F, Hokkanen HMT, Loomans AJM (2006a) Assessing risks of releasing exotic biological control agents of arthropod pests. Annu Rev Entomol 51:609–634 + supplemental material

van Lenteren JC, Cock MJW, Hoffmeister TS, Sands DPA (2006b) Host specificity in arthropod biological control, methods for testing and interpretation of the data. In: Bigler F, Babendreier D, Kuhlmann U (eds) Environmental impact of invertebrates for biological control of arthropods: methods and risk assessment. CAB Int Wallingford, pp 38–63

Vasil'ev RA (1963) On the bionomics of *Gastrolina thoracica* Baly (Coleoptera, Chrysomelidae. Entomologiceskoe Obozrenie 42(2):264–272 (in Russian)

Wajnberg E, Scott JC, Quimby PC (2000) Evaluating indirect ecological effects of biological control. CAB Int Wallingford

Ware RL, Majerus MEN (2007) Intraguild predation of immature stages of British and Japanese coccinellids by the invasive ladybird *Harmonia axyridis*. BioControl (this issue). doi:10.1007/s10526-007-9135-8

Harmonia axyridis in Great Britain: analysis of the spread and distribution of a non-native coccinellid

Peter Michael James Brown · Helen E. Roy · Peter Rothery · David B. Roy · Remy L. Ware · Michael E. N. Majerus

Abstract *Harmonia axyridis* (Pallas) (Coleoptera: Coccinellidae) is native to Asia, and was widely introduced as a biocontrol agent of aphids and coccids in Europe and North America. In Europe *H. axyridis* is considered to be an invasive alien species. Although not known to have been deliberately introduced to Great Britain, it was first recorded there in 2004, in south-east England. *Harmonia axyridis* arrived in Great Britain by various means, primarily from mainland Europe, but also from Canada. Extensive national and local media coverage, and a dedicated website (http://www.harlequin-survey.org), facilitated public involvement in recording *H. axyridis* in Great Britain; in excess of 4,000 verified records of the species were received between 2004 and 2006. Through detailed mapping, the objective of our study was to quantify and analyse the spread of *H. axyridis* in its early stages of invasion in Great Britain. Our data shows that between 2004 and 2006, the species spread north through Great Britain at the rate of 58 km year^{-1} and west at the rate

P. M. J. Brown (✉) · H. E. Roy · P. Rothery · D. B. Roy
NERC Centre for Ecology and Hydrology – Monks Wood, Huntingdon, Cambridgeshire PE28 2LS, UK
e-mail: pmb@ceh.ac.uk

P. Rothery
e-mail: pro@wpo.nerc.ac.uk

D. B. Roy
e-mail: dbr@ceh.ac.uk

P. M. J. Brown · H. E. Roy
Department of Life Sciences, Anglia Ruskin University, East Road, Cambridge CB1 1PT, UK
e-mail: hele@ceh.ac.uk

R. L. Ware · M. E. N. Majerus
Department of Genetics, University of Cambridge, Downing Street, Cambridge CB2 3EH, UK
e-mail: r.ware@gen.cam.ac.uk

M. E. N. Majerus
e-mail: m.majerus@gen.cam.ac.uk

Biological Control to Invasion: Ladybird Harmonia axyridis. H.E. Roy, E. Wajnberg (eds.),
doi: 10.1007/978-1-4020-6939-0_5, © IOBC 2007

of 144.5 km year^{-1}. In England *H. axyridis* spread north to Yorkshire and west to Cornwall by the end of 2006, and also reached Wales. Recorded occurrence (of one or more *H. axyridis* individuals at larval, pupal and/or adult stage) in 10 km squares in Great Britain was: 2004—51; 2005—149; 2006—447. Records of juvenile *H. axyridis* extend through most of the recorded British range of the species, and we present evidence of bi-voltinism in the population in 2006.

Keywords Alien species · Biological control · Coccinellidae · Harlequin ladybird · *Harmonia axyridis* · Invasive species · Non-native species

Introduction

The recording of biological events in Great Britain has a long tradition; detailed recording of many taxa has been carried out since the nineteenth century (Harding 1990). Biological records for Britain and Ireland are integrated and distributed through the National Biodiversity Network (NBN) gateway (http://www.searchnbn.net), an interactive mapping website holding in excess of 27 million records for over 8,500 species. The biological recording infrastructure in Great Britain provides an effective tool for monitoring species undergoing rapid range expansions. Analysis of long-term datasets showed that various taxa (e.g. dragonflies, butterflies, spiders and fish), are moving northwards in Great Britain in response to climate change (Hickling et al. 2006). Invasive species are regarded as one of the most serious threats to biodiversity (Glowka et al. 1994). The monitoring of biological invasions is vital, in order to understand the population dynamics, habitat tolerance and impact on native taxa, of the invading species. In Great Britain, monitoring projects for invasive insects include the horse chestnut leafminer *Cameraria ohridella* Deschka and Dimic (Lepidoptera: Gracillariidae) (http://www.forestresearch.gov.uk/leafminer), and the rosemary beetle *Chrysolina americana* L. (Coleoptera: Chrysomelidae) (http://www.rhs.org.uk/research/projects/rosemary_beetle.asp).

The harlequin ladybird or multicolored Asian lady beetle, *Harmonia axyridis* (Pallas) (Coleoptera: Coccinellidae), is native to Asia (e.g. Dobzhansky 1933; Kuznetsov 1997) and was introduced to North America and Europe to control pest aphid and coccid species on a range of crops (e.g. Adriaens et al. 2003; Koch 2003). It established in North America in 1988 (Chapin and Brou 1991), spread very rapidly (Koch 2003), and became the dominant coccinellid in many areas (Colunga-Garcia and Gage 1998; Smith et al. 1996; Tedders and Schaefer 1994). In mainland Europe, *H. axyridis* was first marketed as a biocontrol agent in 1995 (Coutanceau 2006), established in the late 1990s and expanded its range rapidly, especially from 2002 (Brown et al. 2007). It is not known to have been deliberately introduced to Great Britain, but arrived by various means, including flight and on produce from mainland Europe and in packing cases from Canada (Majerus et al. 2006a; Roy et al. 2005). There is no doubt that *H. axyridis* is an effective aphid predator, and its presence may be welcomed by some. However, there is concern in Europe about the rapid increase in distribution and abundance of *H. axyridis*, and its impact on non-target species (Adriaens et al. 2003; Majerus et al. 2006b; Roy et al. 2006). Although most biocontrol companies have stopped selling *H. axyridis* in Europe, a non-flying variety is still available from Biotop (http://www.biotop.fr).

Harmonia axyridis has the potential to contribute to biotic homogenization (McKinney and Lockwood 1999) and to negatively impact on up to 1,000 species in Great Britain (Majerus 2007). These are primarily insects (including non-target Homoptera, alternative prey, members of aphidophagous and coccidophagous guilds) and parasites, parasitoids, pathogens and symbionts of theses species (Majerus 2007). In the USA, *H. axyridis* has been declared a potential pest of fruit production and processing, as the species sometimes feeds on the juices of ripe fruits, such as apples, pears and grapes, in autumn (Koch 2003). *Harmonia axyridis* may become seriously pestilent to the native fauna of Great Britain (Majerus et al. 2006b). This pest status may extend directly to humans, as *H. axyridis* tends to form very large overwintering aggregations of thousands of individuals, often on or in buildings (Adriaens et al. 2003; Kidd et al. 1995). Some such aggregations were observed in England in 2006 (Brown and Roy 2007).

Uniquely, the early detection of *H. axyridis* in Britain presented the opportunity to study the spread of an invasive animal from the year of its arrival (Majerus et al. 2006b). There is no evidence to suggest that *H. axyridis* was established in Great Britain before 2004 and circumstantial evidence that argues against its presence prior to that year (Majerus and Roy 2005). Through detailed mapping of adult and juvenile stages, the objective of our study was to quantify and analyse the spread of *H. axyridis* in its early stages of invasion in Great Britain. A further objective was to investigate the voltinism of *H. axyridis* in Great Britain, to determine whether there was evidence of multiple generations per year.

Material and methods

Data collection

The spread of *H. axyridis* was monitored by utilising the extensive biological recording community in Great Britain, coupled with engagement of the general public. Recent advances in technology, and high levels of public access to the internet and digital photography, enabled a web-based biological survey to be set up. The Harlequin Ladybird Survey (http://www.harlequin-survey.org) was one of the first online surveys of its kind in Great Britain. It was launched in March 2005 in response to the first report of *H. axyridis* in Great Britain, in September 2004 (Majerus 2004). The survey benefited from high levels of media interest, including the front page of The Times on 15 March 2005. Members of the British public showed great willingness to look for *H. axyridis*, and to register their sightings with the survey.

The dataset presented here comprises 4,117 species records of *H. axyridis* in Great Britain between 2004 and 2006. Each record represents a verified sighting of *H. axyridis* on a given date, and comprises one or more individual ladybirds observed from one or more life stages (larva, pupa and adult; records of eggs were not included). 76% of these records were received from members of the British public by post, or entered online at the Harlequin Ladybird Survey website or UK Ladybird Survey website (http://www.ladybird-survey.org); the latter was set up to encourage the recording of native ladybirds. Verification of the records was made by the authors after receipt of either a specimen or photograph. Most photographs were received as digital images attached to emails. The remaining 24% of records were received from coleopterists and other naturalists, and in particular from the London and Essex Ladybird Survey. Such records were not verified by the authors but are regarded as accurate and so have been included in the dataset. A further 4,316 online records were received that remain unverified (i.e. no photograph or specimen was sent), or were verified as another species, and so are not included in the analyses. Verified records were regularly uploaded to the database of the National Biodiversity

Network. There they could be viewed via online maps, which helped to encourage further recording.

Each species record includes the following data: recorder name; location of sighting (the grid reference of the Ordinance Survey British national grid reference system); locality/site name (not included for all records); date of sighting; life stage observed (larva, pupa, adult); number of each life stage observed (assumed to equal 1, if not specified); and determiner name. Additional optional attributes include the number of each colour form of any adults observed (included for 82% of records), and supplementary comments, including habitat and/or behavioural information.

Three main colour forms of adult *H. axyridis* have been found in Great Britain: f. *succinea*, f. *spectabilis* and f. *conspicua*. The *succinea* complex has been divided into many subforms (e.g. *siccoma*—0 spots; *frigida*—6 spots; *novemdecimsignata*—19 spots) (Dobzhansky 1933), but is treated as a single form here. Thus f. *succinea* has elytra with a ground colour of yellow, orange, or red, and 0–21 black spots, which may or may not be fused. *Forma spectabilis* has black elytra with four yellow, orange, or red spots or other shaped markings, which sometimes contain a central black spot. *Forma conspicua* is as f. *spectabilis*, but with only two spots.

The spatial resolution of the records is variable, and while approximately 20% include a grid reference, enabling resolution to 100 m, the other approximate 80% of records were derived at 1 km resolution from a UK postal code (UK Government Data Standards Catalogue, http://www.govtalk.gov.uk/gdsc/html/frames/PostCode.htm). The option on the online recording form to enter the location via a UK postal code was provided to make the entry of records easier for members of the public unfamiliar with the grid reference system. Whilst the resolution is thus reduced for these records, the reduction in user error (e.g. the problem of grid reference eastings and northings being transposed), is an advantage (Majerus et al. 1990). The postal code method was applicable for sightings of *H. axyridis* made within 200 m of a specified postal code, so could not be used for a minority of records where the ladybird was seen in a semi-natural habitat.

Variability in recording effort (both temporally and spatially) is clearly an issue when analysing a dataset of the kind presented here. Other factors being equal, more records will come from areas with a higher density of recorders. Across Great Britain there were a number of particularly active local groups or individuals, which contributed hotspots of recorder activity, potentially biasing the results for certain areas, e.g. London, because of the high activity of the London and Essex Ladybird Survey. The Harlequin Ladybird Survey benefited from a high profile and enjoyed frequent local and national media attention, thus potentially increasing the volume of records received at certain times. To minimise these effects, the data have mostly been analysed in terms of the presence of *H. axyridis* in 10 km squares (1 km squares for juveniles) by year. Thus, whether the species was recorded just once or many times in any given square in a year, is not reflected in the analyses. To many recorders, juvenile stages were less noticeable and more difficult to identify than the adult stage, thus generally limiting the recording of juveniles.

The possibility of a reporting bias towards sightings early in the season also existed (i.e. some recorders may have reported their first sighting of *H. axyridis*, but not subsequent sightings). In order to minimise this effect, the importance of recording multiple sightings was stressed to recorders. The peaks in record numbers observed late in each year also suggest that any effect of this potential bias was minor.

Rate of spread

The location of the northern range margin of *H. axyridis* in Great Britain was measured by calculating the mean northing of the ten most northerly 10 km squares occupied each year (Hickling et al. 2006). The location of the range margins in the westerly and north-westerly directions were calculated each year using the same method. There is a very strong assumption that *H. axyridis* colonised Great Britain primarily from the coastal regions of northern France, Belgium and the Netherlands (Majerus et al. 2006a), hence the inclusion of the north-westerly direction.

Seasonal pattern

The seasonal pattern of *H. axyridis* sightings in 2005 and 2006 were examined. *Harmonia axyridis* was first recorded in Great Britain in July 2004, so that year was excluded from the analyses because of incomplete data. The data are weekly counts of the occurrence of *H. axyridis* in grid squares. For adult *H. axyridis* 10 km squares were used, but for juveniles (i.e. larvae and pupae), because of a lower number of records, 1 km squares were used.

Seasonal patterns were analysed using a Generalized Additive Model (GAM) in which the counts varied randomly about a smooth trend described by a cubic smoothing spline (Green and Silverman 1994). The degrees of freedom for the fitted GAM were determined using cross-validation (leaving out each data point in turn) and minimising the cross-validation Poisson deviance. A GAM was fitted separately to data from 2005 and 2006 to describe the seasonal pattern of observations in each year. The number of sightings was generally higher throughout 2006 compared to 2005, but the GAM curves allow comparison of the shape of the seasonal pattern each year. The null hypothesis of equality of shape in 2005 and 2006 implies that curves differ by some constant factor, i.e. parallel on a log scale. The alternative hypothesis allows different curves in each year. The test-statistic is $F = \{(D_{null} - D_{alt})/(df_{null} - df_{alt})\}/D_{alt}/df_{alt}$, where D denotes the residual deviance and df is the corresponding degrees of freedom. On the null hypothesis, the statistic follows an F distribution with $(df_{null} - df_{alt})$ and df_{alt} degrees of freedom. The analysis was performed using the statistical package Genstat 6 (Payne et al. 2002).

Results

Abundance and rate of spread

Harmonia axyridis has spread rapidly in Great Britain (Fig. 1a). Recorded occurrence (of one or more *H. axyridis* individuals at larval, pupal and/or adult stage) in 10 km squares in Great Britain was: 51 in 2004; 149 in 2005; 447 in 2006; representing a 2.9-fold increase from 2004 to 2005 and a 3.0-fold increase from 2005 to 2006. There was a 2.7-fold increase in the number of adult *H. axyridis* records, and a 2.9-fold increase in the number of individual adults recorded, from 2005 to 2006 (Table 1).

The number of 10 km squares with records of juvenile *H. axyridis* increased from 8 in 2004, to 51 in 2005, and 75 in 2006 (Fig. 1b). In 2004, evidence of *H. axyridis* breeding was restricted to London and East Anglia, with juveniles recorded in a wider range of counties in 2005 (notably Kent and Derbyshire). Further evidence of breeding was

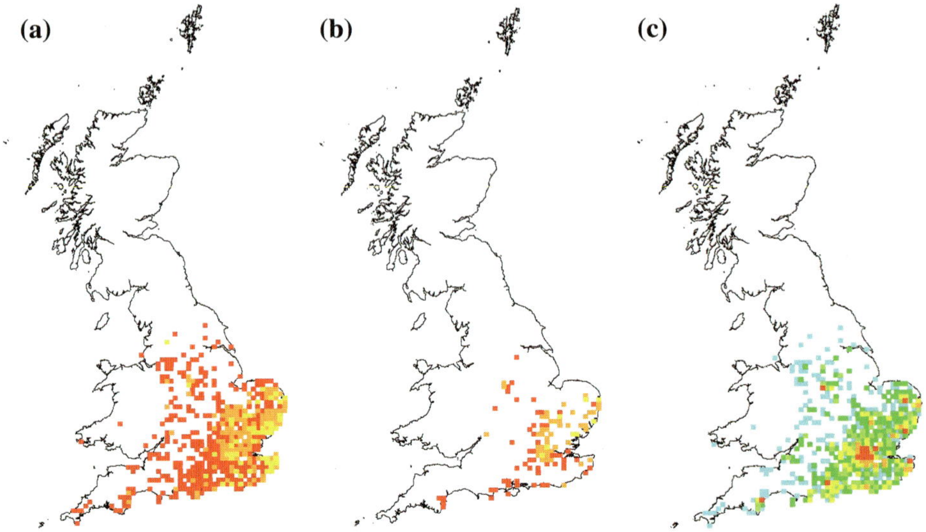

Fig. 1 *Harmonia axyridis* occurrence in 10 km squares in Great Britain from 2004 to 2006. For (**a**) and (**b**), where a square has been recorded in more than one year, occurrence in the earliest year is shown (yellow = 2004; orange = 2005; red = 2006). (**a**) Adults and juveniles; (**b**) Juveniles only; (**c**) Density of records of adults and juveniles; (red = 50+ records per square; orange = 20–49; yellow = 10–19; pale green = 6–9; dark green = 2–5; blue = 1)

recorded across south-east England, and west as far as Devon, and north as far as Yorkshire, in 2006 (Fig. 1b).

Northerly spread (Fig. 2a) distances per year were: 19 km between 2004 and 2005; 97 km between 2005 and 2006; mean—58 km $year^{-1}$. Westerly spread (Fig. 2b) distances per year were: 107 km between 2004 and 2005; 182 km between 2005 and 2006; mean—144.5 km $year^{-1}$. North-westerly spread (Fig. 2c) distances per year were: 81.2 km between 2004 and 2005; 107.5 km between 2005 and 2006; mean—94.3 km $year^{-1}$.

Table 1 Summary of Harlequin Ladybird Survey data from Great Britain (Note—records of large aggregations without a specific number of ladybirds recorded have been excluded)

Life stage	Year	No. verified records	Total no. *H. axyridis* recorded	Mean no. *H. axyridis* per record	Maximum no. *H. axyridis* recorded
Larva	2004	13	54	4.2	10
	2005	182	1,817	10.0	266
	2006	145	1,410	9.7	200
Pupa	2004	5	27	5.4	10
	2005	81	1,194	14.7	227
	2006	23	231	10.0	53
Adult	2004	119	344	2.9	25
	2005	1,045	6,180	5.9	399
	2006	2,825	17,641	6.2	689

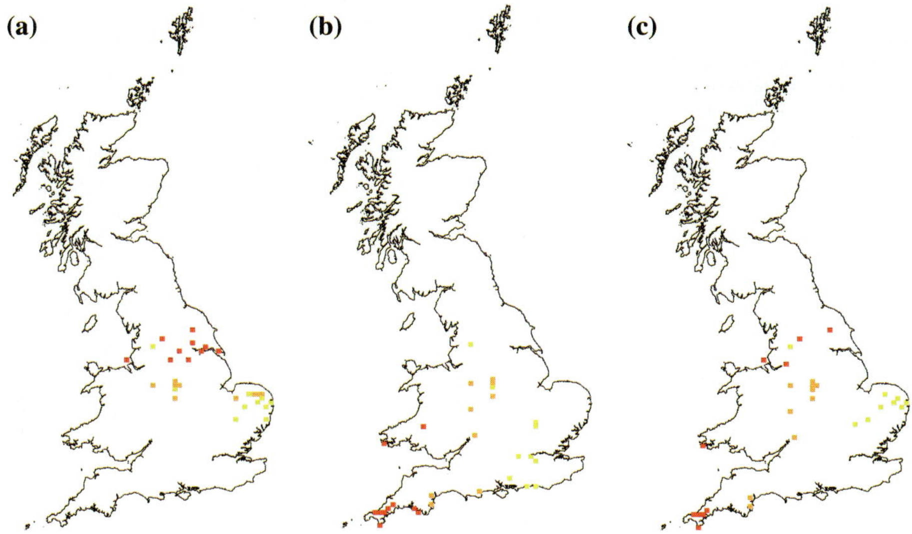

Fig. 2 Range margin of *H. axyridis* in 10 km squares in Great Britain from 2004 to 2006. Where a square has been recorded in more than one year, occurrence in the earliest year is shown (yellow = 2004; orange = 2005; red = 2006). (**a**) Ten most northerly squares per year; (**b**) Ten most westerly squares per year; (**c**) Ten most north-westerly squares per year

The density of records received was highest in and around London, and was also high in parts of East Anglia and along the south coast of England, particularly in Kent, Sussex, Hampshire and Devon (Fig. 1c). There is also a notable hotspot in Derby, the only location in central-north England with a large number of verified records (Fig. 1c).

The mean number of adults per record increased year on year, from 2.9 in 2004, to 5.9 in 2005, and 6.2 in 2006 (Table 1). The trend was different for the juvenile stages, which appear to peak in 2005 (however, see discussion). A similar pattern was observed for the maximum counts observed in each year (Table 1).

Seasonal pattern

Adults

The seasonal patterns of spread of adult *H. axyridis* in 2005 and 2006 were significantly different ($F_{12,78} = 2.52$, $P = 0.007$). The main difference in the smoothed curves was during the period 23–34 weeks, where there was a seasonal peak observed in 2006, which was not observed in 2005 (Fig. 3a and b).

Juveniles

The seasonal patterns of spread of juvenile *H. axyridis* in 2005 and 2006 were significantly different ($F_{6,89} = 3.57$, $P = <0.001$). The main difference is the extra peak around week

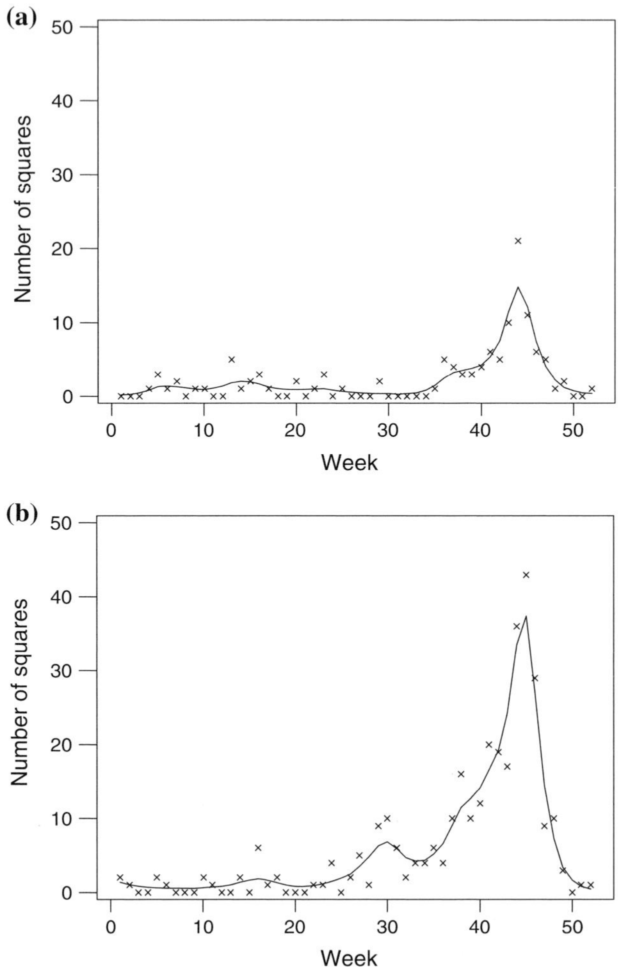

Fig. 3 Adult *H. axyridis* occurrence in 10 km squares in Great Britain with fitted GAMs and degrees of freedom (*df*). (**a**) 2005 (*df* = 12); (**b**) 2006 (*df* = 13)

25 in 2006, and the earlier date of the autumn peak in 2006. For the fitted GAMs, the main peaks occur at week 44 (2005) and week 40 (2006) (Fig. 4a and b).

Discussion

Rate of spread

The rate of spread and increase in abundance of *H. axyridis* in Great Britain has been dramatic. Westerly spread was faster than northerly spread by a factor of approximately 2.5, and this may partly be explained by the likelihood of repeated recruitment of *H. axyridis* from mainland Europe. Indeed, in 2006, eight of the ten most westerly 10 km squares occupied were in the extreme south-west of England and close to the coast, possibly representing new recruits from abroad, rather than spread from the existing British range. This factor is presumed not to apply to the most northerly squares occupied, because

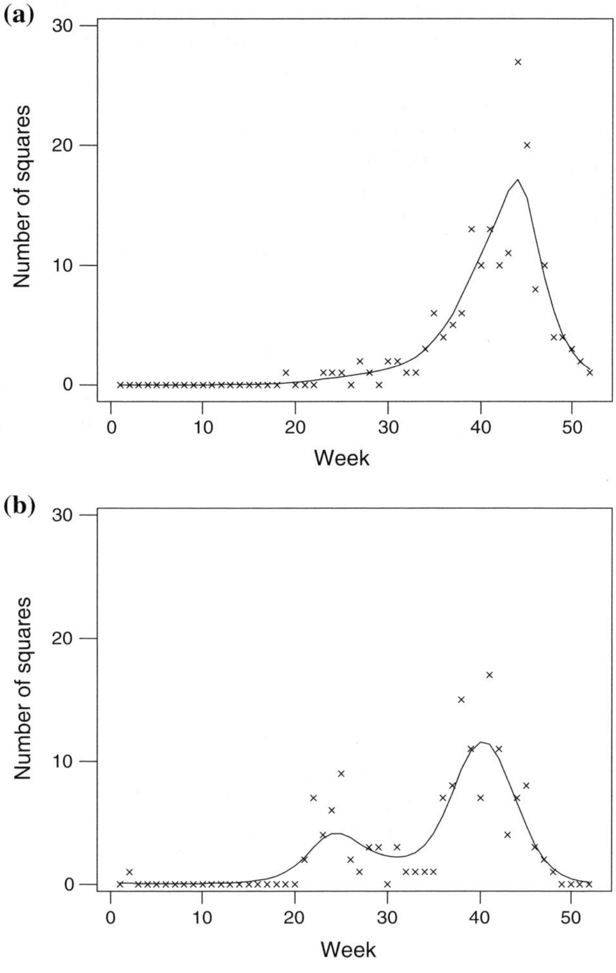

Fig. 4 Juvenile *H. axyridis* occurrence in 1 km squares in Great Britain with fitted GAMs and degrees of freedom (*df*). (**a**) 2005 (*df* = 6); (**b**) 2006 (*df* = 6)

of their greater distance from mainland Europe and, in most cases, their lack of proximity to the British coast. Ecological factors such as climate and habitat type may also have influenced the faster westerly spread. In late October and early November 2006 there was an influx of reports of large aggregations of *H. axyridis* adults, mainly at locations close to the south coast of England. Aggregations of hundreds and in a few cases, thousands, of *H. axyridis* were observed (Brown and Roy 2007). Whether these large aggregations represented new recruits from mainland Europe has yet to be determined.

The most closely related species to *H. axyridis* in Great Britain is *Harmonia quadripunctata* (Pontoppidan) (Coleoptera: Coccinellidae). This species is also a relatively recent arrival to Britain, but is now regarded as a native species, having expanded its native range northwards within Europe, and is assumed to have reached Britain by natural means (Majerus and Kearns 1989; Majerus 1994). It is not considered invasive. Although *H. quadripunctata* is far more habitat-specific than *H. axyridis*, and the mechanisms of spread may not be the same, a comparison of the rate of spread of the two species is enlightening. Majerus and Kearns (1989) outline the spread of *H. quadripunctata*. It was first recorded in Great Britain in West Suffolk (East Anglia) in 1937, and based on earliest records for each vice county, took fifty years to spread west as far as Devon, in south-west England. Our data show that *H. axyridis* took just two years to spread to Devon from a similar starting point in East Anglia.

In Great Britain, the northerly range expansion of some insect groups was calculated by Hickling et al. (2006), giving figures of up to 4.2 km $year^{-1}$ (over a 25 year period) for the most dispersive taxa (dragonflies); but these range expansions were for native species responding to climate change, rather than for invasive species. There is limited quantitative data on the spread of invasive insects in Europe. The spread of the horse chestnut leafminer *C. ohridella* in Germany in the late 1990s was modelled, and the rate of spread calculated to vary between 54 km $year^{-1}$ and 330 km $year^{-1}$ (Gilbert et al. 2004), figures of the same order of magnitude as our calculations for *H. axyridis* in Britain. In eastern North America the rate of range expansion of *H. axyridis* was estimated at 442 km $year^{-1}$ (McCorquodale 1998). As it was probably confounded by many intentional releases at various locations (McCorquodale 1998), this very fast expansion is not a fair comparison with that observed in Great Britain. Expansion was thought to be slower in other parts of North America (Koch et al. 2006).

Abundance and mechanism of spread

Abundance of *H. axyridis* in Great Britain increased rapidly from 2004 to 2006, and this trend is predicted to continue. There is very strong demographic evidence that the spread of *H. axyridis* in Great Britain did not originate from a single immigration event, and that the species arrived at different locations at various times, and by various means. Apart from the spread (by flight) from continental Europe, some specimens of *H. axyridis* are known to have arrived in England on imported flowers from the Netherlands, and in packing cases from Canada (Majerus et al. 2006a). By 2004, *H. axyridis* had sufficient time to adapt to conditions in mainland Europe and was building-up in number and spreading fast. Crossing the English Channel to reach Great Britain was not a major step. It is a coincidence that specimens from Canada were found in Great Britain for the first time in the same year, and it is possible that it had arrived by similar mechanisms in earlier years, but failed to be noticed and failed to establish.

A separate population of *H. axyridis* was evident in Derby (central England) in 2004, which is thought to have originated from specimens arriving with produce to a supermarket in the city (W. Grange, personal communication). Derby was not the most northerly verified location of *H. axyridis* in 2004 (a single specimen was recorded in Lancashire, but with no evidence of subsequent establishment), but is the only place outside of the southeast with multiple verified records of *H. axyridis* in each of 2004, 2005 and 2006. Large numbers of *H. axyridis* (three records each of 50+ individuals) were recorded in Derby in 2005. The colour form profile of the Derby population almost totally lacked melanics in 2004 and 2005 (99.6% f. *succinea*; $n = 495$), in contrast to other parts of the country, where melanic specimens formed a significant proportion of the population (a mean of 20% melanic specimens were recorded in 2005). The clear difference in colour form profile strongly suggests a genetically distinct population, and further work is in progress to examine this.

The observed reduction in both the number of juvenile records, and mean number of juveniles per record, from 2005 to 2006, is not believed to reflect a true reduction in breeding activity. Rather, this is likely to be the effect of reduced recorder effort. In particular, recorders with the London and Essex Ladybird Survey worked diligently in 2005 to gather juvenile records, and provided 46% of all such records; this effort was reduced to 9% of the total in 2006. A 47% increase in the number of 10 km squares with juvenile records from 2005 to 2006 suggests that there was in fact an increase in breeding activity.

Examples of invasive species in Great Britain exhibiting a 'boom-and-bust' population cycle are unusual, although this scenario did apply to the rhododendron lacebug *Stephanitis rhododendri* Horvath (Hemiptera: Tingidae) (Williamson 1996). The evidence from almost 20 years as an established introduced species in North America suggests that this will not apply to *H. axyridis* on either side of the Atlantic.

Evidence of bi-voltinism

Some of the common and widespread coccinellid species in Britain, e.g. *Coccinella septempunctata* L. (Coleoptera: Coccinellidae), are limited by the requirement of a winter dormancy period before they can reproduce (Majerus 1994). *Harmonia axyridis* does not have this limit to population growth and may have several generations per year, five having been recorded in Asia (Wang 1986) and four in southern Europe (Katsoyannos et al. 1997). In the cool and damp maritime climate of Great Britain, we predict *H. axyridis* will achieve two, or possibly three, generations per year. There is evidence of a second generation in 2006 in the data presented here, with a summer peak (week 24) and an autumn peak (week 40), in juvenile records. Production of two generations would help to explain the very rapid spread of *H. axyridis* observed.

Habitat use

The dataset reflects a somewhat urban distribution. Although some of this effect may be biased by higher recorder effort in urban areas, it is assumed to be the result of genuine habitat preference by *H. axyridis*. Lime trees, *Tilia* sp. (Tiliaceae) and sycamore trees, *Acer* sp. (Aceraceae) were recorded as favoured habitats, and these trees are generally found in Britain in urban locations, such as parks, gardens, lining roads and churchyards. Whether *H. axyridis* will thrive in semi-natural habitats is a question that remains to be answered, and an important one in terms of the potential impact of the species on native coccinellids and other insects. *Harmonia axyridis* has been recorded in various semi-natural habitats in Great Britain, including reedbeds, and on planted and self-seeded *Pinus sylvestris* L. (Pinaceae) in East Anglia. Mature *P. sylvestris* is an important habitat for several native coccinellids, including *Anatis ocellata* L. (Coleoptera: Coccinellidae), *Myzia oblongoguttata* L. (Coleoptera: Coccinellidae) and *Myrrha octodecimguttata* L. (Coleoptera: Coccinellidae) (Majerus 1994). Evidence of the strength of intraguild predation by *H. axyridis* (Pell et al. 2007; Roy et al. 2007; Ware and Majerus 2007; Ware et al. 2007) deepens concern that the species will have a serious negative impact on native coccinellids.

Conclusion

We predict that the spread of *H. axyridis* in Great Britain will continue at a rapid pace, and that the species will become established in Scotland by the end of 2008. To determine the impact of *H. axyridis* on native species, monitoring of ladybird populations is required in semi-natural habitats and urban habitats, both of which can support high abundance and species richness. Quantitative research at appropriate spatial and temporal scales is essential, if we are to objectively assess the ecological impact of *H. axyridis*. Field studies need to focus on habitat use by this species and its interactions within the large guild of associated aphidophagous species. *Harmonia axyridis* has many traits that have ensured its status as a successful invasive alien species. The continued population expansion within and beyond its invaded range seems inevitable.

Acknowledgements We extend our thanks to all of the following, who have helped to make the Harlequin Ladybird Survey a success: the thousands of naturalists and members of the British public who have provided records; Department for Environment Food and Rural Affairs, National Biodiversity Network Trust, Joint Nature Conservation Committee, Natural Environment Research Council, University of Cambridge and Anglia Ruskin University for funding and support; Paul Mabbott and recorders from the London and Essex Ladybird Survey; Trevor James, Jim Munford and Andrew Brewer of the National Biodiversity Network Trust; Mark Hill, Cassie Hoyland and Henry Arnold of the Biological Records Centre; Ian Wright, Laura-Jane Michie, James Waters and Francis Rowland of University of Cambridge; Julian Doberski of Anglia Ruskin University; and finally Robert Frost and William Grange for records and valuable contributions.

References

Adriaens T, Branquart E, Maes D (2003) The multicoloured Asian ladybird *Harmonia axyridis* Pallas (Coleoptera: Coccinellidae), a threat for native aphid predators in Belgium? Belg J Zool 133:195–196

Brown PMJ, Roy HE (2007) The spread of the harlequin ladybird *Harmonia axyridis* in late 2006. Atropos 31:32–33

Brown PMJ, Adriaens T, Bathon H, Cuppen J, Goldarazena A, Hägg T, Kenis M, Klausnitzer BEM, Kovar I, Loomans AJM, Majerus MEN, Nedved O, Pedersen J, Rabitsch W, Roy HE, Ternois V, Zakharov IA, Roy DB (2007) *Harmonia axyridis* in Europe: spread and distribution of a non-native coccinellid. BioControl (this issue). doi:10.1007/s10526-007-9132-y

Chapin JB, Brou VA (1991) *Harmonia axyridis* (Pallas), the 3rd species of the genus to be found in the United States (Coleoptera, Coccinellidae). Proc Entomol Soc Wash 93:630–635

Colunga-Garcia M, Gage SH (1998) Arrival, establishment, and habitat use of the multicolored Asian lady beetle (Coleoptera: Coccinellidae) in a Michigan landscape. Environ Entomol 27:1574–1580

Coutanceau J-P (2006) *Harmonia axyridis* (Pallas, 1773): une coccinelle asiatique introduite, acclimatée et en extension en France. Bulletin de la Société Entomologique de France 111:395–401

Dobzhansky T (1933) Geographical variation in ladybeetles. Am Nat 67:97–126

Gilbert M, Grégoire J-C, Freise JF, Heitland W (2004) Long-distance dispersal and human population density allow the prediction of invasive patterns in the horse chestnut leafminer *Cameraria ohridella*. J Anim Ecol 73:459–468

Glowka L, Burhenne-Guilmin F, Synge H, McNeely J, G'ndling L (1994) A guide to the convention on biological diversity (Environmental Policy and Law paper no. 30). International Union for the Conservation of Nature, Gland

Green P, Silverman BW (1994) Non-parametric regression and generalized additive models. Chapman and Hall, London

Harding PT (1990) National species distribution surveys. In: Goldsmith FB (ed) Monitoring for conservation and ecology. Chapman and Hall, London

Hickling R, Roy DB, Hill JK, Fox R, Thomas CD (2006) The distributions of a wide range of taxonomic groups are expanding polewards. Global Change Biology 12:450–455

Katsoyannos P, Kontodimas DC, Stathas GJ, Tsartsalis CT (1997) Establishment of *Harmonia axyridis* on citrus and some data on its phenology in Greece. Phytoparasitica 25:183–191

Kidd KA, Nalepa CA, Day ER, Waldvogel MG (1995) Distribution of *Harmonia axyridis* (Pallas) (Coleoptera, Coccinellidae) in North Carolina and Virginia. Proc Entomol Soc Wash 97:729–731

Koch RL (2003) The multicolored Asian lady beetle, *Harmonia axyridis*: a review of its biology, uses in biological control, and non-target impacts. J Insect Sci 3:1–16

Koch RL, Venette RC, Hutchison WD (2006) Invasions by *Harmonia axyridis* (Pallas) (Coleoptera: Coccinellidae) in the western hemisphere: implications for South America. Neotrop Entomol 35:421–434

Kuznetsov VN (1997) Lady beetles of Russian far east. The Sandhill Crane Press, Gainesville
Majerus MEN (1994) Ladybirds. No. 81, New Naturalist series. HarperCollins, London
Majerus MEN (2004) The ladybird has landed. University of Cambridge press release, 4 October 2004
Majerus MEN (2007) Generalist and specialist ladybirds. British Naturalists' Association, in press
Majerus MEN, Kearns P (1989) Ladybirds. Richmond Publishing, Slough
Majerus MEN, Roy HE (2005) Scientific opportunities presented by the arrival of the harlequin ladybird, *Harmonia axyridis*, in Britain. Antenna 29:196–208
Majerus MEN, Forge H, Walker L (1990) The geographical distributions of ladybirds in Britain (1984–1989). Br J Entomol Nat Hist 3:153–166
Majerus MEN, Mabbott P, Rowland F, Roy HE (2006a) The harlequin ladybird, *Harmonia axyridis*, arrives in Britain. Entom Mon Mag 142:87–92
Majerus M, Strawson V, Roy H (2006b) The potential impacts of the arrival of the harlequin ladybird, *Harmonia axyridis* (Pallas) (Coleoptera: Coccinellidae), in Britain. Ecol Entomol 31:207–215
McCorquodale DB (1998) Adventive lady beetles (Coleoptera: Coccinellidae) in eastern Nova Scotia, Canada. Entomol News 109:15–20
McKinney ML, Lockwood JL (1999) Biotic homogenization: a few winners replacing many losers in the next mass extinction. Trends Ecol Evol 14:450–453
Payne RW (2002) Genstat 6 release 1 reference manual. Oxford University Press, Oxford
Pell JK, Baverstock J, Roy HE, Ware RL, Majerus MEN (2007) Intraguild predation involving *Harmonia axyridis*: a review of current knowledge and future perspectives. BioControl (this issue). doi: 10.1007/s10526-007-9125-x
Roy HE, Rowland F, Brown PMJ, Majerus MEN (2005) Ecology of the harlequin ladybird: a new invasive species. Br Wildl 16:403–407
Roy HE, Brown P, Majerus MEN (2006) *Harmonia axyridis*: a successful biocontrol agent or an invasive threat? In: Eilenberg J, Hokkanen H (eds) An ecological and societal approach to biological control. Kluwer Academic Publishers, Dordrecht
Roy HE, Baverstock J, Ware RL, Clark SJ, Majerus MEN, Baverstock KE, Pell JK (2007) Intraguild predation of the aphid pathogenic fungus *Pandora neoaphidis* by the invasive coccinellid *Harmonia axyridis*. Ecological Entomology, in press
Smith WM, Arnold DC, Eikenbary RD, Rice NR, Shiferaw A, Cheary BS, Carroll BL (1996) Influence of ground cover on beneficial arthropods in pecan. Biol Control 6:164–176
Tedders WL, Schaefer PW (1994) Release and establishment of *Harmonia axyridis* (Coleoptera, Coccinellidae) in the southeastern United States. Entomol News 105:228–243
Wang LY (1986) Mass rearing and utilization in biological control of the lady beetle *Leis axyridis* (Pallas). Acta Entomologica Sinica 29:104
Ware RL, Majerus MEN (2007) Intraguild predation of immature stages of British and Japanese coccinellids by the invasive ladybird *Harmonia axyridis*. BioControl (this issue). doi:10.1007/s10526-007-9135-8
Ware RL, Evans N, O'Farrell K, Majerus MEN, Malpas L, Michie LJ (2007) Intraguild predation by the invasive ladybird *Harmonia axyridis*: 1: British and Japanese coccinellid eggs. Neobiota, in press
Williamson M (1996) Biological invasions. Chapman and Hall, London

Invasion history, habitat preferences and phenology of the invasive ladybird *Harmonia axyridis* in Belgium

Tim Adriaens · Gilles San Martin y Gomez · Dirk Maes

Abstract *Harmonia axyridis* Pallas (Coleoptera: Coccinellidae) is native to Asia, and was widely introduced as a biocontrol agent of aphids and coccids in Europe and North America. In Europe, *H. axyridis* is considered to be an invasive alien species because of its potential to disrupt native ladybird communities. Since 1999, the Belgian Ladybird Working Group mapped all Belgian Coccinellidae and recorded data on substratum plants and habitat. The first feral *H. axyridis* population in Belgium was recorded in 2001, but the expansion rate is decreasing because it now colonised the whole country. Recorded occupancy in Belgium showed an average rate of increase of 189% between 2002 and 2006. In Belgium, *H. axyridis* occurred in a wide range of habitats, including those of high conservation value. However, habitat and land cover analysis showed that *H. axyridis* is more frequently found in urbanised landscapes than in semi-natural landscapes. A niche overlap analysis based on plant use data showed that the potential to affect native species is higher for generalist, deciduous and coniferous tree ladybird species than for heathland and wetland specialist species. Phenology data showed that *H. axyridis* is able to reproduce later in the year than native species. Based on recorded distribution, ecology and phenology, we discuss the potential of *H. axyridis* to disrupt native ladybird assemblages in Belgium.

Keywords Biological control · Coccinellidae · Harlequin ladybird · *Harmonia axyridis* · Invasive species · Multicoloured Asian ladybird · Non-native species · Phenology · Invasion history · Habitat · Niche overlap

---

T. Adriaens (✉) · D. Maes
Research Institute for Nature and Forest (INBO), Kliniekstraat 25, 1070 Brussels, Belgium
e-mail: tim.adriaens@inbo.be

G. San Martin y Gomez
Catholic University of Louvain-la-Neuve (UCL), Ecology and Biogeography Unit, Croix du Sud, 4-5 (Carnoy), 1348 Louvain-la-Neuve, Belgium

Biological Control to Invasion: Ladybird Harmonia axyridis. H.E. Roy, E. Wajnberg (eds.), 69
doi: 10.1007/978-1-4020-6939-0_6, © IOBC 2007

Introduction

Harmonia axyridis Pallas (Coleoptera: Coccinellidae) is used as a biocontrol agent against aphid populations because its larvae are very voracious, polyphagous and easy to rear (Ferran et al. 1996; Maignet 2002; Koch 2003). The species has been widely used in greenhouses, orchards and gardens in North America since 1916 (Gordon 1985) and in Western Europe since 1982 (Iperti and Bertrand 2001), where it was sold by different private companies (Adriaens et al. 2003; Poutsma et al. 2007). *Harmonia axyridis*, among other predatory insects and parasitoids, was ranked as a high-risk species in an environmental risk assessment of exotic natural enemies used in inundative biological control (van Lenteren et al. 2003; van Lenteren et al. 2007). *Harmonia axyridis* is known to have strong dispersal capacities (Koch 2003; Osawa 2000) and studies in North America have shown that it can rapidly colonise large areas (Tedders and Schaefer 1994). In Europe, it has spread very rapidly, particularly since 2002, and the species now exists as feral populations in 13 European countries (Brown et al. 2007a).

Laboratory experiments showed that *H. axyridis* is frequently involved in intraguild interactions with other aphidophagous species, such as the ladybird species *Adalia bipunctata* (L.) and *Coccinella septempunctata* L. (Coleoptera: Coccinellidae), both native to Western Europe (Hironori and Katsuhiro 1997; Cottrell and Yeargan 1998; Yasuda and Ohnuma 1999; Yasuda et al. 2001). Petri dish experiments showed that 4th instar *H. axyridis* larvae easily attack *A. bipunctata* larvae in the absence of other prey (Hautier 2003a, b, 2006). Furthermore, *H. axyridis* larvae showed aggressive behaviour towards the first three larval stages of *A. bipunctata*, resulting in over 80% mortality after 30 min and 100% mortality for all three larval stages after 24 h (Hautier 2003a). In laboratory experiments, the presence of aphids as alternative prey did not alter the mortality figures of *A. bipunctata*. These results showed that *A. bipunctata* is a potential prey for *H. axyridis* under limited food conditions. Therefore, *H. axyridis* can be regarded as an important source of mortality for native ladybird species. Moreover, native species rarely consumed eggs and larvae of *H. axyridis*. Observations in potato fields showed that oviposition of *H. axyridis* usually occur late during aphid development and is potentially accompanied by strong intraguild predation against native ladybird species (Hautier 2006). It was shown that *H. axyridis* can reproduce on a pollen diet, when insect prey is scarce, which gives the species a competitive advantage over indigenous predatory ladybird species that are less capable of doing so (Berkvens et al. 2007). However, there are indications that *A. bipunctata* is also able to compensate for a suboptimal diet of animal prey by supplementary feeding on flower pollen (De Clercq et al. 2005).

Harmonia axyridis can induce dominance shifts in ladybird communities and often becomes the predominant species at the expense of native species (Lamana and Miller 1996; Brown and Miller 1998; Colunga-Garcia and Gage 1998; Burgio et al. 2002; Michaud 2002, 2004; Alyokhin and Sewell 2004). In Belgium, modifications in ladybird communities have been reported through detailed monitoring studies in Brussels (San Martin 2003; Ottart 2005). Prior to the invasion of *H. axyridis*, *A. bipunctata* was the dominant lady beetle on lime (*Tilia* sp.) and maple (*Acer* sp.) in this urban environment and was co-dominant with the native congeneric ladybird *Harmonia quadripunctata* (Pontoppidan) (Coleoptera: Coccinellidae) on Austrian pine (*Pinus nigra* Arn.). Only 2 years after its establishment, *H. axyridis* quickly became the predominant ladybird species in all habitats monitored (Ottart 2005). In addition, a significant decline in the abundance of two native species *A. bipunctata* and *Adalia decempunctata* (L.) (Coleoptera: Coccinellidae) was recorded between 2003 and 2005, while a simultaneous increase of the *H. axyridis* population was observed (Ottart 2005).

Due to its voracity and wide trophic niche, it was believed that *H. axyridis* would harm native aphidophagous guilds. Despite the relatively large body of knowledge on *H. axyridis*, our knowledge of the potential adverse impacts and other non-target effects needs to be improved (Koch 2003). Studies have shown declines in densities of native ladybird species after the arrival of *H. axyridis*, e.g. in orchards (Tedders and Schaefer 1994), potato fields (Alyokhin and Sewell 2004) and other agricultural ecosystems (Colunga-Garcia and Gage 1998). However, very few articles deal with *H. axyridis* population dynamics in (semi)-natural ecosystems.

The knowledge of processes and factors explaining the invasion success is still rudimentary (Williamson 2006). The recent establishment of *H. axyridis* in Belgium and elsewhere in Europe provides a unique opportunity to study invasion mechanisms and the impact of an exotic predator on native organisms. Explaining and predicting the success of an invasive species like *H. axyridis* requires information on microhabitat and phenology.

In this article, we analyse distributional and phenological data collected during a large-scale mapping scheme in Belgium to detect whether *H. axyridis* is able to modify native ladybird communities. First, we investigated differences in habitat selection between *H. axyridis* and native species. Second, we analysed the occurrence of *H. axyridis* in areas of conservation concern, using biological valuation and land cover maps. Finally, we discuss the potential impact on native species based on a niche overlap analysis.

Material and methods

The Belgian ladybird mapping project

In 1999, the Belgian ladybird working group *Coccinula* launched a large-scale field survey on 40 native ladybird species (Coccinellinae, Chilocorinae and Epilachninae) and to date has 527 volunteers providing distribution data. They actively search for ladybirds in a variety of habitats using sweep nets, beating trays, visual search, light trapping, pitfall traps and other sampling methods. Distribution, habitat and substrate plant information is also noted on a standard recording form. The working group maintains a database of observations, literature and collection data of Coccinellidae from 1887 onwards. Preliminary atlases have been published (Branquart et al. 1999; Adriaens and Maes 2004) and updated distribution maps for the Belgian territory are available in Dutch and French online (http://www.inbo.be/docupload/2680.pdf).

The first feral field populations of *H. axyridis* in Belgium were discovered in 2001 (Adriaens et al. 2003). Since then, the Belgian Ladybird Working Group collected data on the spread and distribution of the species and also recorded data on its ecology in Belgium (Adriaens and Gysels 2002; San Martin et al. 2005). Collection events minimally consisted of species, number of individuals, stage (adults, larvae and pupae), observation date, observer and location. Locations were attributed to 1 × 1-km^2 or 5 × 5-km^2 grid cells of the UTM grid (Universal Transverse Mercator). At present, the database contains 53,458 records, of which 15% come from museum collections and literature data. A large part of the Belgian territory has now been surveyed for ladybirds: the database contains records for 85% of all 5 × 5-km UTM grid cells ($N = 1{,}376$) in Belgium. Additionally, information on substratum plants, height in the vegetation, sampling method, habitat type, surrounding landscape, slope orientation, soil type, humidity, vegetation cover and behaviour was noted on the recording form. Observers were also asked to record *H. axyridis* morph type following nomenclature based on literature (Komai 1956;

Iablokoff-Khnzorian 1982; Serpa et al. 2003; Osawa and Nishida 1992): *f. succinea*, *f. conspicua* or *f. spectabilis*. The same nomenclature was applied in the UK's Harlequin Ladybird Survey (Brown et al. 2007b) and The Netherlands' Monitoring Project (Cuppen et al. 2004a, b). A small group of observers have performed detailed counts of morph type frequencies on different locations since the arrival of *H. axyridis* in Belgium in 2001.

Analysis

Distribution and invasion of H. axyridis *in Belgium*

To describe the distribution pattern, we used the Belgian ecoregions (Fig. 1) and calculated the number of 1-km^2 grid cells with and without presence of *H. axyridis* per ecoregion using 2001–2006 data. We used only well-prospected grid cells (i.e. with at least five native species recorded) and compared the presence/absence pattern in each ecoregion using χ^2 tests.

We calculated the percentage of *H. axyridis* observations and the percentage of collection events expressed as the unique combination of [date × 1-km^2 grid cell × observer] with *H. axyridis* in the database per year (2001–2006). We only used collection events with more than one species observed to account for accidental observations. Geographic spread was analysed by calculating mean and maximal distances between centroids of grid cells with *H. axyridis* observations for the period 2001–2006 and taking 2001 as a starting point, using the find distance tool in ArcGis 3.2a Spatial Analist extension (ESRI). We used the Home Range Extension for ArcGIS (Rodgers and Carr 1998) to calculate the

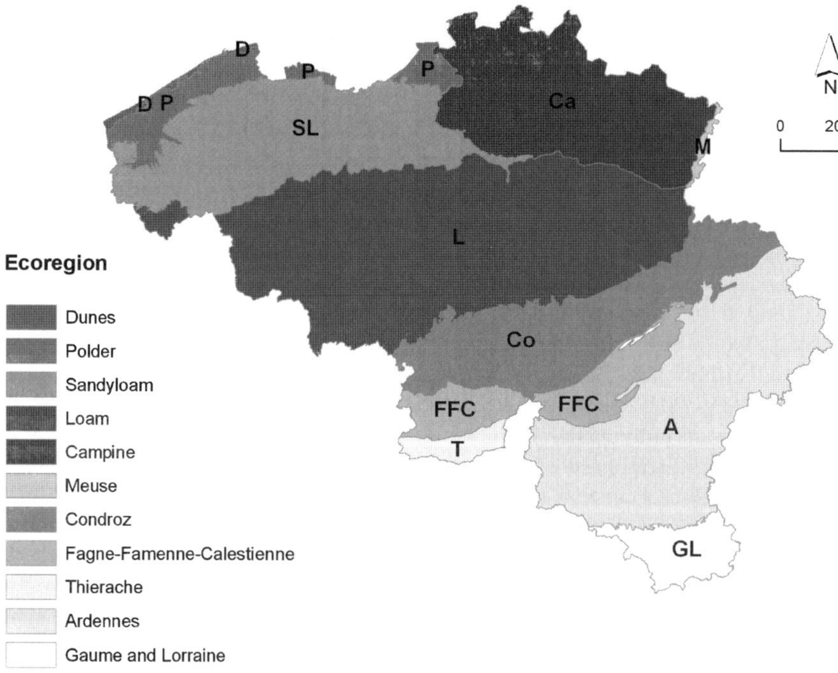

Fig. 1 Location of the Belgian ecoregions

surface area of minimum convex polygons around 1-km^2 grid cell *H. axyridis* data (2001–2006) as a measure of invaded area.

Habitat preference

Absolute and relative number of *H. axyridis* observations (expressed as the unique combination of [date × 1-km^2 grid cell]) on different plant species and genera were calculated based on 1,349 data with substratum plant information. Based on post-2001 substratum plant species data, we also calculated the relative number of *H. axyridis* observations on different vegetation strata (trees, shrubs, dwarf-shrubs, herbs) and habitat types and compared that with patterns in native species using χ^2-tests. For this analysis, we only included habitat types with at least 10 *H. axyridis* observations based on 2001–2006 data. Statistical analyses were performed using Statistica version 6.0 (StatSoft).

Potential interactions with native species

Three niche overlap like indices were calculated: two are based on the use of plant genera and one is based on spatio-temporal co-occurrence.

For each native ladybird species (38 species belonging to the subfamilies Coccinellinae, Chilocorinae, Epilachninae and Coccidulinae), we only used plants with at least two observations and we calculated plant indices only for ladybird species with at least 20 plant data (see Table 3).

We used Czekanowski index as an index of host plant use similarity (Feinsinger et al. 1981; Hurlbert 1978):

$$\text{Plant use similarity} = 1 - 0.5\left(\sum_i |p_{xi} - p_{yi}|\right) = \sum_i \min(p_{xi}, p_{yi})$$

where p_{xi} and p_{yi} are the relative occurrences of species x and y on plant i. This index range from 0 (i.e. no resource shared) to 1 (i.e. all resources shared in the same proportion).

We also calculated Lloyd's interspecific crowding index (Lloyd 1967):

$$\text{Mean crowding of species } y \text{ on species } x = \frac{\sum_i (x_i y_i)}{X}$$

where x_i and y_i are the absolute number of occurrences on plant i and X is the total number of occurrences of species x. This is a nonreciprocal index that gives a relative measure of the degree to which species y impinges on species x by the use of shared resources (Hurlbert 1978). This index can only be positive and a null value indicates no crowding at all. We calculated the degree to which *H. axyridis* impinges on native species and the degree to which native species impinge on *H. axyridis*.

For the spatio-temporal co-occurrence index, we used only collection events (expressed as the unique combination [date × 1-km^2 grid cell]) with at least two ladybird species observed. We considered only collecting events since 2004 when *H. axyridis* was already well established in Belgium. An estimate of probability of co-occurrence is given by the number of collecting events where the species x has been found with *H. axyridis* divided by total number of collecting events with species x.

Ecotope and landscape level analysis

We investigated the occurrence of *H. axyridis* in areas of conservation concern both on the ecotope and the landscape level. Therefore, we compared the average area of very valuable, valuable and less valuable land according to the Biological Valuation Map (BVM, version 2.0) for Flanders (north Belgium; Wils et al. 2006) in 1-km^2 grid cells with and without *H. axyridis* using a two-way ANOVA. We only used well-prospected grid cells i.e. cells in which at least five native species were recorded. The BVM is a uniform field-driven survey of the land cover and vegetation in north Belgium. This survey is translated into a biological valuation, largely based on plant species and vegetations. The biological value of legend units is fixed and determined by a number of ecological criteria: rarity of the biotope, presence or absence of certain species, biodiversity of the biotope, vulnerability and 'replaceability' of the biotope. Areas of very valuable, valuable and less valuable land were calculated using ArcMap (ESRI). The same analysis was performed for Belgium using CORINE Land Cover (Nunes de Lima 2005), using label 1 Land Cover classes (agricultural areas, artificial areas, forests and semi-natural areas, water bodies and wetlands) and label 3 Land Cover classes (beaches/dunes/sands, broad-leaved forest, complex cultivation patterns, coniferous forest, (dis)continuous urban fabric, estuaries, fruit trees and berry plantations, green urban areas, inland marshes, intertidal flats, land principally occupied by agriculture with significant areas of natural vegetation, mixed forest, moors and heathland, natural grasslands, pastures, peat bogs, road and rail networks and associated land, woodland-shrub, water bodies and water courses).

Phenology

The phenology of adults, larvae and pupae was investigated by calculating the number of observations (date $\times$ 1-km^2 grid cell) of *H. axyridis* (2,474 adult, 329 larval/pupal observations of *H. axyridis*) and other native species (Coccinellinae, Coccidulinae, Chilocorinae, Epilachninae) per 15-day period. Given the fact that recording efforts are fairly evenly spread over the different months of the investigated years, we do not expect a bias in phenology due to recording effort. Due to the limited number of data, observations of larvae and pupae were pooled. For native species, we used data from 1999 onward (22,790 adult, 918 larval/pupal observations), as this was the official starting date of the mapping project.

Morph types

We used a log-linear analysis of frequency tables to check for differences in two *H. axyridis* morph types (melanic = *f. spectabilis* and *f. conspicua* versus non-melanic = *f. succinea*) frequency among three consecutive years (2004–2006).

Results

Invasion and distribution in Belgium

Figure 2 shows the trend in the number of field observations of *H. axyridis* in Belgium until 2007. Although *H. axyridis* has been sold in Belgium by Koppert since 1996

(Poutsma et al. 2007) and was commonly used for biological control in Belgium since 1997 (B. Adam, pers. comm.), no observations in the wild were reported until September 2001 when the species was discovered in the cities of Ghent (28/09/2001) and Brussels (single adult on 22/10/2001). In Ghent, 27 adults (of which several freshly emerged individuals), three larvae and one pupa were discovered beating *Acer platanoides* L. and sweep netting *Solidago canadensis* L. vegetation in an urban park.

Since then, the number of *H. axyridis* observations increased (Fig. 2) and the species gradually expanded its range in Belgium (Fig. 3). The relative number of *H. axyridis* observations increased from 0.26% in 2001 to 23% in 2007 (Fig. 2). Also, in 2007, the relative number of ladybird collection events with *H. axyridis* has increased to 64%, compared to 7% in 2002 (Fig. 2).

The species rapidly spread to the east, the west and the south of Belgium (Fig. 3). In 2002, numerous feral populations were discovered in all but the westernmost Belgian provinces and in south Belgium. By the end of 2004, the species had colonised the whole Belgian territory with the exception of the southernmost parts of the country. By that time, *H. axyridis* had also started to spread in The Netherlands (Cuppen et al. 2004a, b), the UK (Brown et al. 2007a) and northern France (Coutanceau 2006).

In 2005 and 2006, the species was recorded in the southernmost parts of Belgium (Fig. 3). Initially, most observations in Belgium originated from cities and anthropogenic sites, and some could be attributed to escapes from biological control in nearby greenhouses. However, individuals were subsequently found in (semi-) natural habitats such as forests, wetlands, meadows and heathlands. Observations of eggs, larvae and pupae became numerous. Since the end of October 2002, relatively small overwintering aggregations (10–30 individuals) were observed in houses. The first large overwintering group was found in November 2002 in a concrete bunker of an old fortress in the surroundings of Antwerp. This aggregation consisted of several thousands of *H. axyridis* mixed with a few *A. bipunctata* individuals. Meanwhile, the phenomenon of October swarming and overwintering aggregates became quite common in Belgium. However, findings of very large overwintering groups, as reported in North America with estimates of hundreds of thousands individuals (Kidd et al. 1995), remained scarce. Only 17 observations were reported of overwintering aggregations of more than 500 individuals. Most aggregations were found inside private houses and concerned groups of 10–500 individuals.

Mean spreading distance increased over the period 2001–2006, which indicates that the species is still colonising new areas in Belgium (Fig. 4). The largest maximum

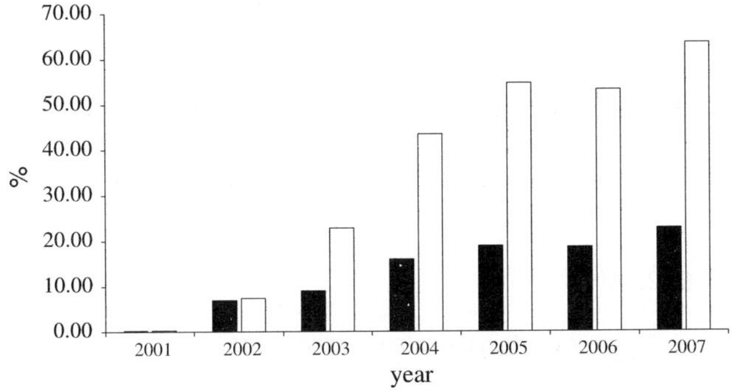

Fig. 2 Relative number of *H. axyridis* data (black bars) and collection events (unique combination of [date × 1-km^2 grid cell × observer]) with *H. axyridis* observed (empty bars) in Belgium per year

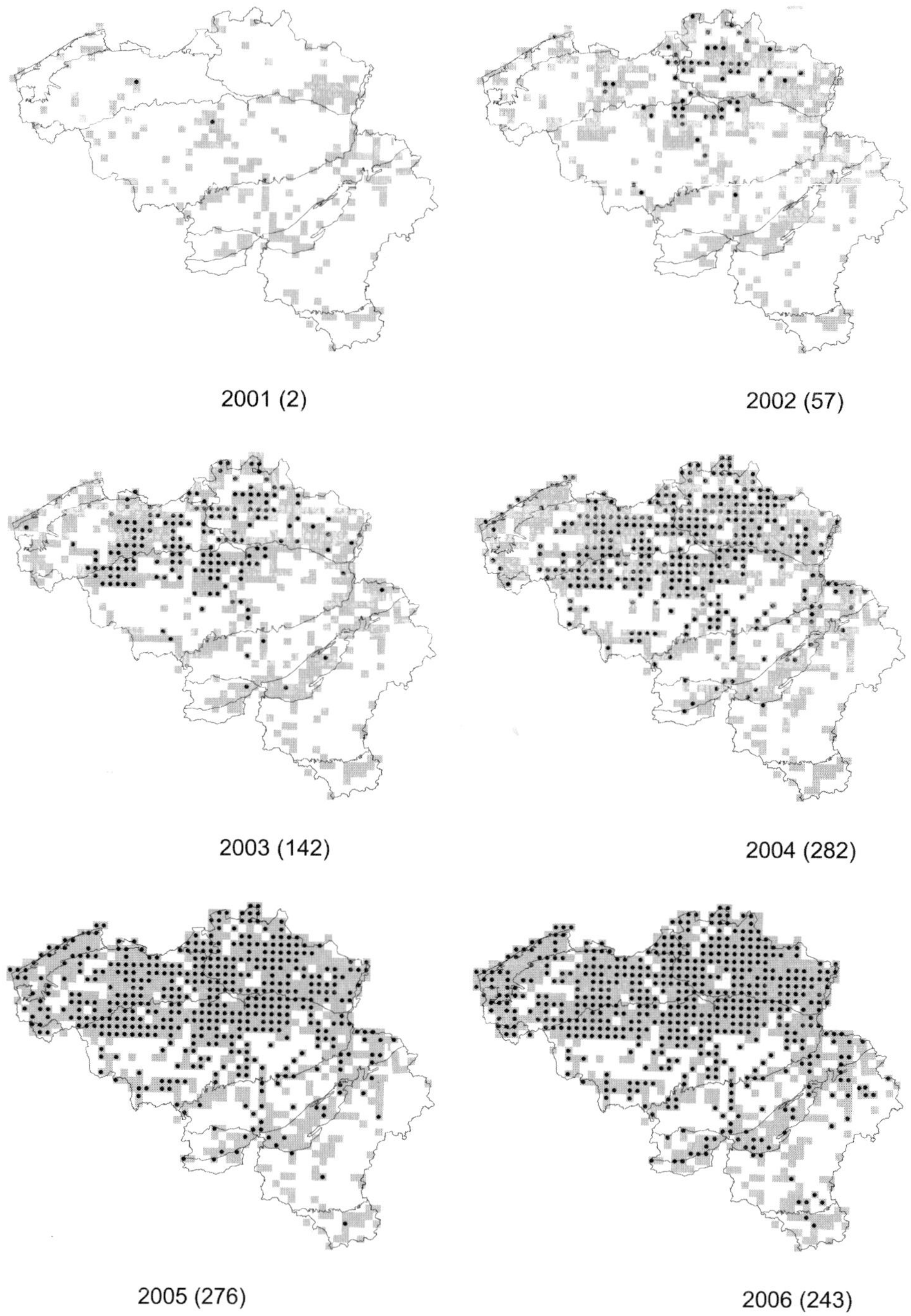

Fig. 3 Distribution of *H. axyridis* in Belgium in the period 2001–2006. Sampled grid cells are marked in grey. Recorded occupancy of the species in 5 × 5-km^2 grid cells is given between brackets

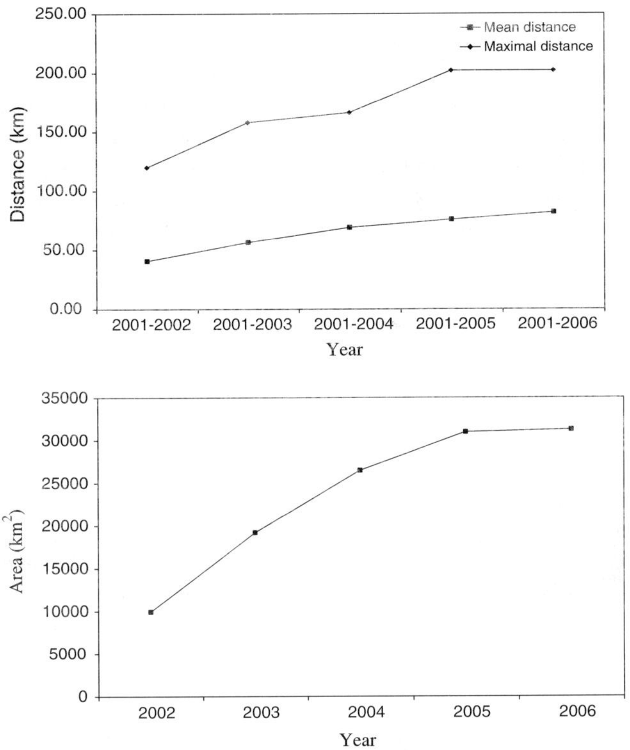

Fig. 4 Mean and maximal distances between centroids of grid cells with *H. axyridis* observations for the period 2001–2006 (top) and (bottom) area of minimum convex polygons around 1 × 1-km^2 grid cell with *H. axyridis* data (2002–2006)

colonisation distance was recorded between 2004 and 2005. Mean distances per year decreased, indicating a slow-down of the expansion (Fig. 4). The area of minimum convex polygons around occupied 1-km^2 grid cells by *H. axyridis* increased from almost 10,000-km^2 in 2002 to 31,000-km^2 in 2006, and inclined towards the total area of Belgium (32,545-km^2—Fig. 4). The number of grid cells occupied by *H. axyridis* was significantly higher in the sandy and loamy ecoregions Campine ($\chi^2_1 = 124.53$, $P < 0.001$), Loam ($\chi^2_1 = 9.24$, $P = 0.024$) and Sandyloam ($\chi^2_1 = 95.98$, $P < 0.001$) and was lower in the four southernmost, less urbanised Belgian ecoregions: the hilly loess-soil ecoregion Condroz ($\chi^2_1 = 198.65$, $P < 0.001$), the Fagne-Famenne-Calestienne ($\chi^2_1 = 85.68$, $P < 0.001$), a region with a lot of chalk hills and calcareous grasslands, the Ardennes, the highest, wet and cold Belgian region with a lot of spruce stands and some moorlands ($\chi^2_1 = 124.53$, $P < 0.001$) and the warm, calcareous southernmost ecoregion Gaume and Lorraine ($\chi^2_1 = 48.4$, $P < 0.001$).

Habitat preference

A total of 1,688 *H. axyridis* observations (40%, 24,543 individuals) had associated plant use data. In total, *H. axyridis* was found on 159 plant species, belonging to 139 genera. The number of observations of adult *H. axyridis* was highest on nettle (*Urtica dioica* L.) and on deciduous trees such as maple (*Acer* sp.), willow (*Salix* sp.), lime (*Tilia* sp.), oak (*Quercus* sp.) and birch (*Betula* sp.) but the number of observations was also high on pine tree (*Pinus* sp.), hawthorn (*Crataegus* sp.) and on a number of herbs such as reed *Phragmites australis* Steud (Table 1). Fewer data were available for *H. axyridis* larvae and pupae compared to adults but preliminary analysis showed the aforementioned tree species to be the most important plants for larvae and pupae as well.

The relative occurrence of *H. axyridis* on different vegetation strata was comparable to that of native species (Table 2) indicating that *H. axyridis* occurrence on different vegetation strata is proportional to the sampling effort. More than half of the *H. axyridis* observations (52%) were done on trees, about one-third (34%) was done on herbs and 14% on shrubs. The use of dwarf shrubs (such as *Calluna* and *Vaccinium*) seems to be lower for *H. axyridis* (0.1%) than for other native species (2.4%).

Habitat data on *H. axyridis* were available for a total of 2,005 records. The species occurred in a variety of habitat types, both in anthropogenic habitats (parks and gardens, road verges, arable fields, pastures, orchards, fallow land, abandoned railways, forest clearings, coal mine spoil piles) as in more natural habitats with conservation value (forest, wood fringes, river banks/lake shores, brushwood, scrub/hedgerows, heathlands, swamps, meadows, marches, marshland, dunes, reed-lands, poor dry grasslands and calcareous grasslands). The relative number of *H. axyridis* observations was significantly higher than native species in parks and gardens ($\chi^2_1 = 22.44$, $P < 0.001$), arable fields ($\chi^2_1 = 23.74$,

Table 1 Absolute and relative number of *H. axyridis* observations on different plant genera in Belgium

Genus	*H. axyridis* observations	%
Urtica spp.	185	11.0
Acer spp.	130	7.7
Salix spp.	112	6.6
Tilia spp.	108	6.4
Quercus spp.	91	5.4
Betula spp.	89	5.3
Pinus spp.	80	4.7
Crataegus spp.	68	4.0
Prunus spp.	54	3.2
Phragmites spp.	49	2.9
Cirsium spp.	46	2.7
Corylus spp.	46	2.7
Picea spp.	41	2.4
Alnus spp.	39	2.3
Tanacetum spp.	35	2.1

The list is restricted to genera with at least 30 *H. axyridis* observations

Table 2 Relative number, proportional to the sampling effort, of *H. axyridis* observations on different vegetation strata compared to other species

	H. axyridis	Other species ($n = 37$)
Trees (29 genera, 71 species)	51.8	48.8
Shrubs (32 genera, 43 species)	13.7	14.6
Dwarf-shrubs (3 genera, 4 species)	0.1	2.4
Herbs (268 genera, 387 species)	34.4	34.2

The list is restricted to genera with at least 20 observations

$P < 0.001$), orchards ($\chi^2_1 = 68.95$, $P < 0.001$), brushwood ($\chi^2_1 = 6.55$, $P < 0.011$) and reed-lands ($\chi^2_1 = 7.15$, $P < 0.008$) and significantly lower in heathlands ($\chi^2_1 = 31.95$, $P < 0.001$), pioneer vegetation ($\chi^2_1 = 8.96$, $P < 0.003$), poor dry grasslands ($\chi^2_1 = 8.87$, $P < 0.003$), dunes ($\chi^2_1 = 11.55$, $P < 0.001$) and calcareous grasslands ($\chi^2_1 = 18.26$, $P < 0.001$).

Potential interactions with native ladybirds

The plant use and co-occurrence indices are given in Table 3 along with the number of data used for the calculations and the number of plant genera with at least two observations for each ladybird species. The native ladybirds have been grouped according to the main biotopes used in Belgium (Branquart et al. 1999; Baugnée et al. 2001; Adriaens and Maes 2004).

The plant use similarity was highest with the four generalist species (45–77%) and particularly with *A. bipunctata* and *Propylea quatuordecimpunctata* (L.) (Coleoptera: Coccinellidae). The two other species, *C. septempunctata* and *Psyllobora vigintiduo-punctata* (L.) (Coleoptera: Coccinellidae) are generalist species that are regularly found in all vegetation strata but show a clear preference for the herb layer. Most of the trees and deciduous trees ladybird species also showed quite high similarity values (around 50%). The plant use similarity was generally lower for the species living on coniferous trees, in heathlands, or in the herb layer of diverse biotopes.

The interspecific crowding was highly asymmetric: *H. axyridis* impinged potentially more on native species than the reverse, according to plant use. The only native species that showed higher interspecific crowding on *H. axyridis* were three generalist ladybirds: *P. quatuordecimpunctata, A. bipunctata* and *C. septempunctata*. The indices of interspecific crowding of *H. axyridis* on native species were highest for ladybirds living on coniferous trees and particularly for the three pine trees specialists: *Myzia oblongoguttata* (L.), *H. quadripunctata* and *Myrrha octodecimguttata* (L.) (Coleoptera: Coccinellidae). The potential interactions were also particularly high for generalist and deciduous trees ladybirds and for the rare myrmecophilous specialist *Coccinella magnifica* Redtenbacher (Coleoptera: Coccinellidae). The potential negative effects of *H. axyridis* on native ladybirds seemed to be lower on some species relatively to other native ladybirds: *Exochomus nigromaculatus* (Goeze) (a rare dry heathland specialist), *Tytthaspis sedecimpunctata* (L.) (a mycophagous herb layer species living mainly on grasses) and *Epilachna argus* (Geoffroy) (a phytophagous species living only on *Bryonia dioica* Jacq.) (Coleoptera: Coccinellidae).

The spatio-temporal co-occurrence with *H. axyridis* was very high for almost all species: most native ladybirds were found together with *H. axyridis* in 40–65% of the collecting events. There were no clear differences among the groups of species living in different biotopes. Only two species were found with *H. axyridis* in less than 30% of the observations: *Coccinella hieroglyphica* L. (a heathland specialist: 20% of the observations) (Coleoptera: Coccinellidae) and *E. argus* (27%).

Land cover analysis

Grid cells with *H. axyridis* had more area of less valuable biotopes and less area of very valuable biotopes according to the BVM than grid cells where the species was not observed ($F_{(2,\ 1887)} = 10.901$, $P < 0.001$—Fig. 5). The area of agricultural area ($F_1 = 6.485$,

Table 3 Plant use and co-occurrence indices

Main biotope used	Ladybird species	*N* data	*N* plant genus with at least two observations	Plant use similarity with *H. axyridis*	Plant use interspecific crowding of *H. axyridis* on other species	Plant use interspecific crowding of other species on *H. axyridis*	*N* collecting events with at least two sp. since 2004	Spatio-temporal cooccurence with *H. axyridis* (%)
Generalist	*Harmonia axyridis*	1,349	84	1.00	58.19	58.19	1,371	100.0
Generalist	*Adalia bipunctata*	1,800	97	0.77	49.75	66.39	739	61.2
	Coccinella septempunctata	2,303	164	0.60	43.25	73.84	1,817	53.3
	Propylea quatuordecimpunctata	1,635	117	0.72	45.72	55.42	1,115	52.2
	Psyllobora vigintiduopunctata	867	80	0.45	34.75	22.34	545	55.6
Trees	*Exochomus quadripustulatus*	766	42	0.51	54.12	30.73	393	57.3
	Rhyzobius chrysomeloides	100	14	0.26	42.40	3.14	96	58.3
Deciduous trees	*Adalia decempunctata*	676	37	0.61	56.45	28.29	293	63.8
	Calvia decemguttata	421	28	0.51	60.21	18.79	186	67.7
	Calvia quatuordecimguttata	839	43	0.63	58.76	36.55	340	60.6
	Chilocorus renipustulatus	264	24	0.44	52.03	10.18	90	33.3
	Halyzia sedecimguttata	718	39	0.53	49.36	26.27	356	56.5
	Oenopia conglobata	192	23	0.51	61.91	8.81	94	75.5
	Oenopia impustulata	6	2	–	–	–	2	100.0
	Vibidia duodecimguttata	24	6	0.23	50.25	0.89	8	37.5
Coniferous trees	*Anatis ocellata*	269	22	0.35	59.28	11.82	128	50.8
	Aphidecta obliterata	538	22	0.28	44.99	17.94	134	56.0
	Harmonia quadripunctata	371	14	0.16	67.80	18.64	193	53.9
	Myrrha octodecimguttata	151	3	0.09	71.62	8.02	62	45.2
	Myzia oblongoguttata	53	3	0.12	65.13	2.56	33	39.4
Heathlands	*Chilocorus bipustulatus*	42	7	0.26	38.26	1.19	32	59.4
	Coccinella hieroglyphica	100	12	0.22	19.66	1.46	59	20.3
	Exochomus nigromaculatus	65	4	0.10	10.78	0.52	57	40.4

Table 3 continued

Main biotope used	Ladybird species	*N* data	*N* plant genus with at least two observations	Plant use similarity with *H. axyridis*	Plant use interspecific crowding of *H. axyridis* on other species	Plant use interspecific crowding of other species on *H. axyridis*	*N* collecting events with at least two sp. since 2004	Spatio-temporal cooccurence with *H. axyridis* (%)
Herb layer	*Coccinella quinquepunctata*	308	42	0.42	37.64	8.59	354	49.2
	Coccinella undecimpunctata	140	29	0.38	39.36	4.09	164	46.3
	Coccinula quatuordecimpustulata	44	12	0.18	30.82	1.01	121	48.8
	Hippodamia variegata	158	32	0.22	17.78	2.08	226	46.5
	Rhyzobius litura	16	4				58	46.6
	Subcoccinella vigintiquatuorpunctata	58	7	0.12	17.26	0.74	53	35.8
	Tytthaspis sedecimpunctata	126	28	0.11	8.79	0.82	415	42.4
Hygrophylous herb layer	*Anisosticta novemdecimpunctata*	177	16	0.11	21.72	2.85	120	49.2
	Coccidula rufa	49	10	0.04	7.80	0.28	60	38.3
	Coccidula scutelata	12	2				22	40.9
	Cynegetis impunctata	30	3	0.10	17.90	0.40	18	44.4
	Hippodamia septemmaculata	0	0				0	0.0
	Hippodamia tredecimpunctata	22	6	0.15	26.55	0.43	22	31.8
Myrmecophilous	*Coccinella magnifica*	44	11	0.32	57.50	1.88	52	48.1
	Platynaspis luteorubra	20	6	0.18	35.55	0.53	29	44.8
Phyto-monophagous	*Epilachna argus*	122	2	0.00	1.97	0.18	44	27.3
	Total	14,875	213	0.73	47.84	527.15	2699	50.8

"*N* plant genus data" is the number of data for which plant information is available. *N* plant genus with at least two observations" is a number of taxa (e.g. plant genus) for which we have at least two observations

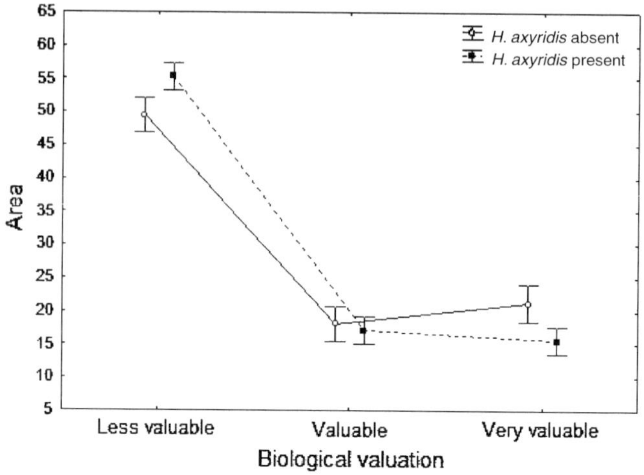

Fig. 5 Mean area (in Km^2) of less valuable, valuable and very valuable land in grid cells with and without *H. axyridis*. Error bars are SE

$P = 0.011$), artificial surfaces ($F_1 = 3.857$, $P = 0.050$), forest and semi-natural areas ($F_1 = 7.391$, $P = 0.007$) was significantly larger in grid cells with *H. axyridis* than in grid cells without the species. Using label 3 CORINE land cover classes, grid cells with *H. axyridis* had a significantly larger area of urban fabric ($F_1 = 6.625$, $P = 0.01$), estuaries ($F_1 = 4.700$, $P = 0.030$) and land principally occupied by agriculture with significant areas of natural vegetation ($F_1 = 9.122$, $P = 0.003$). Grid cells with *H. axyridis* had a significantly smaller area of broad-leaved forest ($F_1 = 10.04$, $P = 0.002$), mixed forest ($F_1 = 21.403$, $P < 0.001$), moors and heathland ($F_1 = 6.881$, $P_1 = 0.009$), peat bogs ($F_1 = 6.528$, $P = 0.011$) and water courses ($F_1 = 4.468$, $P = 0.035$).

Phenology

Our field data showed that *H. axyridis* adults were found throughout the summer, from the second half of April–first half of May to September, with a peak in the first half of August and two smaller peaks in June and September. A second peak for *H. axyridis* adults appeared in the second half of October, about 2 months later than native species and was probably related to swarming behaviour. A peak of immature stages of *H. axyridis* appeared in between the two summer peaks of adults, around the second half of June, and again in the first half of October. By May–June 2007 there were indeed several records of large groups of *H. axyridis* larvae crawling onto sunny garden walls, walls of houses, churches and cars to pupate, causing (aesthetic) nuisance to people. Adults of native species showed only two peaks, the first half of May and the first half of August, with a single peak of immature stages in between around the first half of July (Fig. 6).

Morph types

A total of 748 records (out of 5,164) of *H. axyridis* had information on morph type. The form *succinea* was the most commonly encountered morph type (72%), followed by

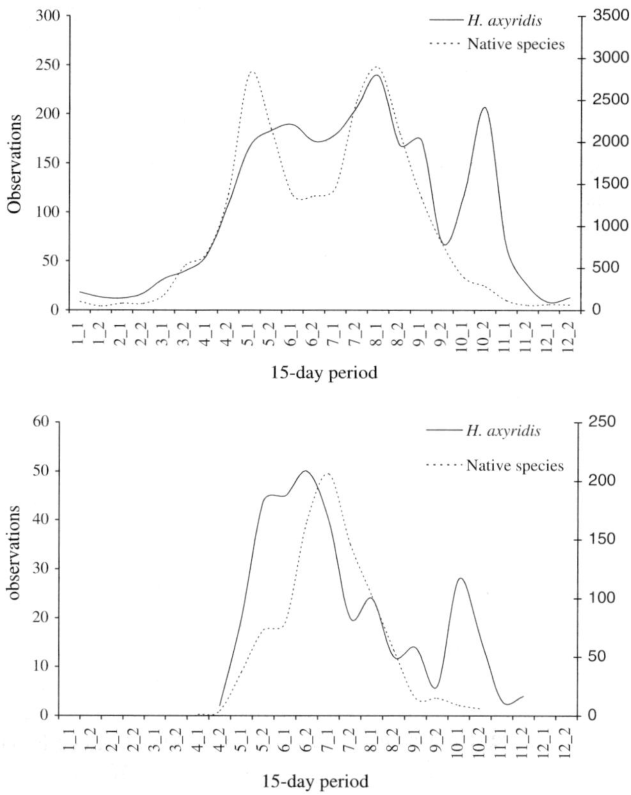

Fig. 6 Phenology of adults (top) and immature stages (bottom) of *H. axyridis* (left axis) and native species (right axis) based on the number of field observations. Numbers on *x*-axis refer to 15 day-periods throughout the year (1_1 = first half of January, 1_2 = second half of January, etc.)

spectabilis (19%) and *conspicua* (5%). About 4% belonged to one of two melanic morphs *conspicua/spectabilis*. These relative frequencies were comparable to the frequencies in counted overwintering groups ($N = 1{,}838$): 70% *succinea*, 18% *spectabilis*, 8% *conspicua* and 5% *conspicua/spectabilis* and to the frequencies in various single populations. The ratio of melanic and non-melanic *H. axyridis* did not differ among years (2004–2006) (Log linear analysis $\chi^2_3 = 0.488$, $P = 0.92$).

Discussion

About 4 years after its discovery in 2001 in the north of Belgium, *H. axyridis* had already spread throughout the Belgian territory. Based on its current distribution in Belgium, *H. axyridis* can clearly be considered as a eurytopic species with large amplitude for geomorphologic and climatic conditions in Belgium. Analysis of occupied and well-prospected unoccupied grid cells however, showed that the species is less common in the southern parts of Belgium. This pattern might be influenced by lower sampling effort in the southern part of Belgium. However, we believe it is the result of invasion history of the species since the species is less common on typical ecotopes present in these regions e.g. calcareous grasslands, moorlands and forests. The species showed a clear preference for the sandy and loamy regions in the north of Belgium. These are also the regions that are most species-rich in native coccinellids (Adriaens and Maes 2004). Hotspots for other arthropod groups, such as butterflies, are also located here (Maes et al. 2005).

The results on substratum plants were only partly consistent with literature. *H. axyridis* is generally regarded a (semi-)arboreal species, occurring mostly on deciduous trees (e.g. Hodek 1973; Iablokoff-Khnzorian 1982) but also in various herbaceous habitats (Koch et al. 2006). In Belgium, *H. axyridis* was indeed observed mostly on deciduous trees, but was also frequently found on pine trees and in the herb layer. Iablokoff-Khnzorian (1982) mentioned *H. axyridis* on pine trees in China, but added that this might in fact concern *H. yedoensis* (Takizawa) (Coleoptera: Coccinellidae). Our results clearly showed that, at least in Belgium, *H. axyridis* frequently uses pine trees. However, these results have to be interpreted with care. As aphids are more scattered in the herb layer an effect of dilution might occur that could explain the lower number of individuals on herbs. Moreover, sweeping, generally, does not allow the observer to identify individual plant species in the herb layer and observers were probably more able to determine tree species than herb species.

Plant use and co-occurrence indices only give a coarse idea of the potential impact of *H. axyridis* on native species because they do not take into account the real trophic resources (aphids or other prey that are not necessarily limiting) and intra-guild predation events (Wissinger 1992). But co-occurrence and shared habitat are preliminary conditions for competition and intra-guild predation (Wissinger 1992). Based on the niche overlap analyses, generalist and deciduous tree ladybirds are probably the most negatively affected native species. Moreover, our results showed that pine tree specialist ladybirds, although they have a quite different niche, could also potentially be affected by *H. axyridis* interactions because they are highly specialised on a resource frequently used by *H. axyridis*. It seems that the most threatened and localised species, heathland and wetland specialists, would be less affected relative to other species. This is consistent with the results on habitat preference and the area of different land cover classes according to the biological valuation and CORINE land cover maps. Few methods are presently available to control *H. axyridis* in natural environments (Kenis et al. 2007). Our data suggest that control measures should focus on managing *H. axyridis* populations outside semi-natural areas (e.g. mechanical methods to control aggregation) to prevent the species spreading into more natural areas of conservation concern.

Habitat data showed that *H. axyridis* is frequently found in a broad range of semi-natural biotopes. However, biotope and landscape level data and data on land cover showed that the species is more frequently found in more urbanised and anthropogenic landscapes and less frequent in landscapes with forests and other natural elements (moors and heathland, peat bogs, water courses). This could explain the apparent slower colonisation rate of the four southernmost natural ecoregions that are less urbanised and much more wooded.

Harmonia axyridis is considered bivoltine in most of its range. In central Japan, the species had a bivoltine cycle (Sakurai et al. 1988). In its invasive range, *H. axyridis* was also reported as bivoltine, e.g. in a Minnesota agricultural landscape (Koch and Hutchison 2003) and in Oregon (Lamana and Miller 1996). However, up to four or five generations per year have been observed, for example in Italy (Bazzocchi et al. 2004) and Greece (Katsoyannos et al. 1997). In southern France, two generations were reported (Ongagna et al. 1993) and evidence was presented of bi-voltinism in the UK population (Brown et al. 2007b). The same seems to be true for the Belgian *H. axyridis* population. Despite species-specific variations, the main pupal eclosion period for most native ladybird species is from mid-July to mid-August (Majerus 1994). The adult phenology of native species, based on the number of observations, is consistent with this pattern, clearly showing two adult peaks in May and August and a July peak of immature stages in between. *Harmonia axyridis* adults and immature stages, however, peak again in October, very late in the season and about 2 months later than native species. This might indicate that *H. axyridis* is more able to exploit alternative resources and reproduce in the absence of aphid prey. Berkvens et al. (2007) also suggested that *H. axyridis* might have a competitive advantage over indigenous species because it can exploit pollen when insect prey is scarce. Immature stages of first generation *H. axyridis* showed an overlap with native species, indicating a high potential for intraguild predation (Pell et al. 2007).

The extreme phenotypic variability in colour morphs of *H. axyridis* appears to have a genetic basis (for a review see Komai 1956). The occurrence of different colour morphs is subject to spatial variation (Dobzhansky 1933) and could give information on the origin of *H. axyridis* in Belgium. The observed frequencies of the various colour morphs in Belgium were largely the same as in the rest of Europe, with a mixture of the three morphs present and with *f. succinea* predominating (Brown et al. 2007a; Majerus and Roy 2005). This might provide evidence that introduced *H. axyridis* in Belgium came from a population with a homogenous colour morph profile. We could find no evidence of a decline in the percentage of melanic forms, as has been reported in the UK where it has been related to a separate introduction of a population with a different colour morph profile (Brown et al. 2007a). The *equicolor* morph type, observed several times in The Netherlands, has not yet been recorded in Belgium.

We conclude that large-scale field data provide valuable information on the invasion process and the potential negative impact of *H. axyridis* on native ladybird species. Future work on feral *H. axyridis* populations in Belgium is needed to estimate its impact on the indigenous fauna, aphidophagous or other, and to acquire insights in the underlying mechanisms. Until now, monitoring the abundance of *H. axyridis* and native ladybirds has been restricted to the urban environment of Brussels in Belgium. More quantitative data are, however, needed to determine the exact impact of *H. axyridis* on the native fauna in the invasion range. A more extensive monitoring scheme comprising different habitat types, including those with a high conservation value, should be applied to assess the impact of this invasive alien.

Acknowledgements We are very grateful to the numerous volunteers of the Belgian Ladybird Working Group for providing us with the field data on *H. axyridis* in Belgium. We thank three anonymous referees for useful comments on the manuscript. We are grateful to Helen Roy for correcting the English and to Dirk Bauwens for help in statistical analyses.

References

Adriaens T, Gysels J (2002) Veelkeurig aziatisch lieveheersbeestje, van biologische bestrijder tot pestsoort? Natuur Focus 1(4):148–152

Adriaens T, Maes D (2004) Voorlopige verspreidingsatlas van lieveheersbeestjes in vlaanderen, resultaten van het lieveheersbeestjesproject van de jeugdbonden. Bertram 2(1bis):1–69

Adriaens T, Branquart E, Maes D (2003) The multicoloured Asian ladybird *Harmonia axyridis* Pallas (Coleoptera : Coccinellidae), a threat for native aphid predators in Belgium? Belg J Zool 133(2): 195–196

Alyokhin A, Sewell G (2004) Changes in a lady beetle community following the establishment of three alien species. Biol Invasions 6:163–471

Baugnée J-Y, Branquart E, Maes D (2001) Velddeterminatietabel voor de lieveheersbeestjes van België (Chilocorinae, Coccinellinae en Epilachninae). Jeugdbond voor Natuurstudie en Milieubescherming (Gent), Jeunes & Nature asbl (Wavre) i.s.m. Instituut voor Natuurbehoud (Brussel)

Bazzocchi GG, Lanzoni A, Accinelli G, Burgio G (2004) Overwintering, phenology and fecundity of *Harmonia axyridis* in comparison with native coccinellid species in Italy. BioControl 49(3):245–260

Berkvens N, Bonte J, Berkvens D, Deforce K, Tirry L, De Clercq P (2007) Pollen as an alternative food for *Harmonia axyridis*. BioControl (this issue). doi:10.1007/s10526-007-9128-7

Branquart E, Baugnée J-Y, Mairesse JL, Gaspar C (1999) Inventaire de la faune des coccinelles de Wallonie (Chilocorinae, Coccinellinae et Epilachninae). Rapport final. Gembloux

Brown MW, Miller SS (1998) Coccinellidae (Coleoptera) in apple orchards of Eastern West Virginia and the impact of invasion by *Harmonia axyridis*. Entomological News 109(2):143–151

Brown PMJ, Adriaens T, Bathon H, Cuppen J, Goldarazena A, Hägg T, Kenis M, Klausnitzer BEM, Kovář I, Loomans AJM, Majerus MEN, Nedved O, Pedersen J, Rabitsch W, Roy HE, Ternois V, Zakharov IA, Roy DB (2007a) *Harmonia axyridis* in Europe: spread and distribution of a non-native coccinellid. BioControl (this issue). doi:10.1007/s10526-007-9132-y

Brown PMJ, Roy HE, Rothery P, Roy DB, Ware RL, Majerus MEN (2007b) *Harmonia axyridis* in Great Britain: analysis of the spread and distribution of a non-native coccinellid. BioControl (this issue). doi:10.1007/s10526-007-9124-y

Burgio G, Santi F, Maini S (2002) On intra-guild predation and cannibalism in *Harmonia axyridis* (Pallas) and *Adalia bipunctata* L. (Coleoptera: Coccinellidae). Biol Control 24(2):110–116

Colunga-Garcia M, Gage SH (1998) Arrival, establishment, and habitat use of the multicolored Asian lady beetle (Coleoptera: Coccinellidae) in a Michigan landscape. Environ Entomol 27(6):1574–1580

Cottrell TE, Yeargan KV (1998) Intraguild predation between an introduced lady beetle, *Harmonia axyridis* (Coleoptera : Coccinellidae), and a native lady beetle, *Coleomegilla Maculata* (Coleoptera: Coccinellidae). J Kans Entomol Soc 71(2):159–163

Coutanceau J (2006) *Harmonia axyridis* (Pallas, 1773): une coccinelle asiatique introduite, acclimatée et en extension en France. Bull Soc entomol Fr 111(3):395–401

Cuppen J, Heijerman Th, van Wielinck P, Loomans AJM (2004a) Het lieveheersbeestje *Harmonia axyridis* in Nederland: een aanwinst voor onze fauna of een ongewenste indringer (Coleoptera: Coccinellidae)? Nederlandse Faunistische Mededelingen 20:1–12

Cuppen J, Heijerman Th, van Wielinck P, Loomans AJM (2004b) Veelkleurig aziatisch lieveheersbeestje in opmars. Natura 66(3):3

De Clercq P, Bonte M, Van Speybroeck K, Bolckmans K, Deforce K (2005) Development and reproduction of *Adalia bipunctata* (Coleoptera: Coccinellidae) on eggs of *Ephestia kuehniella* (Lepidoptera: Phycitidae) and pollen. Pest Manag Sci 61(11):1129–1132

Dobzhansky T (1933) Geographical variation in lady beetles. Am Nat 67:97–126

Feinsinger P, Spears E, Poole RW (1981) A simple measure of niche breadth. Ecology 62(1):27–32

Ferran A, Niknam H, Kabiri F, Picart JL, Deherce C, Brun J, Iperti G, Lapchin L (1996) The use of *Harmonia axyridis* larvae (Coleoptera: Coccinellidae) against *Macrosiphum Rosae* (Hemiptera: Sternorrhyncha: Aphididae) on rose bushes. Eur J Entomol 93(1):59–67

Gordon RD (1985) The Coleoptera (Coccinellidae) of America north of Mexico. J New York Entomol Soc (93):1–912

Hautier L (2003a) Impacts sur l'entomofaune indigène d'une coccinelle exotique utilisée en lutte biologique. TFE Université Libre de Bruxelles, IGEAT, Brussels

Hautier L (2003b) Intraguild predation of *Adalia bipunctata* by *Harmonia axyridis*. Presentation. Workshop Ladybirds and biological control in Belgium, Brussels, November 2003

Hautier L (2006) Intraguild predation by *Harmonia axyridis*. In: Branquart E, Baus E, Pieret N, Vanderhoeven S, Desmet P (eds) SOS Invasions—uitheemse invasieve soorten in België, Brussels, March 9, 2006

Hironori Y, Katsuhiro S (1997) Cannibalism and interspecific predation in two predatory ladybirds in relation to prey abundance in the field. Entomophaga 42(1–2):153–163

Hodek I (1973) Life history and biological properties. In: Hodek I (ed) Biology of Coccinellidae. Dr. W. Junk, The Hague

Hurlbert SH (1978) The measurement of niche overlap and some relatives. Ecology 59(1):67–77

Iablokoff-Khnzorian SM (1982) Les coccinelles (Coleoptera: Coccinellidae). Tribu Coccinellini des régions Paléarctique et Orientale. Boubée, Paris

Iperti G, Bertrand E (2001) Hibernation of *Harmonia axyridis* (Coleoptera: Coccinellidae) in South-Eastern France. Acta Soc Zool Bohem (65):207–210

Katsoyannos P, Kontodimas DC, Stathas GJ et al (1997) Establishment of *Harmonia Axyridis* on citrus and some data on its phenology in Greece. Phytoparasitica 25(3):183–191

Kenis M, Roy HE, Zindel R, Majerus MEN (2007) Current and potential management strategies against *Harmonia axyridis*. BioControl (this issue). doi:10.1007/s10526-007-9136-7

Kidd KA, Nalepa CA, Day ER, Waldvogel MG (1995) Distribution of *Harmonia Axyridis* (Pallas) (Coleoptera, Coccinellidae) in North-Carolina and Virginia. Proc Entomol Soc Wash 97(3):729–731

Koch RL (2003) The multicolored Asian lady beetle, *Harmonia axyridis*: a review of its biology, uses in biological control, and non-target impacts. J Insect Sci 3(32):1–16

Koch RL, Hutchison WD (2003) Phenology and blacklight trapping of the multicolored Asian lady beetle (Coleoptera: Coccinellidae) in a Minnesota agricultural landscape. J Entomol Sci 38(3):477–480

Koch RL, Venette RC, Hutchison WD (2006) Invasions by *Harmonia axyridis* (Pallas) (Coleoptera: Coccinellidae) in the Western Hemisphere: implications for South America. Neotrop Entomol 35(4):421–434

Komai T (1956) Genetics of ladybeetles. Adv Genet 8:155–188

Lamana ML, Miller JC (1996) Field observations on *Harmonia axyridis* Pallas (Coleoptera: Coccinellidae) in Oregon. Biol Control 6(2):232–237

Lloyd M (1967) 'Mean crowding'. J Anim Ecol 36(1):1–30

Maes D, Bauwens D, De Bruyn L, Anselin A, Vermeersch G, Van Landuyt W, De Knijf G, Gilbert M (2005) Species richness coincidence: conservation strategies based on predictive modelling. Biodivers Conserv 14(6):1345–1364

Maignet P (2002) Bilan de l'introduction en France de la coccinelle *Harmonia axyridis* Pallas en lutte biologique contre les pucerons. Deuxième Conférence Internationale sur les Moyens Alternatifs de Lutte contre les Organismes Nuisibles aux Végétaux, March 4, 2002

Majerus MEN (1994) Lady birds. Harper-Collins, London

Majerus MEN, Roy DB (2005) Scientific opportunities presented by the arrival of the Harlequin ladybird, *Harmonia axyridis*, in Britain. Antenna 29:196–208

Michaud JP (2002) Invasion of the Florida citrus ecosystem by *Harmonia axyridis* (Coleoptera: Coccinellidae) and asymmetric competition with a native species, *Cycloneda Sanguinea*. Environ Entomol 31(5):827–835

Michaud JP (2004) Natural mortality of Asian citrus psyllid (Homoptera: Psyllidae) in central Florida. Biol Control 29(2):260–269

Nunes de Lima MV (2005) CORINE land cover (updating for the year 2000). European Commission, Italy

Ongagna P, Giuge L, Iperti G, Ferran A (1993) Life-cycle of *Harmonia axyridis* (Col, Coccinellidae) in its area of introduction—South-Eastern France. Entomophaga 38(1):125–128

Osawa N (2000) Population field studies on the Aphidophagous ladybird beetle *Harmonia axyridis* (Coleoptera: Coccinellidae): resource tracking and population characteristics. Popul Ecol 42(2): 115–127

Osawa N, Nishida T (1992) Seasonal-variation in elytral color polymorphism in *Harmonia axyridis* (the ladybird beetle)—the role of nonrandom mating. Heredity 69:297–307

Ottart N (2005) Impacts de la coccinelle invasive *Harmonia axyridis* sur les populations de coccinelles indigènes à Bruxelles. TFE Université Libre de Bruxelles, Ecole Interfacultaire de BioIngénieur, Brussels

Pell JK, Baverstock J, Roy HE, Ware RL, Majerus MEN (2007) Intraguild predation involving *Harmonia axyridis*: a review of current knowledge and future perspectives. BioControl (this issue). doi:10.1007/s10526-007-9125-x

Poutsma J, Loomans AJM, Aukema B, Heijerman T (2007) Predicting the potential geographical distribution of the harlequin ladybird, *Harmonia axyridis* using the CLIMEX model. BioControl (this issue). doi:10.1007/s10526-007-9140-y

Rodgers AR, Carr AP (1998) HRE: The Home Range Extension for ArcView. Thunder Bay, Ontario, Canada, Ontario Ministry of Natural Resources, Centre for Northern Forest Ecosystem Research

Sakurai H, Takeda S, Kawai T (1988) Diapause regulation in the lady beetle, *Harmonia axyridis*. In: Niemczyk E, Dixon AFG (eds) Ecology and effectiveness of Aphidophaga, S.P.B. Academic Publishing, The Hague

San Martin G. (2003) Etude de l'impact de l'urbanisation sur les populations de coccinelles à Bruxelles. TFE Université Libre de Bruxelles, Faculté des Sciences, Service d'Eco-Ethologie Evolutive

San Martin G, Adriaens T, Hautier L, Ottart N (2005) La coccinelle asiatique *Harmonia axyridis*. Insectes 136(1):7–11

Serpa L, Schanderl H, Brito C, Soares AO (2003) Temporal and geographic variations in elytral colour of the polymorphic ladybird beetle *Harmonia axyridis* Pallas (Coleoptera:Coccinellidae): the role of fitness and genetic transmission. In: Proceedings of the 8th international symposium on ecology of Aphidophaga

Tedders WL, Schaefer PW (1994) Release and establishment of *Harmonia axyridis* (Coleoptera, Coccinellidae) in the Southeastern United States. Entomol News 105(4):228–243

van Lenteren JC, Babendreier D, Bigler F, Burgio G, Hokkanen HTM, Kuske S, Loomans AJM, Menzler-Hokkanen I, van Rijn PCJ, Thomas MB, Tommasini MG, Zeng Q-Q (2003) Environmental risk assessment of exotic natural enemies used in inundative biological control. BioControl 48(1):3–38

van Lenteren JC, Loomans AJM, Babendreier D, Bigler F (2007) *Harmonia axyridis*: an environmental risk assessment for Northwest Europe. BioControl (this issue). doi:10.1007/s10526-007-9120-2

Williamson M (2006) Explaining and predicting the success of invading species at different stages of invasion. Biol Invasions 8(7):1561–1568

Wils C, Paelinckx D, Adams Y, Berten B, Bosch H, De Knijf G, De Saeger S, Demolder H, Duelinckx R, Lust P, Oosterlynck P, Scheldeman K, t'Jollyn F, Van Hove M, Vandebussche V, Vriens L (2006) Biologische Waarderingskaart van het Vlaamse Gewest. Instituut voor Natuur- en Bosonderzoek (INBO), Brussels

Wissinger SA (1992) Niche overlap and the potential for competition and intraguild predation between size-structured populations. Ecology 73:1431–1444

Yasuda H, Ohnuma N (1999) Effect of cannibalism and predation on the larval performance of two ladybird beetles. Entomol Exp Appl 93(1):63–67

Yasuda H, Kikuchi T, Kindlmann P, Sato S (2001) Relationships between attack and escape rates, cannibalism, and intraguild predation in larvae of two predatory ladybirds. J Insect Behavior 14(3):373–384

Phenotypic variation in invasive and biocontrol populations of the harlequin ladybird, *Harmonia axyridis*

Eric Lombaert · Thibaut Malausa · Rémi Devred · Arnaud Estoup

Abstract Despite numerous releases for biological control purposes during more than 20 years in Europe, *Harmonia axyridis* failed to become established until the beginning of the 21st century. Its status as invasive alien species is now widely recognised. Theory suggests that invasive populations should evolve toward greater phenotypic plasticity because they encounter differing environments during the invasion process. On the contrary, populations used for biological control have been maintained under artificial rearing conditions for many generations; they are hence expected to become specialised on a narrow range of environments and show lower phenotypic plasticity. Here we compared phenotypic traits and the extent of adaptive phenotypic plasticity in two invasive populations and two populations commercialized for biological control by (i) measuring six phenotypic traits related to fitness (eggs hatching rate, larval survival rate, development time, sex ratio, fecundity over 6 weeks and survival time of starving adults) at three temperatures (18, 24 and 30°C), (ii) recording the survival rate and quiescence aggregation behaviour when exposed to low temperatures (5, 10 and 15°C), and (iii) studying the cannibalistic behaviour of populations in the absence of food. Invasive and biocontrol populations displayed significantly different responses to temperature variation for a composite fitness index computed from the traits measured at 18, 24 and 30°C, but not for any of those traits considered independently. The plasticity measured on the same fitness index was higher in the two invasive populations, but this difference was not statistically significant. On the other hand, invasive populations displayed significantly higher survival

E. Lombaert (✉)
Centre de Recherches de Sophia Antipolis, Unité de Lutte Biologique, INRA, 400 Route des Chappes, BP 167, 06903 Sophia Antipolis Cedex, France
e-mail: lombaert@sophia.inra.fr

T. Malausa · R. Devred
INRA, UMR ROSE (INRA/UNSA), 06903 Sophia-Antipolis Cedex, France
e-mails: tmalausa@sophia.inra.fr; devred@sophia.inra.fr

A. Estoup
INRA, UMR CBGP (INRA/IRD/Cirad/Montpellier SupAgro), Campus international de Baillarguet, CS 30016, 34988 Montferrier-sur-Lez Cedex, France
e-mail: estoup@supagro.inra.fr

Biological Control to Invasion: Ladybird Harmonia axyridis. H.E. Roy, E. Wajnberg (eds.), doi: 10.1007/978-1-4020-6939-0_7, © IOBC 2007 and higher phenotypic plasticity when entering into quiescence at low temperatures. In addition, one invasive population displayed a singular cannibalistic behaviour. Our results hence only partly support the expectation of increased adaptive phenotypic plasticity of European invasive populations of *H. axyridis*, and stress the importance of the choice of the environmental parameters to be manipulated for assessing phenotypic plasticity variation among populations.

Keywords Adaptive phenotypic plasticity · Alien species · Biological control · *Harmonia axyridis* · Biological invasion

Introduction

The Asian ladybird beetle *Harmonia axyridis* (Pallas) was first brought into Europe in 1982 (Coutanceau 2006). The species was studied in southern France in the laboratory and in experimental greenhouses during the eighties with a view to using this coccinellid as a biological control agent of pest aphids and scale insects. Large experimental as well as commercial releases *in natura* were then performed until 2003 in many European countries (Coutanceau 2006). Despite those numerous releases during more than 20 years, the species failed to become established until 2000–2001 when it started to be observed and subsequently spread into Germany and Belgium. It is now present in many European countries from southern France to Denmark (Brown et al. 2007a). Its status as invasive species is now widely recognised for a number of reasons including its impact on functional biodiversity (van Lenteren et al. 2007). Despite some differences, the European situation parallels the North American experience where *H. axyridis* was first released in 1916 but first established populations were not observed until 1988 after which there was a very rapid spread across the continent (Koch 2003). In both cases, whether the invasive populations resulted from intentional introductions, accidental migrants or both remains unknown. Therefore, the chronology of *H. axyridis* invasions is symptomatic of a general recurrent question around invasion biology: why now and not before?

Understanding the factors driving biological invasions has become of major interest within the past few decades. This is because the recent rise of human activities has greatly accelerated the invasion rate of non-native species and some of these invasions have dramatic economical, ecological or human-health consequences (Mack et al. 2000). However, among the species which arrive in a new location, only few persist and even less spread (Williamson and Fitter 1996). The main reason for that is the unsuitability of the site and/or the environmental stochasticity which promote local extinction of non-adapted populations. Therefore, of particular interest are (i) whether key-characteristics which predispose a species to successful establishment exist and (ii) whether those characteristics evolve during the geographical spread following the establishment phase (Kolar and Lodge 2001; Lee et al. 2007). One mechanism that is frequently suggested in this context is adaptive phenotypic plasticity (Agrawal 2001; Kaufman and Smouse 2001; Yeh and Price 2004; Richards et al. 2006; Geng et al. 2007; Ghalambor et al. 2007).

Adaptive phenotypic plasticity can be defined as a set of processes historically selected to produce the highest fitness among different environments by means of various plastic traits (Debat and David 2001). The plasticity of a trait can be assessed by determining the pattern of its phenotypic expression in different environments (called a reaction norm). Absolute adaptive phenotypic plasticity should lead to a flat fitness reaction norm (i.e. fitness homeostasis, Scheiner 1993; Richards et al. 2006). Theory suggests that invasive populations are expected to evolve toward greater phenotypic plasticity because of the wide range of environments encountered during the invasion process. However, genetic assimilation, the evolutionary loss of plasticity after successful colonization of a novel environment, should be taken into account as an alternative scenario (West-Eberhard 2003). Some studies that have previously addressed the role of plasticity in invasions have reported increased levels of plasticity in invasive species or populations. The overall evidence remains however limited so that it is premature to draw any firm and general conclusions from these results (reviewed in Richards et al. 2006). Moreover, there is likely to be a bias towards publishing positive results. *Harmonia axyridis* is a suitable biological model to test such predictions because its invasion has been far from instantaneous, despite the numerous intentional releases, and variation in level of plasticity has already been described in this species (Grill et al. 1997; Preziosi et al. 1999). While invasive populations of *H. axyridis* are expected to show high adaptive phenotypic plasticity, biocontrol populations which have long failed to invade are expected to display low phenotypic plasticity. This hypothesis rests on the low variability of the artificial rearing conditions which should lead to the loss of adaptive plasticity (Masel et al. 2007).

In this paper, we compare the adaptive phenotypic plasticity displayed by two invasive (from England and southern France) and two biocontrol populations of *H. axyridis*. In a first experiment, we measured six phenotypic traits related to fitness (eggs hatching rate, larval survival rate, development time, sex ratio, fecundity over 6 weeks and survival time of starving adults) at three temperatures (18, 24 and 30°C). In a second experiment, we recorded the survival rate and quiescence aggregation behaviour when exposed to low temperatures (5, 10 and 15°C). Finally, we studied the cannibalistic behaviour of populations in the absence of food. The implications of our results in relation to the choice of the environmental parameters to be manipulated for assessing phenotypic plasticity variation among *H. axyridis* populations are discussed.

Material and methods

Population sampling and rearing

Four populations were used in this study. Two strains maintained in the laboratory for several years and used as biological control agents were provided by the firm BIOTOP (Valbonne, France): the biocontrol strain, commercialized between the years 1995 and 1999 all over Europe (hereafter referred to as population Biocontrol 1) and the so called flightless strain, selected in the late 1990's from the Biocontrol 1 strain for its incapacity to fly and disperse (Tourniaire et al. 2000a, b) and commercialised since 2000 in Europe (hereafter referred to as population Biocontrol 2). Although the biocontrol strains 1 and 2 evolved separately for 50–100 generations and phenotypic traits are supposed to evolve quickly, they cannot be considered as strictly independent evolutionary replicates of biocontrol populations. Two other samples were collected in the wild from two invasive populations in Europe. The first one, referred to as population London, was collected on September 2006 in London, England (51°28′44″ North; 00°09′02″ West) where *H. axyridis* has been reported since 2004 (Majerus et al. 2006; Brown et al. 2007b). The second one, referred to as population Roquefort, was collected on October 2006 in Roquefort-les-Pins, Southern France (43°40′44″ North; 07°02′26″ East), where it has been observed for at least 3 years including 2007 (Christine Delclos, Pers. Com. and Pers. Observation).

Before the experiment started, we maintained all four populations in the lab for two generations, under strict control conditions, in order to avoid bias due to maternal effects. During these two generations, populations were exclusively fed with ionized *Ephestia kuehniella* (Lepidoptera: Pyralidae) eggs and reared at constant environmental conditions (20°C; 60% HR; L:D 16:8). At generation F2, males and females were separated immediately after emergence to avoid any mating event. They were then maintained in the same environmental conditions for 2 weeks in order to insure their reproductive maturity at the beginning of the experiment. Fifty families of each population were then randomly created by pooling one male and one female in a cylindrical box (height = 3 cm; diameter = 8.5 cm) and temperature was increased to 24°C. Eggs produced by these families were then used to start the experiments. We further used 30 randomly chosen families from the 50 initially created.

Experiment 1: life history traits and phenotypic plasticity

The protocol used for this experiment is summarized in Fig. 1. At the beginning of the experiment, 45 eggs (3 × 15) of each family were equally distributed in three different rectangular boxes (length = 25 cm; width = 12 cm; height = 8 cm). The three boxes were placed in three separate rooms differing by their temperature: 18, 24 and 30°C. Relative humidity was maintained at ~60% in all rooms. After hatching, larvae were fed to excess every 2 days with fresh ionized eggs of *E. kuehniella* until adulthood. Several traits were measured for each box: (i) the number of hatched eggs among the 15 initially placed in each box (egg survival), (ii) the number of individual reaching adulthood (larval survival), (iii) the total development time (from egg laying to adult emergence) of each individual, and (iv) the sex of each adult. The boxes were then discarded after individuals were picked for subsequent experiments.

One adult female was picked from each box to measure fecundity. Each female was placed in a cylindrical box (see above) with one male from another box of the same population and the same temperature treatment. Eggs were counted and removed every two days during 6 weeks (42 days).

We also measured the lifespan of starving adults (male or female) at each temperature (one individual per family for each temperature) by placing each individual just after emergence in a small cylindrical box (height = 2 cm; diameter = 5 cm) with a damp piece of cotton wool. These small boxes were monitored daily and the date of death of each individual was recorded.

A global fitness index (w) for each family in each environment was calculated from four of the above traits using the following equation:

$$w = P_h * N_l * (1 - S_r) * F_{tot} \qquad (1)$$

where P_h is the proportion of hatched eggs, N_l the number of individuals reaching adulthood, S_r the sex ratio (expressed as the proportion of males) and F_{tot} the total fecundity of the female after 6 weeks of adulthood.

For this fitness index, the adaptive phenotypic plasticity was quantified by computing the relative distance plasticity index (RDPI) proposed by Valladares et al. (2006). For each population, the RDPI was computed using the following equation:

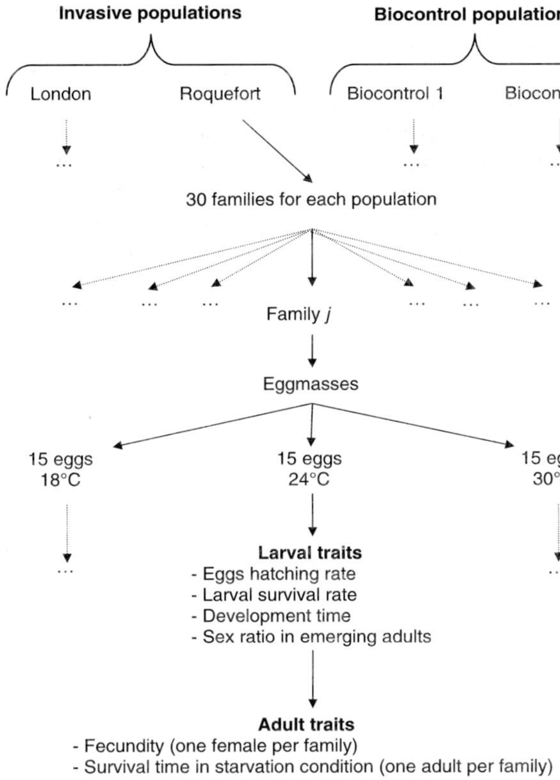

Fig. 1 Protocol design of experiment 1

$$RDPI_w = \frac{1}{n} \sum_j \sum_{i,i'} \frac{|w_{ij} - w_{i'j}|}{w_{ij} + w_{i'j}} \tag{2}$$

where j is the family index, i and i' are temperature indexes ($i \neq i'$) and n is the total number of relative distances. The $RDPI_w$ ranges from 0 to 1, and a value close to 0 means that the fitness is well canalised among environments, and thus that adaptive phenotypic plasticity is potentially high.

Experiment 2: quiescence

About 15 new-laid eggs of the 30 couples of each F2 population were randomly pooled in five rectangular boxes (length = 25 cm; width = 25 cm; height = 8 cm). Indeed, contrary to experiment 1, we could not use family structure for practical reasons (low number of individuals per family available at this stage of the experiment and reduced space available in environmental test chambers). Individuals were raised until adulthood in constant abiotic conditions (24°C; 60% HR; L:D 16:8) and fed with fresh ionized eggs of *E. kuehniella*. Temperature was then lowered to 18°C for 1 month. Twelve groups of 14 individuals (seven males and seven females) of each population were then put into rectangular boxes (length = 25 cm; width = 12 cm; height = 8 cm) with a damp piece of cotton wool, but no food. The bottom of the box was covered with a piece of corrugated cardboard (length = 25 cm; width = 12 cm). For each population, four boxes were then placed in a climatic chamber. Three climatic chambers were used in order to test three temperatures (5, 10 and 15°C; 60% HR; L:D 12:12). After 5 weeks, we measured at each temperature (i) the number of live individuals in each box and (ii) the proportion of live individuals that aggregated under the cardboard, revealing quiescence aggregation-like behaviour.

Because we did not use a family structure here as in experiment 1, we could not calculate a RDPI parameter. Hence we calculated a coefficient of variation (CV) for each population with the mean number of survivors at each temperature to roughly evaluate fitness canalisation from our quiescence data. In this case, low CV indicates strong adaptive phenotypic plasticity.

Experiment 3: cannibalism

Depending on the population, from 13 to 18 females were randomly collected from fecundity measures of experiment 1 at 24°C. Each female was put alone in a small cylindrical box (height = 3 cm; diameter = 8.5 cm) with no food except 20 of its own eggs and 20 eggs laid by a randomly chosen female from one of the three other populations. Eggs were all laid in the preceding 12 h on small pieces of drawing paper which were marked in order to discriminate between the origin of the different egg patches. Monitoring was performed after 24 h and 48 h by counting eggs eaten and identifying their origin.

Data analysis

In experiment 1, we used generalized linear models to assess the effect on each of the six studied phenotypic traits and on the fitness index of the temperature, the population status (either invasive or biocontrol), the population nested within status and the two interactions involving the temperature. A binomial probability distribution and a logit link function were used for rate data (i.e. hatching rate of eggs, larval survival rate and sex-ratio). A Gamma probability distribution and an inverse link function were used for temporal data (i.e. family mean development time and lifespan of starving adult). Finally, a Poisson probability distribution and a log link function were used for count data (i.e. 6 weeks total fecundity of each female and fitness index as the latter was expressed as an indirect count of descendants). The effect of the population on the RDPI was tested with a non-parametric Kruskal–Wallis test using family scores as replicate units within population.

In experiment 2, a generalized linear model with a binomial probability distribution and logit link function was used to test the effect of the temperature, the population status, the population nested within status and the two interactions involving the temperature on the survival rate of individuals. For each population at each temperature, we investigated the aggregation behaviour by testing deviation from the null hypothesis of a 1:1 ratio of individuals under and over the cardboard using a χ^2 test. Because in standard rearing conditions, individuals are generally active and patrol all over the cardboard surface available and the surface under and over the cardboard was the same, we have considered that a 1:1 ratio corresponded the null hypothesis of random (i.e. non-aggregative) distribution of the beetles in the box.

In experiment 3, generalized linear models with a binomial probability distribution and logit link function were used to test the effect of the population on (i) the proportion of eggs eaten among the 40 after 24 h and after 48 h (cannibalism rates at T + 24 h and T + 48 h, respectively) and (ii) the proportion of own-laid eaten eggs among the total number of eaten eggs after 24 h (self-cannibalism rate). Using observation records at T + 24 h and T + 48 h, we also assessed for each population which type of egg patch (own-laid eggs versus eggs laid by a randomly chosen female from one of the three other populations) was consumed first by using a sign test.

All statistical analyses were performed with SAS software version 8.1 (SAS Institute Inc. 1999).

Results

Phenotypic plasticity of life history traits and fitness

We found a significant effect of the temperature for every trait (Table 1; Fig. 2). The population status (invasive or biocontrol) had a significant effect for three traits (larval survival rate, development time, and fecundity over 6 weeks) and for the fitness index. Most importantly, the interaction between the population status and the temperature, which reflects a potentially different response to temperature of invasive and biocontrol populations, was significant for the composite fitness index computed from the traits measured at 18, 24 and 30°C ($\chi^2 = 7.58$; df = 2; $P < 0.05$), but not for any of those traits considered independently. The population nested within status had a significant effect on all traits including fitness index (excepted on the fecundity over 6 weeks). No significant effect of the interaction between the latter factor and the temperature was detected for most traits (except for larval survival rate and sex ratio). Therefore, populations of same status displayed different traits values in each treatment but responded to variation of temperature in approximately the same way. Fitness was higher at 24°C for all populations, and biocontrol populations were globally more efficient in our experimental conditions (Fig. 2g). This trend was observed for most traits, including the total development time (not incorporated in the fitness index) for which the population London took slightly longer time to reach adulthood than the three others (Fig. 2c). This feature was less clear for the starving adult survival time (not incorporated in the fitness index) for which invasive populations lived in some cases longer than biocontrol populations (Fig. 2f).

The RDPI of fitness index, which is inversely proportional to the extent of adaptive phenotypic plasticity, was on average higher for the biocontrol populations ($RDPI_w$_Biocontrol 1 = 0.56; $RDPI_w$_Biocontrol 2 = 0.51) than for the invasive populations ($RDPI_w$_London = 0.42; $RDPI_w$_Roquefort = 0.5), but this difference was not statistically significant ($\chi^2 = 2.42$; df = 3; $P = 0.49$).

Quiescence

In experiment 2, the number of survivors after 5 weeks without any food was strongly explained by the temperature ($\chi^2 = 126.7$; df = 2; $P < 10^{-4}$), the population status ($\chi^2 = 148.29$; df = 1; $P < 10^{-4}$) and the interaction between both factors ($\chi^2 = 18.72$; df = 2; $P < 10^{-4}$) (Fig. 3). Population nested within status had a significant effect as well ($\chi^2 = 28.22$; df = 2; $P < 10^{-4}$), but the effect of the interaction between the latter factor

Table 1 Summary of statistical results using the generalized linear models for each traits of experiment 1, and the composite fitness index computed from these traits

	Effects				
	Temperature	Status	Temperature × status	Population (status)	Temperature × population (status)
Trait A: egg hatching rate	$\chi^2 = 83.31$	$\chi^2 = 0.17$	$\chi^2 = 0.18$	$\chi^2 = 36.6$	$\chi^2 = 4.51$
	df = 2	df = 1	df = 2	df = 2	df = 4
	$\boldsymbol{P < 10^{-4}}$	$P = 0.68$	$P = 0.91$	$\boldsymbol{P < 10^{-4}}$	$P = 0.34$
Trait B: larval survival rate	$\chi^2 = 20.82$	$\chi^2 = 4.83$	$\chi^2 = 1.44$	$\chi^2 = 8.37$	$\chi^2 = 14.23$
	df = 2	df = 1	df = 2	df = 2	df = 4
	$\boldsymbol{P < 10^{-4}}$	$\boldsymbol{P < 0.05}$	$P = 0.49$	$\boldsymbol{P < 0.05}$	$\boldsymbol{P < 0.01}$
Trait C: development time	$\chi^2 = 9183.92$	$\chi^2 = 10.91$	$\chi^2 = 5.51$	$\chi^2 = 10.11$	$\chi^2 = 1.6$
	df = 2	df = 1	df = 2	df = 2	df = 4
	$\boldsymbol{P < 10^{-4}}$	$\boldsymbol{P < 10^{-3}}$	$P = 0.06$	$\boldsymbol{P < 0.01}$	$P = 0.81$
Trait D: sex ratio	$\chi^2 = 63.48$	$\chi^2 = 1.08$	$\chi^2 = 3.74$	$\chi^2 = 27.57$	$\chi^2 = 21.74$
	df = 2	df = 1	df = 2	df = 2	df = 4
	$\boldsymbol{P < 10^{-4}}$	$P = 0.3$	$P = 0.15$	$\boldsymbol{P < 10^{-4}}$	$\boldsymbol{P < 10^{-3}}$
Trait E: fecundity over 6 weeks	$\chi^2 = 31.21$	$\chi^2 = 9.16$	$\chi^2 = 4.76$	$\chi^2 = 4.37$	$\chi^2 = 0.52$
	df = 2	df = 1	df = 2	df = 2	df = 4
	$\boldsymbol{P < 10^{-4}}$	$\boldsymbol{P < 0.01}$	$P = 0.09$	$P = 0.11$	$P = 0.97$
Trait F: starving adult survival	$\chi^2 = 841.58$	$\chi^2 = 0.89$	$\chi^2 = 4.91$	$\chi^2 = 6.72$	$\chi^2 = 3.72$
	df = 2	df = 1	df = 2	df = 2	df = 4
	$\boldsymbol{P < 10^{-4}}$	$P = 0.35$	$P = 0.86$	$\boldsymbol{P < 0.05}$	$P = 0.45$
Trait G: fitness index	$\chi^2 = 55.22$	$\chi^2 = 7.16$	$\chi^2 = 7.58$	$\chi^2 = 11.29$	$\chi^2 = 3.05$
	df = 2	df = 1	df = 2	df = 2	df = 4
	$\boldsymbol{P < 10^{-4}}$	$\boldsymbol{P < 0.01}$	$\boldsymbol{P < 0.05}$	$\boldsymbol{P < 0.01}$	$P = 0.55$

Significant *P*-values at the 5% threshold level are in bold characters. Status = invasive or biocontrol population

and temperature was not significant ($\chi^2 = 2.65$; df = 4; $P = 0.62$). Therefore, invasive populations always had higher survival rates than the biocontrol populations, with the population London showing the lowest mortality. At 5°C, 86% of the population London survived versus 54% for the population Roquefort and 41% for both biocontrol populations. The coefficients of variation (CV) calculated from the mean number of survivors at each temperature were substantially higher for the biocontrol populations (CV_Biocontrol 1 = 0.85; CV_Biocontrol 2 = 0.75) than for the invasive populations (CV_London = 0.16; CV_Roquefort = 0.33). Whereas at 15°C no population displayed a significant trend for aggregation under the cardboard, significant aggregation behaviour was observed for all populations at 5°C. At 10°C, the population London was the only one to display significant aggregation behaviour (Table 2).

Cannibalism

The factor population significantly explained the cannibalism at T + 24 h ($\chi^2 = 29.37$; df = 3; $P < 10^{-4}$) and T + 48 h ($\chi^2 = 12.61$; df = 3; $P < 0.01$) (Fig. 4a), as well as

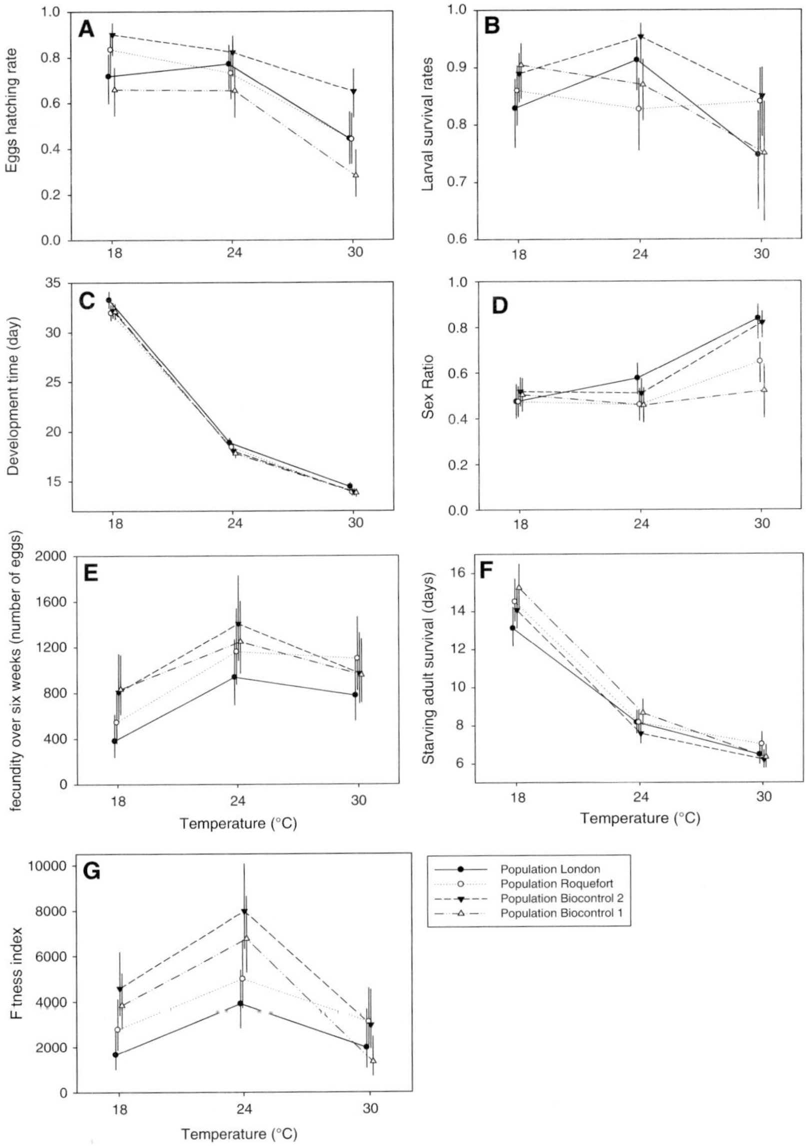

Fig. 2 Reaction norms to temperature (18, 24 and 30°C) of traits measured in experiment 1: (**a**) eggs hatching rates, (**b**) larval survival (from eggs until adult emergence), (**c**) family mean development time (until adult emergence), (**d**) sex ratio (proportion of males) of emerging adults, (**e**) fecundity over 6 weeks, (**f**) survival time in starvation conditions and (**g**) composite fitness index (see "Materials and methods" section for details). For each population and each temperature, vertical lines correspond to 95% confidence interval. The *P*-values associated to the effects of "temperature", "status", "temperature × status", "population" (nested within status) and "temperature × population" of the generalized linear models are given for each trait in Table 1. Status = invasive or biocontrol population

differences in self-cannibalism rates ($\chi^2 = 40.57$; df = 3; $P < 10^{-4}$) (Fig. 4b). The population London was mostly responsible for this effect. At T + 24 h, *H. axyridis* from population London had eaten only 8% of the total number of eggs versus 61–85% for the

Table 2 Proportion of individuals which aggregated under the cardboard (P_u) in experiment 2

Population	5°C			10°C			15°C		
	n	P_u	*P*-value	*n*	P_u	*P*-value	n	P_u	*P*-value
London	48	**0.96**	**$<10^{-4}$**	53	**0.79**	**<0.01**	38	0.42	0.49
Roquefort	30	**0.90**	**$<10^{-3}$**	45	0.58	0.46	24	0.29	0.14
Biocontrol 1	23	**1.00**	**$<10^{-4}$**	32	0.53	0.80	1	NC	0.41
Biocontrol 2	23	**1.00**	**$<10^{-4}$**	30	0.53	0.79	3	NC	0.68

Significant *P*-values at the 5% threshold level are in bold characters
n is the total number of surviving individuals at the time of the P_u measurement
P-values are obtained from χ^2 test
NC = χ^2 test not computed due to low sample size ($n < 5$)

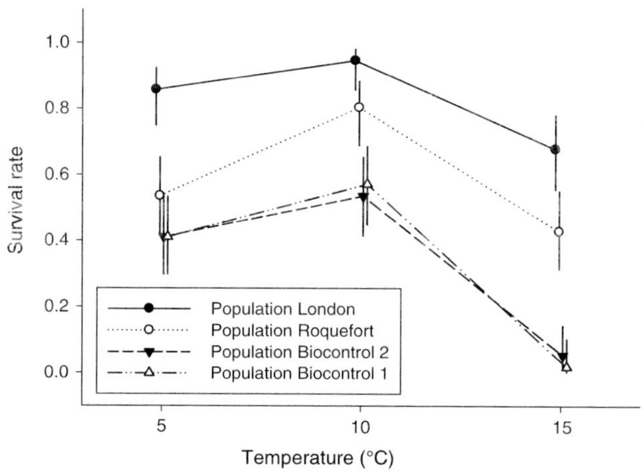

Fig. 3 Survival of adults after 5 weeks at low temperature (5, 10 and 15°C) and without food. Vertical lines correspond to 95% confidence interval

three other populations. After 48 h, more than 75% of all the eggs were eaten in all populations. The population London was the only one for which the eggs originating from other populations were eaten first (signed test; $M = 4$; $P = 0.0386$). Indeed, only 2% of the eggs eaten after 24 h by females from the population London were their own eggs (versus ~50% for the three other populations).

Discussion

Results from experiment 1 show that invasive and biocontrol populations display significantly different responses to temperature variation for the composite fitness index computed from the traits measured at 18, 24 and 30°C, but not for any of those traits considered independently. The adaptive plasticity measured from the RPDI of the fitness index was higher in the two invasive populations than in the two biocontrol populations. However, this difference was far from being statistically significant. Thus, the results of experiment 1 suggest only minor differences in adaptive phenotypic plasticity between

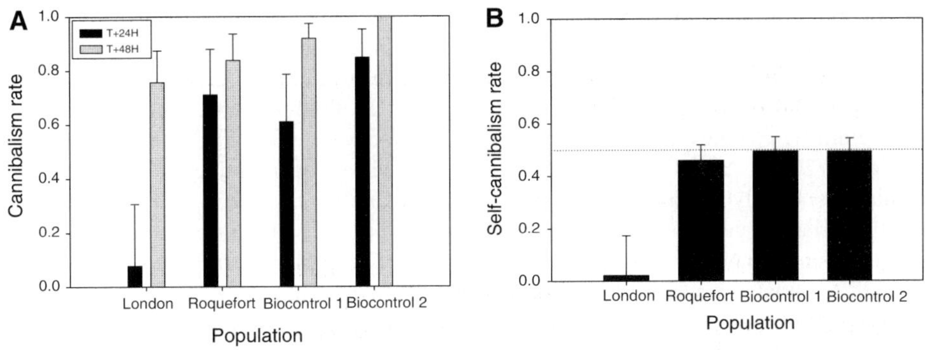

Fig. 4 Cannibalism in *H. axyridis*. (**a**) mean proportion of eggs eaten after 24 and 48 h (T + 24 h and T + 48 h, respectively) for each population. (**b**) Self-cannibalism = mean proportion of self-eaten eggs among eaten eggs after 24 h. Vertical lines correspond to 95% confidence interval

populations and hence do not strongly support the expectation of increased phenotypic plasticity in invasive populations of *H. axyridis*.

This conclusion should, however, be mitigated for at least three reasons. First, some other environmental parameters may be more suitable than temperature to detect phenotypic plasticity (Stillwell et al. 2007). For example, food may be a better environmental parameter to test for phenotypic plasticity in *H. axyridis* which is known to be polyphagous (e.g. Preziosi et al. 1999; Specty et al. 2003; Berkvens et al. 2007) and encounter a taxonomically diverse range of phytophagous insects associated with various vegetation communities. Second, our estimation of fitness is likely to be a poor representation of fitness in the wild. In particular, the "flightless" population Biocontrol 2 had the highest fitness index in our experiment, but probably suffers from low or null fitness in the wild because of its incapacity to disperse and migrate to aggregation sites (Tourniaire et al. 2000a). Third, the use of *E. kuehniella* eggs in our experiments may have distorted at least some of our results, as this food is likely to favour the biocontrol populations that have been fed this way for at least 20 years (Schanderl et al. 1988). This probably increased artificially the extent of adaptive phenotypic plasticity in biocontrol populations and may have prevented us from detecting differences between invasive and biocontrol populations. Indeed, high resource availability and high resource acquisition capacity are expected to mask resource allocation strategies in response to environmental variations (e.g. Malausa et al. 2005). In other words, the fact that biocontrol populations consumed more food than the other populations probably allowed them to canalize their fitness better as they could allocate increased resources to the expression of every phenotypic trait whatever the environmental conditions.

In contrast to the estimation of reaction norms to temperature ranging from 18 to 30°C (experiment 1), measures of survival during quiescence (experiment 2) clearly showed higher fitness and adaptive plasticity of invasive over biocontrol populations. Invasive populations and especially the population London suffered far lower mortality than biocontrol populations, the latter showing a poor ability to enter into quiescence. The response to low temperature variations assessed from the three tested temperatures was significantly different between invasive and biocontrol populations and CV were substantially lower for invasive populations. Such low coefficients of variation indicate fitness homeostasis through adaptive phenotypic plasticity. The problem of entering into quiescence experienced by the biocontrol populations may explain, at least partly, why the species failed to become established for around 10 years despite numerous intentional releases in the 1990's of individuals originating from such populations. It is worth noting here that this result should not be taken as an argument that the present invasive populations in Europe do not originate from those biocontrol populations. Adaptive evolutionary change can indeed be very rapid, and this might be particularly important in biological invasions, which often involve drastic changes in selective regimes (e.g. Stockwell et al. 2003; Lambrinos 2004; Roy et al. 2008). Yet the origin of genetic variance at quantitative traits in invasive populations largely remains a mystery. In particular, the respective roles of ancestral genetic variation and in situ creation of new genotypes by mutation or recombination or hybridization events due to multiple introductions of individuals originating from genetically differentiated populations remain unclear. To tackle this question, we have started research actions based on genetic markers to elucidate pathways of introduction of invasive *H. axyridis* populations as well as their level of genetic variation relatively to native and biocontrol populations both in Europe and in America.

Cannibalism may also be an important trait in an invasion context. Our results highlight strong differences in cannibalistic behaviour of the invasive population London compared to the three others. First, the population London displayed a significantly lower degree of cannibalism after 1 day of starvation. Cannibalism can either be globally beneficial or costly depending on the ecological context (Polis 1981; Osawa 1992; Pervez et al. 2006; Williams and Hernandez 2006). The potential benefit of delaying cannibalistic behaviour during invasion remains unknown. Second, the population London clearly avoided self-cannibalism whereas the three other populations did not discriminate. This feature parallels cannibalism results obtained previously in a more standard kin selection context in other coccinellid species (Agarwala and Dixon 1993; Pervez et al. 2005). Selective cannibalism might be a determinant trait in an invasion context as it could be linked to associated behaviours such as inter-specific predation.

In conclusion, our results indicate that, despite globally weak differences in responses to temperature variation between invasive and biocontrol populations, phenotypic plasticity and its evolution may still play a role in determining the success of invasive populations in some extreme and/or ecologically relevant environmental conditions. Our results also highlight the fact that the traits to be measured and environments to be tested must be chosen carefully when attempting to detect variation of adaptive phenotypic plasticity among populations. In the case of *H. axyridis*, traits relative to activity regulation (ability to enter into quiescence during periods of low resource availability) and ability to forage for a variety of different food sources (including through cannibalism) appear to be of particular interest. In a more general perspective, a comparison based on those traits of invasive populations with populations from the native range would be of great interest to assess the evolution of phenotypic plasticity in *H. axyridis* during the invasion process.

Acknowledgments We thank Michael Majerus, the firm Biotop and Christine and Aurélia Delclos who kindly provided the population London, the biocontrol populations and the population Roquefort, respectively. We also thank André Ferran and Ludovic Giuge for advice about the *Harmonia axyridis* nursery stock, Helen Roy for comments and help in correcting our English writing, Thomas Guillemaud for helpful discussions, and three anonymous reviewers for useful comments. This work was supported by a grant from the French "Agence Nationale de la Recherche", project BioInv-4I.

References

Agarwala BK, Dixon AFG (1993) Kin recognition: egg and larval cannibalism in *Adalia bipunctata* (Coleoptera: Coccinellidae). Eur J Entomol 90:45–50

Agrawal AA (2001) Phenotypic plasticity in the interactions and evolution of species. Science 294:321–326

Berkvens N, Bonte J, Berkvens D, Deforce K, Tirry L, De Clercq P (2007) Pollen as an alternative food for *Harmonia axyridis*. BioControl (this issue). doi:10.1007/s10526-007-9128-7

Brown PMJ, Adriaens T, Bathon H, Cuppen J, Goldarazena A, Hägg T, Kenis M, Klausnitzer BEM, Kovář I, Loomans AJM, Majerus MEN, Nedved O, Pedersen J, Rabitsch W, Roy HE, Ternois V, Zakharov IA, Roy DB (2007a) *Harmonia axyridis* in Europe: spread and distribution of a non-native coccinellid. BioControl (this issue). doi:10.1007/s10526-007-9132-y

Brown PMJ, Roy HE, Rothery P, Roy DB, Ware RL, Majerus MEN (2007b) *Harmonia axyridis* in the Great Britain: analysis of the spread and distribution of a non-native coccinellid. BioControl (this issue). doi:10.1007/s10526-007-9124-y

Coutanceau J-P (2006) *Harmonia axyridis* (Pallas, 1773): une Coccinelle asiatique introduite, acclimatée et en extension en France. Bulletin de la société entomologique de France 111:395–401

Debat V, David P (2001) Mapping phenotypes: canalization, plasticity and developmental stability. Trends Ecol Evol 16:555–561

Geng YP, Pan XY, Xu CY, Zhang WJ, Li B, Chen JK, Lu BR, Song ZP (2007) Phenotypic plasticity rather than locally adapted ecotypes allows the invasive alligator weed to colonize a wide range of habitats. Biol Invasions 9:245–256

Ghalambor CK, McKay JK, Carroll SP, Reznick DN (2007) Adaptive versus non-adaptive phenotypic plasticity and the potential for contemporary adaptation in new environments. Funct Ecol 21:394–407

Grill CP, Moore AJ, Brodie ED (1997) The genetics of phenotypic plasticity in a colonizing population of the ladybird beetle, *Harmonia axyridis*. Heredity 78:261–269

Kaufman SR, Smouse PE (2001) Comparing indigenous and introduced populations of *Melaleuca quinquenervia* (Cav.) Blake: response of seedlings to water and pH levels. Oecologia 127:487–494

Koch RL (2003) The multicolored Asian lady beetle, *Harmonia axyridis*: a review of its biology, uses in biological control, and non-target impacts. J Insect Sci 3:1–16

Kolar CS, Lodge DM (2001) Progress in invasion biology: predicting invaders. Trends Ecol Evol 16:199–204

Lambrinos JG (2004) How interactions between ecology and evolution influence contemporary invasion dynamics. Ecology 85:2061–2070

Lee CE, Remfert JL, Chang YM (2007) Response to selection and evolvability of invasive populations. Genetica 129:179–192

Mack RN, Simberloff D, Lonsdale WM, Evans H, Clout M, Bazzaz FA (2000) Biotic invasions: causes, epidemiology, global consequences, and control. Ecol Appl 10:689–710

Majerus MEN, Strawson V, Roy H (2006) The potential impacts of the arrival of the harlequin ladybird, *Harmonia axyridis* (Pallas) (Coleoptera: Coccinellidae), in Britain. Ecol Entomol 31:207–215

Malausa T, Guillemaud T, Lapchin L (2005) Combining genetic variation and phenotypic plasticity in tradeoff modelling. Oikos 110:330–338

Masel J, King OD, Maughan H (2007) The loss of adaptive plasticity during long periods of environmental stasis. Am Nat 169:38–46

Osawa N (1992) Sibling cannibalism in the ladybird beetle *Harmonia axyridis*: fitness consequences for mother and offspring. Res Popul Ecol 34:45–55

Pervez A, Gupta AK, Omkar (2005) Kin recognition and avoidance of kin cannibalism by the larvae of co-occurring ladybirds: a laboratory study. Eur J Entomol 102:513–518

Pervez A, Gupta AK, Omkar (2006) Larval cannibalism in aphidophagous ladybirds: influencing factors, benefits and costs. Biol Control 38:307–313

Polis GA (1981) The evolution and dynamics of intraspecific predation. Annu Rev Ecol Syst 12:225–251

Preziosi RF, Snyder WE, Grill CP, Moore AJ (1999) The fitness of manipulating phenotypes: implications for studies of fluctuating asymmetry and multivariate selection. Evolution 53:1312–1318

Richards CL, Bossdorf O, Muth NZ, Gurevitch J, Pigliucci M (2006) Jack of all trades, master of some? On the role of phenotypic plasticity in plant invasions. Ecol Lett 9:981–993

Roy HE, Baverstock J, Ware RL, Clark SJ, Majerus MEN, Baverstock KE, Pell JK (2008) Intraguild predation of the aphid pathogenic fungus *Pandora neoaphidis* by the invasive coccinellid *Harmonia axyridis*. Ecol Entomol (in press)

SAS Institute Inc. (1999) SAS/Stat user's guide, version 8. SAS Institute, Cary

Schanderl H, Ferran A, Garcia V (1988) Rearing two coccinellids *Harmonia axyridis* and *Semiadalia undecimnotata* on eggs of *Anagasta kuehniella* killed with ultraviolet radiation. Entomol Exp Appl 49:235–244

Scheiner SM (1993) Genetics and evolution of phenotypic plasticity. Annu Rev Ecol Syst 24:35–68

Specty O, Febvay G, Grenier S, Delobel B, Piotte C, Pageaux JF, Ferran A, Guillaud J (2003) Nutritional plasticity of the predatory ladybeetle *Harmonia axyridis* (Coleoptera: Coccinellidae): comparison between natural and substitution prey. Arch Insect Biochem Physiol 52:81–91

Stillwell R, Wallin W, Hitchcock L, Fox C (2007) Phenotypic plasticity in a complex world: interactive effects of food and temperature on fitness components of a seed beetle. Oecologia 153:309–321

Stockwell CA, Hendry AP, Kinnison MT (2003) Contemporary evolution meets conservation biology. Trends Ecol Evol 18:94–101

Tourniaire R, Ferran A, Gambier J, Giuge L, Bouffault F (2000a) Locomotory behavior of flightless *Harmonia axyridis* pallas (Col., Coccinellidae). J Insect Physiol 46:721–726

Tourniaire R, Ferran A, Giuge L, Piotte C, Gambier J (2000b) A natural flightless mutation in the ladybird, *Harmonia axyridis*. Entomol Exp Appl 96:33–38

Valladares F, Sanchez Gomez D, Zavala MA (2006) Quantitative estimation of phenotypic plasticity: bridging the gap between the evolutionary concept and its ecological applications. J Ecol 94:1103–1116

van Lenteren JC, Loomans AJM, Babendreier D, Bigler F (2007) *Harmonia axyridis*: an environmental risk assessment for Northwest Europe. BioControl (this issue). doi:10.1007/s10526-007-9120-2

Williams T, Hernandez O (2006) Costs of cannibalism in the presence of an iridovirus pathogen of *Spodoptera frugiperda*. Ecol Entomol 31:106–113

Williamson M, Fitter A (1996) The varying success of invaders. Ecology 77:1661–1666

Yeh PJ, Price TD (2004) Adaptive phenotypic plasticity and the successful colonization of a novel environment. Am Nat 164:531–542

Predicting the potential geographical distribution of the harlequin ladybird, *Harmonia axyridis*, using the CLIMEX model

J. Poutsma · A. J. M. Loomans · B. Aukema · T. Heijerman

Abstract *Harmonia axyridis* (Pallas, 1773) (Coleoptera: Coccinellidae) is a ladybird beetle native to temperate and subtropical parts of Asia. Since 1916 populations of this species have been introduced throughout the world, either deliberately, or by accident through international transport. *Harmonia axyridis* was originally released as a classical biological control agent of aphid and coccid pests in orchards and forests, but since 1994 it is also available as a commercial product for augmentative control in field and greenhouse crops. It is a very voracious and effective natural enemy of aphids, psyllids and coccids in various agricultural and horticultural habitats and forests. During the past 20 years, however, it has successfully invaded non-target habitats in North America (since 1988), Europe (1999) and South America (2001) respectively in a short period of time, attacking a wide range of non-pest species in different insect orders. Becoming part of the agricultural commercial pathway, it is prone to being introduced into large areas across the world by accident. We use the CLIMEX programme (v2) to predict the potential geographical distribution of *H. axyridis* by means of matching the climate of its region of origin with other regions in the world and taking in account biological characteristics of the species. Establishment and spread seem likely in many regions across the world, including those areas which *H. axyridis* has already invaded (temperate Europe, North America). Based on the CLIMEX prediction a large part of Mediterranean Europe, South America, Africa, Australia and New Zealand seem highly suitable for long-term survival of *H. axyridis* as

J. Poutsma · A. J. M. Loomans (✉) · B. Aukema
Department of Entomology, National Reference Laboratory, Plant Protection Service, P.O. Box 9102, 6700 HC Wageningen, The Netherlands
e-mail: a.j.m.loomans@minlnv.nl

J. Poutsma
e-mail: j.poutsma@minlnv.nl

B. Aukema
e-mail: b.aukema@minlnv.nl

T. Heijerman
Biosystematics Group, Wageningen University, Generaal Foulkesweg 37, 6703 BL Wageningen, The Netherlands
e-mail: theodoor.heijerman@wur.nl

Biological Control to Invasion: Ladybird Harmonia axyridis. H.E. Roy, E. Wajnberg (eds.),
doi: 10.1007/978-1-4020-6939-0_8,

well. In addition we evaluate CLIMEX as a strategic tool for estimating establishment potential as part of an environmental risk assessment procedure for biological control agents we discuss biological and ecological aspects necessary to fine-tune its establishment and spread in areas after it has been introduced.

Keywords *Harmonia axyridis* · Biological control · CLIMEX · Potential distribution · Climate matching

Introduction

The introduction of species into new regions of the world, either deliberately as a biological control agent or accidentally as a pest, can have a huge impact on agriculture and biodiversity. Before the import and release of any natural enemy as a biological control agent an environmental risk assessment can give an estimate of such an impact. van Lenteren et al. (2003) developed a methodology for assessing the risks of introducing biological control agents, which integrates information on the potential of an organism to establish, its dispersal capacities, its host range, and its direct and indirect effect on non-targets. The direct effects can be divided into effects on non-target herbivores, vectoring, and effects on other trophic levels such as intraguild predation, facultative hyperparasitism, and omnivory. The indirect effects can be divided into competition, indirect effects of intraguild predation, effects if the agent is a food source of another natural enemy, genetic changes due to hybridisation, linked trophic interactions, and non-food requirements like protection, pollination and (seed) dispersal. In such an environmental risk assessment procedure about establishment potential of a species introduced in a new region, is an important criterion (van Lenteren et al. 2003). Except using thermal data, such as cold tolerance and survival at low temperatures (Berthiaume et al. 2003; Hatherly et al. 2005), to assess the establishment potential in the area of release, modelling various stress parameters such as done by using the CLIMEX model may provide a reliable estimate of establishment potential for candidate non-native biological control agents outside its area of origin (Sutherst 2003; Sutherst and Maywald 2005).

Harmonia axyridis (Pallas, 1773) (Coleoptera: Coccinellidae), the harlequin ladybird, is a ladybird beetle native to continental, temperate and subtropical parts of east and central Asia (Iablokoff-Khnzorian 1982). Since the beginning of the 20th century this species has been introduced several times as a biological control agent into different parts of the world. In North America, South of Europe and the Soviet Union it has been released in classical biological control programmes against aphid and/or coccid pests. Since the mid 1990s *H. axyridis* has been offered as a commercial biological control agent in Europe for aphid control in greenhouse and urban ecosystems, and multiple releases have been made for a 1–2 decades in several countries. First records of establishment outside its native range took place in North America in 1988 (Chapin and Brou 1991), in Europe in 2000 (Bathon 2002) and South America in 1998 (Saini 2004). Gradually *H. axyridis* has expanded its range across these continents by natural dispersal during fall, spring and summer, or is currently doing so. In North America *H. axyridis* has spread across the east coast within 5–6 years time (Koch et al. 2006). *H. axyridis* is showing two kinds of migratory behaviour: autumn and spring migration to and from overwintering sites, and summer migrations when food sources are scarce. These summer migrations are random in direction, but often wind-driven. The autumn migrations on the other hand are not random, but directional away from the afternoon sunlight (Lo and Liu 1976; Nalepa et al. 2005), i.e. heading to the northeast. Likely the spring migration, when adults leave their overwintering sites and return to foraging areas, is directional as well, but no data are available for this. Besides intentional introductions, accidental introductions may have taken place as well, as a result of increase in international trade. Eggs, pupae, larvae and adults may be carried as a stow-away in ships and airplanes when attached to leafs, branches and trunks of potted plants, trees, and shrubs. On Tenerife, the Canary Islands, a single specimen was found in the autumn of 2003 and again in 2004 (Machado 2006). In the UK specimens have been intercepted from transatlantic produce in 2004 (Majerus et al. 2006), and in Norway on horticultural produce from the Netherlands in 2006 (Arnstein Staverløkk, personal communication 2007).

What characteristics make *H. axyridis* a successful invader? First, it is known from a large native geographical area, with a wide variety of phenotypes, that differ both within populations and geographic location (Mader 1926–1937, Dobzhansky 1933). Grill et al. (1997) showed that *H. axyridis* displays phenotypic plasticity in several characters important for successful colonization of a new area, e.g. the ability to adapt to a wide range of climates. Secondly, *H. axyridis* has a strong potential to spread by flight, by accidental introductions, and through commercial releases. Thirdly, *H. axyridis* has a wide food range and consumes all kind of soft-bodied insects of many taxa besides aphids, its main prey (Iablokoff-Khnzorian 1982). *Harmonia axyridis* reproduces in a wide range of habitats, whereas many native coccinellids tend to be more habitat- and niche-specific (Majerus et al. 2006). It is also able to reproduce during a long time of the year: the adults become active early in the season (April) and larvae and pupae still can be found in November (Majerus et al. 2006). Being a successful invader could also mean that it has an adverse effect on native aphidophagous predators (e.g. Lanzoni et al. 2004). Cannibalism and intraguild predation are common, and several sources (e.g. Brown and Miller 1998; Colunga-Garcia and Gage 1998; Snyder et al. 2004) report a decline of the population of native aphidophagous ladybeetles that can lead to displacement of native species.

In this paper we use climate scenarios from the simulation software CLIMEX 2.0 to predict the establishment potential of *Harmonia axyridis* in new areas of introduction of the world and verify the outcome by visually matching its current presence in newly invaded areas. We also evaluate CLIMEX as a strategic tool for establishment potential and environmental risk assessments of biological control agents and rapid decision making.

Material and methods

The CLIMEX model

In this study we used the CLIMEX model for Windows Version 2 (Sutherst et al. 2004) to develop a model of climatic responses of *H. axyridis* based on its recorded distribution. The best way to make a prediction is to use the native distribution as a basis for fitting the model parameters. CLIMEX uses climatic responses of an organism and meteorological data to predict the potential distribution and relative abundance of poikilothermic organisms under current climatic conditions. Predictions can be made on a world scale or on a regional scale (Sutherst et al. 2004). In CLIMEX over 2000 locations (meteorological stations) worldwide are included. The climatic data associated with these locations span the period 1960–1990 (Müller 1996).

For a simple comparison of meteorological data at different locations, the function 'Match Climates' can be used, e.g. for targeting collection sites of new, unknown biological control agents, adapted to a certain area of release (e.g. performed by Koch et al. (2006) for South America) but it does not take species characteristics into account. Here, we used this function for targeting the collection sites that could have been the source area of releases in Europe and North America. A second function, the species-specific response (CLIMEX) model, however, allows the user to predict the potential geographic distribution of a given species based on its biology and long-term meteorological data ('Compare Locations'). Species parameters are part of the input in the CLIMEX model, and describe a species' response to climate (Sutherst et al. 2004). These parameters are divided into three groups: one group defines seasonal population growth, another defines stresses (cold, hot, wet and dry) that limit the geographical distribution, and the third describes the limiting conditions that may exclude species from some areas. The probability that the populations can survive under unfavourable conditions is described by these parameters. From these parameters the growth index (GI) and stress temperature and moisture indices (TI, MI) are calculated weekly, and can be combined into an overall annual index of climatic suitability, the ecoclimatic index (EI). Scaled from 0 to 100 it is a measure of climatic suitability for the species concerned. A location can only have an EI of 100 when that location provides ideal conditions, without stress, throughout the year. In temperate areas or regions with distinct stressful wet and dry seasons the maximum EI would be around 50. Values of >20 already have been found to support substantial population densities (Sutherst 2003).

A species' behaviour in a CLIMEX simulation depends on the parameter values. The values of the species parameters are based on seasonal distribution data and experimental observations on developmental threshold temperatures and moisture levels. The more values of a parameter are known, the more reliable the picture (map of potential species distribution) in CLIMEX will be. If some values of the species parameter are missing, values of the parameter templates can be used. Once the species' parameter file has been adequately defined as a starting set of parameter values, the 'Compare Locations' function helps the user to fit the parameters. During the iterative adjustments-and-comparison process the parameter values for *H. axyridis* change to visually fit the observed native geographic distribution.

Native geographic distribution

Harmonia axyridis has its origin in the east of the palaearctic region. Its distribution ranges from the Altai mountains in the west to Japan, Yeniseisk (Krasnoyarsk) in the north, Sakhalin and the Kurile Islands in the east and Guanxi (China) in the south (Dobzhanksy 1933; Komai et al. 1950; Iablokoff-Khnzorian 1982; Kuznetkov 1997) (Fig. 1). Several sources indicate its presence in Taiwan (CABI/EPPO 2007; Poorani 2004), but Sasaji (1982) states that the observations and collected specimens did not concern *H. axyridis*, but *Harmonia yedoensis* (Takizawa). The records from the Himalaya and/or Nepal (Iablokoff-Khnzorian 1982; Poorani 2004) need confirmation because these papers are the only, but not original records. Also records from certain areas in Russia, as cited by CABI/EPPO (2007), are not supported by the underlying original papers and would need confirmation.

Fitting parameters

There was a large degree in similarity between the potential distribution generated by the temperate template in the CLIMEX model and the native range of *H. axyridis* in Asia. We

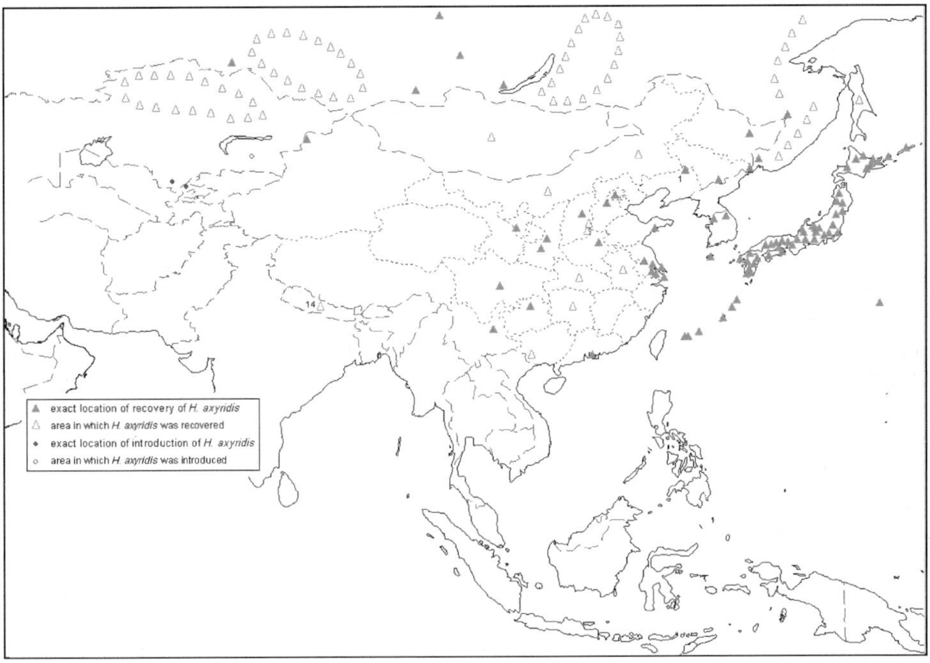

Fig. 1 Distribution of *Harmonia axyridis* in Asia (based on CABI/EPPO 2007; Komai et al. 1950 and others, corrected and modified)

therefore used that template as the starting point for our iteration process. In the next paragraphs we discuss what experimental values we used in the model.

Temperature

DV parameters are the limiting and optimal temperatures for population growth of a species. DV0 and DV3 are the limiting low and high temperatures respectively, at or below and above no population growth takes place. DV1 and DV2 are the lower optimal and upper optimal temperature respectively. Population growth of a species is reduced if the ambient temperature is below DV1. No population growth takes place when the ambient temperature equals or exceeds DV3. DV0 is not exactly equivalent to a developmental threshold for any life stage, but the temperature threshold for population growth, which is at best established under field conditions. Experimental values for DV0 (limiting low temperature) range from 9.3°C (pre-oviposition period) (Soares et al. 2003) to 13.3°C (pupa) (cited in Hodek and Honěk 1996). In this study 10.5°C was adopted for DV0 (Schanderl et al. 1985). Three sources of optimal temperatures have been found. Soares et al. (2003) found 20°C to be the lower optimal temperature DV1, and 25°C for the upper optimal temperature DV2. Barçin et al. (2001) found values for these two parameter of 15°C and 25°C respectively. Yuan et al. (1994) found values for the optimal temperature for adult development of 23–25°C. In this study 15°C is adopted for DV1, and 25°C for DV2. The highest limiting temperature DV3 values found in literature varied between 30°C and 35°C (Barçin et al. 2001; Michaud 2002). In this study DV3 was set to 35°C.

The rigour of an unfavourable season is not the only way in which climate can limit a species' distribution. It may also be necessary to have enough thermal accumulation during the growth season to enable the species to complete a minimum amount of development. PDD is the minimum number of degree-days above DV0, necessary to complete a generation of an organism. If the number of degree-days set for PDD is not reached at a certain location the ecoclimatic index (EI) is set to zero (cross on the maps), and the species is excluded from this location. Thermal constants for a complete life-cycle were not found. Only thermal constants of the egg stage, larval stage and pupal stage were found and ranged from 231.3 (egg-to-adult) (Schanderl et al. 1985) to 267.3 (LaMana and Miller 1998). Because the pre-oviposition period is missing in the calculation of these degree-days, the PDD value was set to 330 to make the life cycle complete.

Moisture

SM parameters are the soil moisture indices for population growth of a species. SM0 and SM3 are the limiting low and high index respectively, at or below and above no population growth takes place. SM1 and SM2 are the lower and upper optimal index respectively. Population growth of a species is reduced if soil moisture is below SM1. No population growth takes place when soil moisture equals or exceeds SM3.Values for soil moisture in relation to the performance of *H. axyridis*, a leaf and branch foraging predator, are of indirect importance through host plant growth and availability of suitable prey. In this study the SM-values from the temperate template were used as the start (SM0: 0.25; SM1: 0.8; SM2: 1.5; SM3: 2.0). Initially, irrigation was not included in the model.

Diapause

During winter, adults of *H. axyridis* exhibit diapause (Nalepa et al. 1996). In late winter or early spring, the beetles switch from diapause to a quiescence state. Preceding diapause the beetles are migrating to their overwintering habitats. In Asia the migration from feeding habitats to overwintering habitats occurs from mid-October to late November. During these migration flights the beetles fly towards prominent, isolated objects such as mountaintops and buildings and land on surfaces, like rocks or walls, that are often white or lightly coloured (Lo and Liu 1976, Nalepa et al. 2005). Little is known about the distance covered between the feeding habitat and the overwintering habitat though observations on winter aggregations suggest they may not travel long distances during that period (Nalepa et al. 1996). Overwintering sites are mountain cleaves, but also houses or other buildings. Like other coccinellids, *H. axyridis* forms mass aggregations at the overwintering sites. During summer beetles may become quiescent and aestivate, although nothing is known about the conditions at which this occurs. DPD0 and DPT0 are the diapause induction day length and the diapause induction temperature respectively. DPT1 is the diapause termination temperature, for winter diapause the minimum temperature that determines it. DPD (diapause development days) is the minimum number of days below DPT0 required for diapause development to be completed. DPSW is an indicator for summer or winter diapause, to be set to 1 or 0 respectively. *Harmonia axyridis* enters diapause in the adult stage. No exact values for the diapause induction day-length and diapause induction temperature have been found in literature. From several sources a value of 12 h, and 16°C and 18°C could be traced for the diapause induction daylength and temperature (Sakurai et al. 1988). In this study a value of 12 is adopted for DPD0 and 16 for DPT0. No general value could be found for the diapause termination temperature. Soares et al. (2001) found different values for the adult lower developmental threshold for the phenotypes 'aulica' and 'nigra': 9.3°C and 10.4°C, respectively. In this study 9.3°C was initially adopted for the DPT1. Values for DPD (diapause development days) were not found in literature. In this study a value of 0 is initially adopted for this parameter.

Light

In the model, LT0 is the day-length above which growth is at a maximum, and LT1 is the day-length below which growth is zero. No data on day-length was found in literature, so LT-values were not used in this study.

Stress

Stress parameters are set to a species' ability to survive adverse seasonal conditions, and so determine its geographical distribution. Stress, leading to negative population growth, is assumed to accumulate outside the values of DV0 and DV3 (temperature limits) and of SM0 and SM3 (moisture limits). Accordingly, there are four kinds of stress: cold, heat, dry and wet stress. Cold and heat stress can occur in two different ways. In the degree-day model, it is assumed that cold stress occurs because the seasonal thermal accumulation is too low to complete the life cycle. In case of heat stress it is assumed that the daily heat load is too high for the essential physiological processes. In the lethal temperature method, stress occurs in response to excessively low minimum temperatures or high maximum temperatures. Both types of stress may occur for the one species. In this study the stress temperature threshold method is used. From survival time data at −5°C to 5°C, Watanabe (2002) found that in Japan the optimum temperature for overwintering is between 0°C and −5°C. Berthiaume et al. (2003) found similar results for Canada, but recorded a significant decrease in survival at −10°C. A value of 0°C was initially adopted for TTCS (cold stress temperature threshold). TTHS (heat stress temperature threshold) was set to 35°C. Infor mation on the rate of heat and cold stress accumulation (THCS, THHS) as well as on dry and wet stress parameters was not found in literature, so values of temperate template were adopted (THCS: 0°C; THHS: 0.005°C). Interactions of stress indices were not found in literature, so these parameters were not used in this study. When no specific value could be found in literature for some parameters of the model, we used values estimated from the temperate template.

Experimental values of temperature-moisture- and diapause-related parameters mentioned above were included in the CLIMEX model and estimated visually by iteration, using the known Asian distribution of *H. axyridis* as a reference. Parameter values used are shown in Table 1. These CLIMEX parameter values were then applied to meteorological data from stations across the world and we visually compared the similarity between the predictions made in CLIMEX and the current distribution of *H. axyridis* in newly invaded areas (North America, Europe, South America). After that we performed a sensitivity test of parameter values for these areas to verify the fit.

Table 1 CLIMEX parameters settings *Harmonia axyridis*

Code	Parameter description	Values	
		Before iteration	After iteration
Temperature			
DV0	Limiting low temperature (°C)	10.5	10.5
DV1	Lower optimal temperature (°C)	15	15
DV2	Upper optimal temperature (°C)	24	25
DV3	Limiting high temperature (°C)	35	35
PDD	Minimum degree-days above DV0 (DD)	330	330
Moisture			
SMO	Limiting low moisture index	0.25	0.1
SM1	Lower optimal moisture index	0.8	0.3
SM2	Upper optimal moisture index	1.5	1.9
SM3	Limiting high moisture index	2.0	2.0
Diapause			
DPD0	Diapause induction day length (h)	12	14
DPT0	Diapause induction temperature (°C)	16	16
DPT1	Diapause termination temperature (°C)	9.3	7.3
DPD	Diapause development days (days)	0	40
DPSW	Summer/winter diapause	0	0
Stress			
TTCS	Cold stress temperature threshold (°C)	0	−5
THCS	Cold stress temperature rate	0	0.00002
DTCS	Cold stress degree-day threshold (DD)	0	0
DHCS	Cold stress degree-day rate	0	0
TTHS	Heat stress temperature threshold (°C)	35	35
THHS	Heat stress temperature rate	0.005	0.005
DTHS	Heat stress degree-day threshold (DD)	0	0
DHHS	Heat stress degree-day rate	0	0
SMDS	Dry stress threshold	0.2	0.01
HDS	Dry stress rate	−0.005	−0.001
SMWS	Wet stress threshold	2	2
HWS	Wet stress rate	0.002	0.01

Results

The iteration process

During the iterative adjustments-and-comparison process for *H. axyridis*, values of some parameters had a large effect on the EI and therefore on the potential distribution. The CLIMEX parameter values that gave the closest visual fit between the known (Fig. 1) and predicted (Fig. 2) distribution in Asia are shown in Table 1. During the iteration process, diapause showed to be the key factor for the absence of positive EI-values of *H. axyridis* in the north and in the south. When iterating the values of the diapause parameters, the diapause termination temperature (DPT1) was the key factor for explaining the absence of

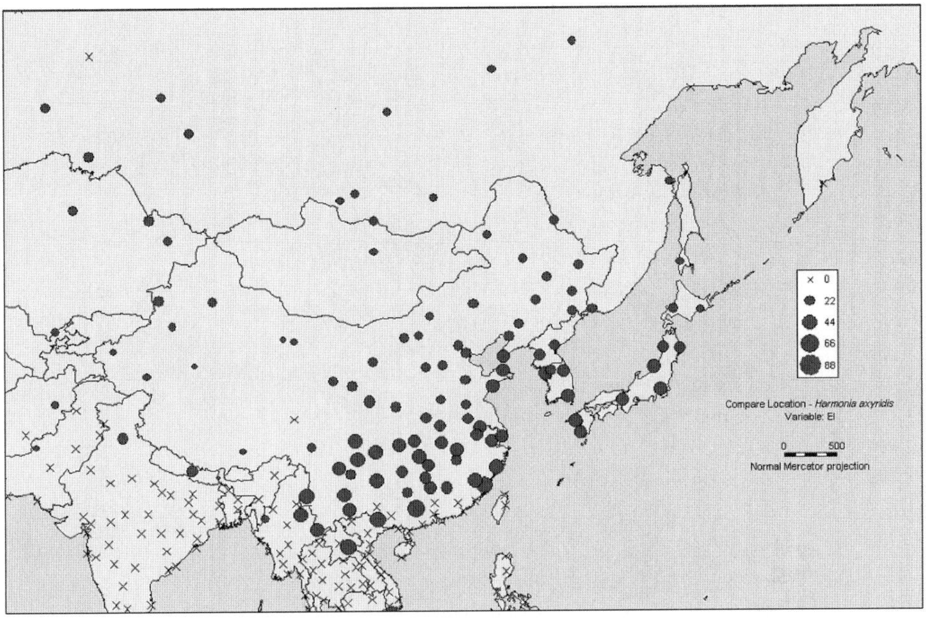

Fig. 2 CLIMEX map of Asia indicating Ecoclimatic Indices (EI). Crosses represent an EI of zero. Spots represents EI's greater than zero. The larger the spot the larger the EI and the more suitable the climate at that station

EI-values in the north. We therefore downgraded the value of this parameter from 9.3°C to 7.3°C to match its presence in South Siberia e.g. Irkutsk, from where *H. axyridis* is known to occur (Fig. 1). On the other hand, the diapause development days (DPD), the minimal number of days necessary to end the state of diapause, showed to be the critical parameter in explaining the southern edge of its native area of distribution. When iterating parameter values ranging between 0 day (no diapause) and 180 days (obligate diapause), a minimal value of 40 day gave the best visual fit, matching south Chinese records in Yunnan (Kuang and Xiao 1995) and in Guang near the Vietnamese border (CABI/EPPO 2007). Temperature and soil moisture parameters largely affected the value of the EI for *H. axyridis*, but was much less limiting in matching its distribution potential as diapause. Stress factors (dry, wet) proved to be of little or no importance on the outcome of the model: cold stress values (CS) remained below a value of 25 (of 100) for populations in the north, but heat stress values increased from Beijing towards the south of China (HS 12–75). Besides climatic suitability, the occurrence of *H. axyridis* depends on humidity through food availability. Its prey, primarily aphids, often occur in areas where irrigation is applied. A 2 mm summer irrigation application per day, after checking several values (0–10 mm), gave the best fit.

When we applied these parameters (Table 1) to meteorological data from stations across the world within CLIMEX, the map of EI's on a worldwide scale suggests that a large part of the global climate is suitable for establishment of *H. axyridis* (Fig. 3). For Europe (Finland) and North America (Alaska) the potential northern distribution limit of *H. axyridis* lies somewhere at 64°N, whereas it most northern record in Siberia (Yeniseisk/Krasnoyarsk) is from 58°N. In North America the eastern part of the continent is suitable for survival on a long-term scale. In the western part of the continent the EI values are

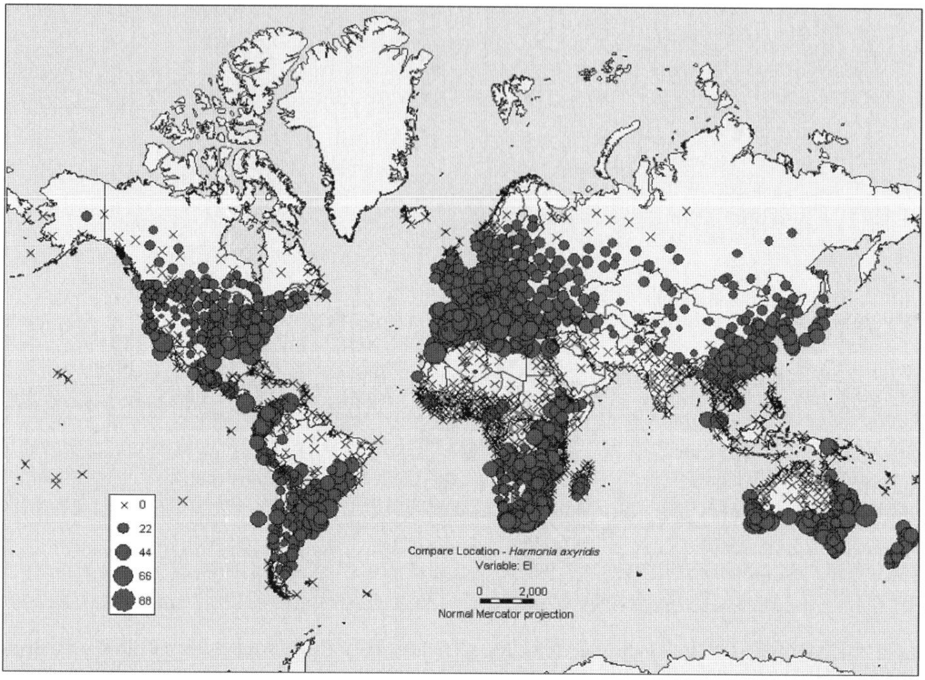

Fig. 3 CLIMEX map of the world indicating Ecoclimatic Indices (EI). Crosses represent an EI of zero. Spots represents EI's greater than zero. The larger the spot the larger the EI and the more suitable the climate at that station

smaller for the inland regions, and higher for the coastal regions (Washington, Oregon and California) (Fig. 4). In Europe concentrations of high EI scores (16–72) are shown for the whole European continent, except for the northern part of Scandinavia and Iceland (Fig. 6). For both continents, dry and wet stress factors were of minor importance, whereas cold and heat stress determined the GI value and therefore the ecoclimatic index EI in the north and south respectively. In particular, changing DPD values (from 40 to 90 days or more) for Europe and North America largely affected the outcome of the EI values in the south and therefore shifted the area of potential distribution towards the north, giving an even better fit.

Match with current geographic distribution

Harmonia axyridis has been introduced several times for classical biological control purposes in North America in the beginning of the 20th century (Table 2). These introductions did not result in any known establishment. From 1978 till 1985, series of releases were made with *H. axyridis* for the control of aphid species like *Monellia caryella* (Fitch) and coccids like *Matsucoccus resinosae* Bean and Godwin (Gordon 1985). Populations from different origins in Asia were released in several states (ARS 2005): Japan (Washington, Delaware), South Korea (Connecticut, Georgia, Maine and Maryland), Russia (Delaware, Maine, Maryland, Mississippi and Nova Scotia-Canada). In addition, releases have been made later on in several areas in Mexico (cited in Koch et al. 2006). The first

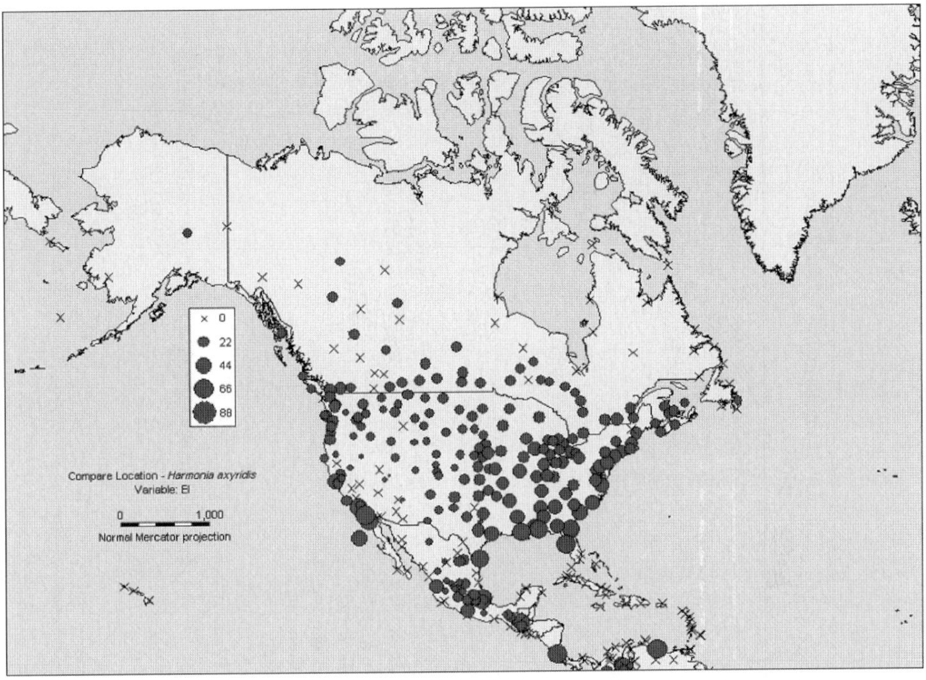

Fig. 4 CLIMEX map of North America indicating Ecoclimatic Indices (EI). Crosses represent an EI of zero. Spots represents EI's greater than zero. The larger the spot the larger the EI and the more suitable the climate at that station

feral record in North America was made in 1988 was by Chapin and Brou (1991). Subsequent records were made at three widely separated locations in northern and southern Mississippi, and in northern Georgia in 1990 (Chapin and Brou 1991; Tedders and Schaefer 1994) and in northern Alabama in 1991 (Day et al. 1994). From there *H. axyridis* has spread throughout Georgia, northern Florida and eastern South Carolina in 1992, reached Virginia in early 1993, and by fall 1993 it was collected farther north, in Delaware, Pennsylvania, and New Jersey (Tedders and Schaefer 1994; Day et al. 1994) and in 1993–1994 all of the states east of the Mississippi were invaded, as well as California, Oregon and Washington on the west coast. From 1994 onwards *H. axyridis* was recorded in Canada as well and is now widely established in North America (Fig. 5) (Koch et al. 2006). When visually comparing the predicted (Fig. 4) and realized (Fig. 5) distribution in North America there is a close fit between both distributions, except for the most northern Canadian stations.

In Europe, *H. axyridis* was first introduced from the Far East (Primorskyi (Vladivostok)) as a classical biological control agent in Georgia in 1927 (Iablokoff-Khnzorian 1982), in the Ukraine from 1964 till 1971 (Koval and Zderchuk 1973), in Kazakhstan in 1968 (Savoiskaya 1970), and in Belarus (Sidlyarevich and Voronin 1973) (Table 3). These introductions, however, did not result in establishment of the species. A second introduction was made by the French National Institute for Agronomical Research (INRA) in 1982 using a population from China (Iperti and Bertrand 2001). It was kept in quarantine till 1990, after which experimental field releases were made in France from 1990 onwards. As far as currently known, most releases in western Europe have been made with offspring

Table 2 Introduction of *Harmonia axyridis* in North America (after Gordon 1985, and ARS 2005)

Place of origin	Year of import	Year of introduction	Place of introduction	Establishment from release	Reference
?	1916		Hawaii	Unknown	Iablokoff-Khnzorian (1982)
China	1927	1927–1928	Florida	No	Watson (1929)
Japan	1916, 1964	1916, 1964, 1965	California	No (Gordon 1985)	Chant (1964, 1965); Essig, 1931 (in Gordon 1985)
Japan (Fukuoka)	1978	1978–1982	Washington, Delaware	Unknown	ARS (2005); J. Coulson, in Gordon (1985)
Japan	1978	>1978	Georgia (Houston: Perry; Peach: Byron)	No (Gordon 1985)	R. Dysart, in Gordon (1985)
Former USSR (Primorskiy Kray: Ussuriysk)	1978	1981	Nova Scotia	Unknown	R. Dysart, in Gordon (1985)
Former USSR (Primorskiy Kray: Ussuriysk)	1978–1984	1978 onwards	Connecticut (New Haven: Mt Carmel-Lockwood), Georgia, Louisiana, Maryland (Prince Georges: Beltsville), Washington D.C., Delaware, Maine (Hancock: Catherine Hill Mtn.), Mississippi (Washington: Leland and Stoneville), Ohio, Pennsylvania	Unknown	ARS (2005); R. Dysart, in Gordon (1985)
Japan	1979	1979, 1980	Louisiana, Mississippi (Washington: Leland and Stoneville)	Unknown	Chapin and Brou (1991); Tedders and Schaefer (1994)
South Korea	1981	1981–1985	Connecticut, Georgia, Maine and Maryland	Unknown	ARS (2005)

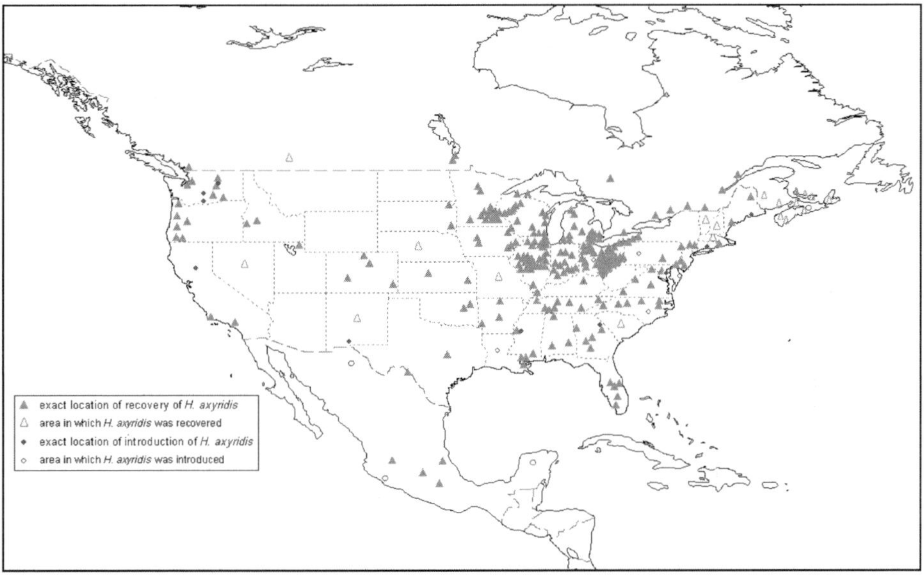

Fig. 5 Distribution of *Harmonia axyridis* in North America (based on CABI/EPPO 2007, and Koch et al. 2006)

of the INRA source population: as a biological control agent of *Toxoptera citrida* in citrus in Portugal (Azores: Garcia 1986) in 1984, and in Greece in 1994 (Katsoyannos et al. 1997), and for aphid control outdoors in France since 1993 (Kabiri 2006). From the mid 1990s onwards biocontrol companies have commercialized *H. axyridis* as a commercial product for aphid control in augmentative biological control programmes indoors. Commercial releases were made in western Europe from 1993 onwards, in France (1993: Kabiri 2006), Belgium (1996: Adriaens et al. 2003), the Netherlands (1995: Cuppen et al. 2004), Italy (1996: Orlandini and Martellucci 1997) and has been offered commercially for a few years or more in most other European countries as well. It has never been sold officially in Germany, UK and Switzerland.

The first feral aggregation in Europe was observed during the winter of 1999–2000 on a balcony in a Frankfurt (Germany) suburb (Bathon 2002). Subsequent feral populations were recorded from Belgium in 2001, the Netherlands in 2002 (Cuppen et al. 2004), France in 2003 (Kabiri 2006), Luxemburg in 2003 (Schneider and Loomans 2006), United Kingdom in 2004 (Majerus et al. 2007), Switzerland in 2004 (Klausnitzer 2004), Austria in 2006 (Rabitsch and Schuh 2006) and Czechia (Brown et al. 2007) (Fig. 7). Kangas (1945) reported *H. axyridis* from Finland, but after studying the specimen (Loomans, personal observation 2007), it turned out to be *Harmonia conformis* (Boisduval). Large numbers are nowadays regularly observed in the Netherlands, Belgium, Luxemburg, Germany, northern France and UK. In South-European countries *H. axyridis* has been recorded, but only incidentally and in low numbers (e.g. Katsoyannos et al. 1997). Other releases in the Westpalaearctic region are known from Egypt and Syria (Abdel-Salam and Abdel-Baky 2001: origin USA; but also China (INRA)) (Table 3). When we compare the potential (Fig. 6) and realized (Fig. 7) distribution in Europe and North Africa there is not as close a fit between both distributions as for North America, but the current area of distribution falls within the predicted range.

Table 3 Introductions of *Harmonia axyridis* in Eurasia

Region of origin	Year of import	Year of introduction	Place of introduction	Establishment from release	Reference
Eastern Maritime Territory (Primorskiy Kray)	1964	1964	Ukraine (Chernovitskaya)	Unknown	Koval and Zderchuk (1973); Sinadskii and Kozarzhevskaya (1980); Voronin (1968)
		1969	Ukraine (Crimea)	Unknown	
Eastern Maritime Territory (Primorskiy Kray)	1969	1969	South-eastern Kazakhstan	Unknown	Savoiskaja (1970)
Belarus	1968	1968–1970	Minsk, glasshouse release	Unknown	Sidlyarevich and Voronin (1973)
China	1982	>1993	France	Unknown	Ongagna et al. (1993); Kabiri (2006)
China[a]	1984	1984	Azores Islands		Garcia (1986)
China[a]	1993	1994	Greece	Yes	Katsoyannos et al. (1997)
China[a]	1995	1996	The Netherlands	Yes	Cuppen et al. (2004)
China[a]	1995	1997	Belgium	Yes	Adriaens et al. (2003)

[a] distributed by INRA and Biotop

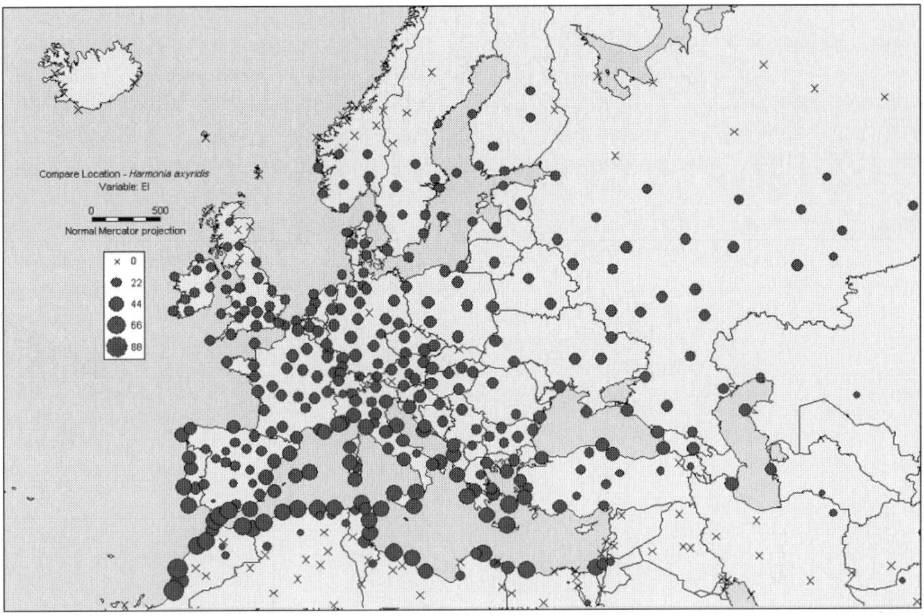

Fig. 6 CLIMEX map of Europe and northern Africa indicating Ecoclimatic Indices (EI). Crosses represent an EI of zero. Spots represents EI's greater than zero. The larger the spot the larger the EI and the more suitable the climate at that station

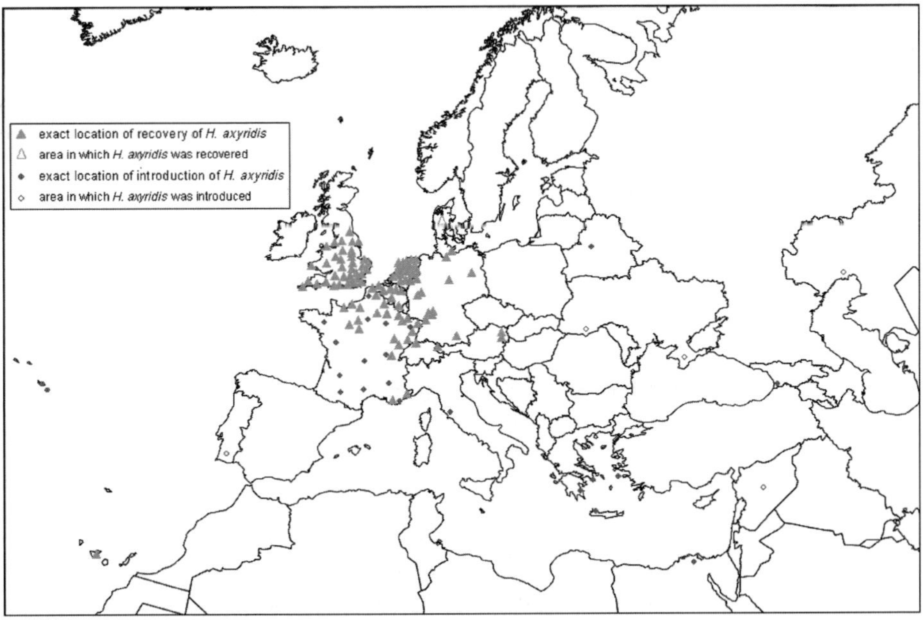

Fig. 7 Distribution of *Harmonia axyridis* in Europe and northern Africa (based on CABI/EPPO 2007, corrected and modified)

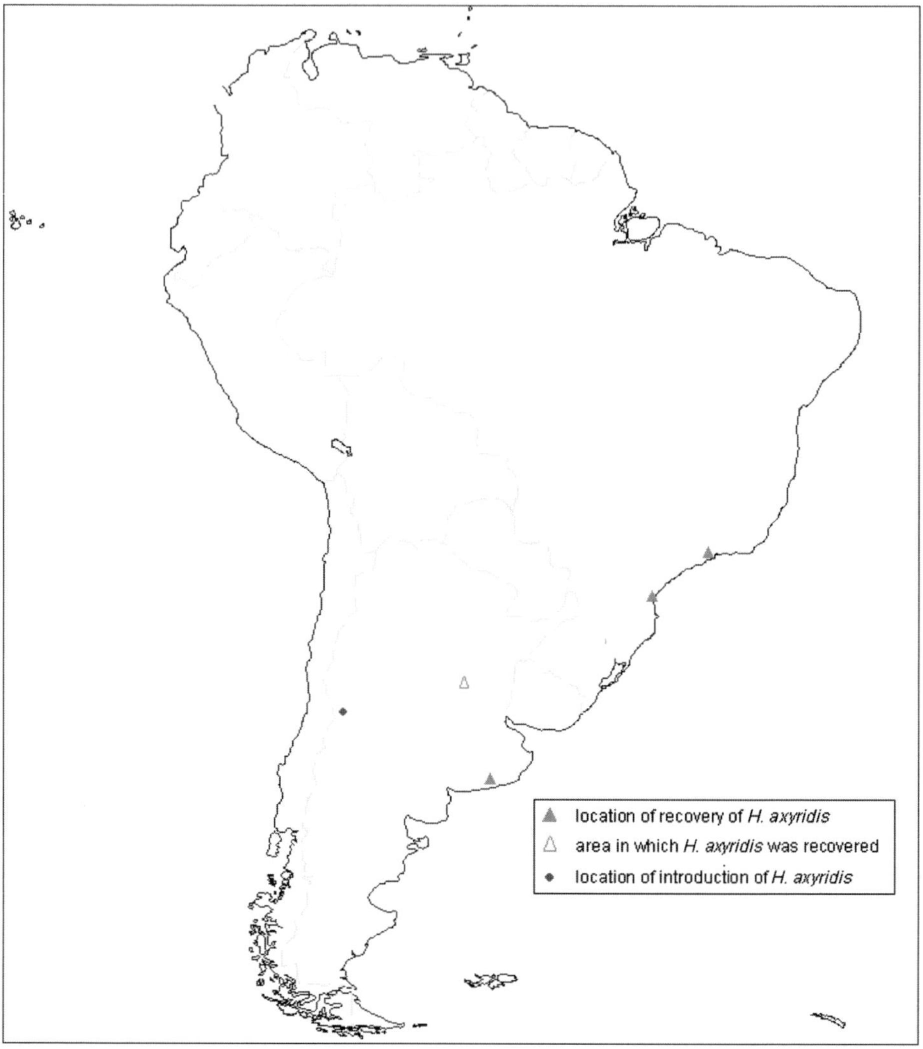

Fig. 8 Distribution of *Harmonia axyridis* in South America (after Saini 2004 and author's information)

In South America *H. axyridis* has been introduced through INRA in Mendoza—Argentina in 1986, 1987 (García et al. 1999) and 1999 (Saini 2004) and in a few years time it has spread into a large part of that country (Fig. 8). Since 2001 it is found in the Buenos Aires area (Saini, 2004) and south of there. In Brazil *H. axyridis* has never been released but was recorded from Curitiba in 2002 (de Almeida and Da Silva 2002) and Sao Paolo in 2007 (Enio Branco, personal communication 2007). Introduction and spread into South America is quite a recent date, and *H. axyridis* may still expand beyond its current range. Other regions on the southern hemisphere, like Africa, Australia and New Zealand, that according to the outcome of the CLIMEX model would be suitable for long-term survival of *H. axyridis* (Fig. 3), would be under threat.

Discussion

The output of a CLIMEX run using several indices for a single location, are integrated into one index (EI) that will indicate the long-term survival of a particular species on that location (Sutherst 2003). The strength of CLIMEX 'Compare Locations' is that it combines the freestanding climatic data, like temperature, precipitation, and relative humidity and integrates these with the biological characteristics of species and shows the indices for long-term survival of all the locations in one glance. CLIMEX has shown to be an effective tool for predicting the distribution and assessing the risks of a large range of species, plant pests as well as natural enemies (Sutherst and Maywald 2005). The current analysis projects the potential geographic distribution of *H. axyridis* in various parts of the world, based on fitting of CLIMEX parameters. Simulated trends show that, for those areas where *H. axyridis* already has been introduced, established and spread, its predicted distribution is largely supported by field data: there is a great consistency between the predicted and realized distribution in North America (Fig. 4, 5) and in Europe (Fig. 6, 7). Predictions for its northern and southern range in North America and Europe are not (yet) completely covered by the actual records. The accuracy of this prediction using the CLIMEX model can be divided in model-based factors and species-based factors. Below we discuss what factors are involved and what experimental data would be needed to verify and specify the model and the accuracy of its use.

Model based factors

Climatic suitability versus actual distribution

As every simulation program for ecological phenomena, CLIMEX has its limitations. Climate models greatly rely on the number and distribution of meteorological stations: the network is not equal per region and is rather low in northern parts of the world, where there are few inhabitants. Areas with a small number of stations may not give a representative view of the climate in that region, because the location of the meteorological stations is frequently unrepresentative for the surrounding area (Bennett et al. 1998). CLIMEX records the presence or absence of a given species, but does not include a measure of abundance, that allows favourable areas to be distinguished from marginal areas. CLIMEX works with climatic data and climate related species characteristics only, and it works with the assumption that the range of a species is solely determined by climate. Within the climatic range of a species also non-climatic factors will of course limit its potential distribution as well, acting individually or in combination (Baker et al. 2000). These physical and biological factors, such as soil type, availability of food and substrates for specific developmental processes (e.g. larval development, overwintering), dispersal capacities of the species, and interactions between species, such as competition, predation and parasitism as well as existing pest management practices are not included in CLIMEX simulations (Baker et al. 2000). Northern regions like central Canada and the Great Plains in the USA are mostly cultivated and have flat terrain and overwintering sites are limited to trees and aggregations are subject to extreme cold when not able to shelter. This could partly explain why these regions have low or no distribution records (Fig. 5), while the model predicts a significant EI value (Fig. 4). Extreme climatic conditions such as cold winters, however, could prevent that *H. axyridis* from establishing.

Species based parameters

Input biological species characteristics

Several factors (species parameters) will affect the potential distribution as indicated by the CLIMEX map, but eventually one of these factors acts as the limiting factor, though another may become limiting as other species parameter values change. CLIMEX predictions are only based on the long-term survival of species on locations during the whole year. To what extent *H. axyridis* is able to build up seasonal populations beyond its overwintering range and can cause severe impact during the growing season, is not clear.

Genetic differences

The observed geographical distribution of a species reflects the sum of all the genetic variation contained in that species (Sutherst et al. 2000). When a species like *H. axyridis*, however, has been introduced into a new area only a small proportion of the gene pool has been used. Consequently, the founder individuals may produce populations that may have characters that differ from the mother population (Samways 2003). Specimens of different populations of a species with a great native distribution range could react differently to climate, caused by the genetic differences between the geographically widely separated populations. The different populations and phenotypes of *H. axyridis* could have different adaptations to climate (Kuang and Xiao 1995) and season (Osawa 2001). Therefore, it is not always safe to assume that a small proportion, which is used in research, is representative for the whole population. Because of this, the species parameter values should be used as an indication, and can be modified if the potential distribution will not fit the actual distribution range. However, the more observations and species parameter values are introduced, the more reliable the picture will be.

Harmonia axyridis is widespread throughout temperate and subtropical Asia and geographical genetic polymorphism could occur in their response to climatic cues. The elytral pattern polymorphism shows a distinct geographical pattern (Mader 1926–1937; Dobzhansky 1933, Komai et al. 1950) and is caused by different factors, including climate, food sources, and genetic components. The macro-geographical variations are mostly related to climatic factors, while micro-geographic variations are mostly related to food availability (Muggleton 1978; Soares et al. 2003). Within populations there are also seasonal variations in the proportions of the light and dark coloured individuals. Research by Osawa and Nishida (1992) showed an increase in the frequency of the 'succinea' phenotype from spring to summer in relation to non-random mating among the different phenotypes. Specimens of different populations of a species with a large distribution range, like *H. axyridis*, could react differently to climate (phenotypic plasticity), caused by the genetic differences between the widely geographically separated populations. Different populations could imply different adaptations to climate. Non-melanic and melanic forms of *H. axyridis* show different responses to climates (Komai 1956) and differences in relative fitness exist between melanics and non-melanics (Soares et al. 2001, 2003) and their phenotypes (Kuang and Xiao 1995). Melanic forms would be in advantage over non-melanic forms in colder climates, melanic forms being able to attain higher body temperatures than non-melanic forms when exposed to sunlight and therefore a greater fitness (Koch et al. 2006).

The individuals introduced by INRA into Europe originate from a Chinese population (Ongagna et al. 1993). It is unknown, however, from what location in China they exactly came from ('Bejing' as cited by García et al. 1999). We have strong indications that the INRA population originates from individuals imported from north-east China because: (1) most individuals found in western Europe belong to the 'succinea' type (~80%) while 20% consists of the melanic types such as 'spectabilis' (~15%) and 'conspicua' (~5%), (2) about 95% of the individuals found in Europe, show an elytral ridge. This fits well the descriptions given by Dobzhansky (1933) and Komai et al. (1950) for north-east China and as well as those examined by I. Zakharov (March 2007, personal communication). Using the 'Match Climates' option in CLIMEX, which is independent of the species requirements, climates of different locations in north-east China (Shenyang, Tianjin, Linxi) and those in Europe show a high level of similarity (0.60 or more) with most European locations.

Compared to the European populations melanic forms are rare in North America, suggesting a different source. From what origin the North American feral populations originate (Krafsur et al. 1997), however, is not exactly known. Releases have been made in North America with offspring of populations from various imports (Korea, USSR, Japan), (ARS 2005; Gordon 1985; Table 2). Instead, because of intentional introductions from 1978 to 1982 (Tedders and Schaefer 1994), Day et al. (1994) presume that, based on its spread, initial establishment was caused by accidental seaport introductions at the port of New Orleans in Louisiana. Also in western Washington, especially near Puget Sound, an inlet used by ships headed to the ports of Seattle and Tacoma, the species is common. There, the first recovery occurred 12 years after the releases in 1981/1982 (ARS 2005). Also other authors (Krafsur et al. 1997; Koch 2003) suggested that, based on the genetic similarities between widely distributed North American populations, feral populations of *H. axyridis* in North America might have come from a single source. Both melanic and non-melanic forms were initially collected and reared in the USDA quarantine facilities, but black colour forms have been found in the Oregon and Washington states only, whereas in other parts of the United States only the 'succinea' form is present (LaMana and Miller 1998; Krafsur et al. 1997). Our predictions from CLIMEX, using the same set of parameters for both continents, however, give an adequate match with the realized distribution in both cases, suggesting that few differences exist. Morphological examination of specimens of both continents by I. Zakharov (personal communication, 2007) also indicated a large similarity. Molecular research will shed more light on the composition and differences between populations in both continents.

Geographical distribution of species

A CLIMEX-prediction is more useful for species with a restricted home range, like *H. axyridis*, than for a more cosmopolitan species. The size of a home-ranged species is clearly restricted. Subsequently it is easier to extrapolate the climate of the home range to other continents. In the case of *H. axyridis*, the situation is ideal because this species has already been introduced and established in some other continents. In this way the parameter values can be easily verified. We can use the observed distribution outside its native range as a check for the parameter values in CLIMEX. Because establishment in North America and Europe are well covered by the EI values higher than zero, and also the establishments South America are covered, it can be concluded that the parameter values in this research were appropriate.

A relatively large discrepancy remains between the current and predicted distribution in Europe. Currently the area of distribution of *H. axyridis* ranges from Denmark in the north, UK in the west, Poland in the east and Northern Italy in the south (see Brown et al. 2007). CLIMEX predicts both temperate and Mediterranean climates as being suitable for establishment of *H. axyridis* (Fig. 6). However, until now no large numbers of *H. axyridis* have been recorded from Southern Europe, in spite of multiple releases made in that area (e.g. Katsoyannos et al. 1997 in citrus). When adapting DPD values from 40 to 90 days or more, the minimum number of days below DPT0 required for diapause to be complete, the prediction in CLIMEX gave a better fit. Is it a matter of time for *H. axyridis* to expand beyond its current range or does its spread come to a halt? Climate, based on our analysis, is not a limiting factor by itself. Seasonality, however, determining certain biological and ecological features of host plants, available prey and unfavourable winter conditions, could play an important role. Though polyphagous in food acceptance, aphids are the primary food source of *H. axyridis*. Aphid pests are characteristic for temperate areas, showing large variety in species in different habitats and seasonal fluctuations. In north and central Europe aphids developing on young plant material can rapidly develop large populations in spring, summer and fall, under warm and moist conditions. In Mediterranean areas there are relatively more evergreen plants such as citrus, and summers are relatively dry, resulting in relatively less vegetative growth and long periods without substantial food sources. A second explanation for the absence of dense populations in Southern Europe is the intensity in which the landscape is used. In north-west Europe the landscape largely consist of intensively used, urban, industrial and cultivated areas. This is expressed in a dominance of annual crops and a diversity of semi-arboreal habitats. Therefore, in addition to CLIMEX, the influence of soil, vegetation type, habitat and land use should be explored for fine-tuning its potential spread.

A third explanation could be that favourable (low temperature) overwintering conditions and sites in the south are limited to certain restricted natural areas, thus limiting the winter survival by *H. axyridis* in large quantities. Adults entering diapause will stay in relative warm conditions over winter in the south of Europe (and in heated buildings in the north). They will be depleted of fat-reserves during the winter period and die before being able to allocate new prey in spring.

In conclusion, we see that according to the CLIMEX predictions, *H. axyridis* will be able to survive and establish in large parts of the world. In North America, the prediction of the model fits the current distribution very well. In Europe, where its spread is still ongoing, the area of recovery largely fits the prediction too, but its limit to the eastern and southern expansion has to be verified. On the southern hemisphere, where *H. axyridis* has recently been introduced intentionally or may arrive by accident, the chances of survival are substantial in a large part. CLIMEX estimates have proven its use as a strategic tool for pest risk assessments (Baker et al. 2000), when time and opportunities are lacking and rapid decisions have to be taken with little experimental data available. For predicting the likelihood and impact of establishment of a biological control agent such as *H. axyridis* (van Lenteren et al. 2003), it can be a first step tool in assessing risks for non-target species.

References

Abdel-Salam AH, Abdel-Baky NF (2001) Life table and biological studies of *Harmonia axyridis* Pallas (Col., Coccinellidae) reared on the grain moth eggs of *Sitotroga cerealella* Olivier (Lep., Gelechiidae). J Appl Entomol 125:455–462

Adriaens T, Branquart E, Maes D (2003) The multicoloured Asian ladybird *Harmonia axyridis* Pallas (Coleoptera: Coccinellidae), a threat for native aphid predators in Belgium? Belgian J Zool 133: 195–196

de Almeida LM, da Silva VB (2002) First record of *Harmonia axyridis* (Pallas) (Coleoptera, Coccinellidae): a lady beetle native to the Palaearctic region. Rev Bras Zool 19:941–944

ARS (2005) National Invertebrate Genetic Resources. ROBO Database Reports Menu—1981–1985. http://www.ars-grin.gov/nigrp/menu.html. Cited 30 Jul 2007

Baker RHA, Sansford CE, Jarvis CH, Cannon RJC, MacLeod A, Walters KFA (2000) The role of climatic mapping in predicting the potential geographical distribution of non-indigenous pests under current and future climates. Agric Ecosyst Env 82:57–71

Barçin EA, Smith BN, Hansen LD, Booth GM (2001) Calorespirometry to determine the effects of temperature on the metabolic efficiency of *Harmonia axyridis* (Col.: Coccinellidae) larvae. http://esa.confex.com/esa/2001/techprogram/paper_3292.htm. Cited 23 Dec 2004

Bathon H (2002) *Harmonia axyridis*, eine invasive Marienkäferart in Mitteleuropa. DGaaE Nachr 16: 109–110

Bennett SJ, Said N, Enneking D (1998) Modelling climatic similarities in Mediterranean areas: a potential tool for genetic resources and breeding programmes. Agric Ecosyst Env 70:129–143

Berthiaume R, Hebert C, Bauce E (2003) Impact of temperature and duration of cold exposure on adult survival of the Asian ladybird beetle, *Harmonia axyridis* (Pallas). Phytoprot 84(2):85–91

Brown PMJ, Adriaens T, Bathon H, Cuppen J, Goldarazena A, Hägg T, Kenis M, Klausnitzer BEM, Kovář I, Loomans AJM, Majerus MEN, Nedved O, Pedersen J, Rabitsch W, Roy HE, Ternois V, Zakharov IA, Roy DB (2007) *Harmonia axyridis* in Europe: spread and distribution of a non-native coccinellid. BioControl (this issue). doi:10.1007/s10526-007-9132-y

Brown MW, Miller SS (1998) Coccinellidae (Coleoptera) in apple orchards of eastern West Virginia and the impact of invasion by *Harmonia axyridis*. Ent News 109:136–142

CABI/EPPO (2007) *Harmonia axyridis*. Distribution Maps of Plant Pests No. 689. CABI Head Office, Wallingford, UK

Chapin JB, Brou VA (1991) *Harmonia axyridis* (Pallas), the third species of the genus to be found in the United States (Coleoptera: Coccinellidae). Proc Entomol Soc Wash 93:630–635

Colunga-Garcia M, Gage SH (1998) Arrival, establishment, and habitat use of the multicolored Asian lady beetle (Coleoptera: Coccinellidae) in a Michigan landscape. Environ Entomol 27:1574–1580

Cuppen J, Heijerman T, van Wielink P, Loomans A (2004) Het lieveheersbeestje *Harmonia axyridis* in Nederland: een aanwinst voor onze fauna of een ongewenste indringer (Coleoptera: Coccinellidae)? Ned Faun Med 20:1–12

Day WH, Prokrym DR, Ellis DR, Chianese RJ (1994) The known distribution of the predator *Propylea quatuordecimpunctata* (Coleoptera: Coccinellidae) in the United States, and thoughts on the origin of this species and five other exotic lady beetles in eastern North America. Ent News 105:224–256

Dobzhansky T (1933) Geographical variation in lady-beetles. Am Nat 67:97–126

Garcia V (1986) Appraoches to integrated control of some citrus pests in the Azores and Algarve (Portugal). In: Cavalloro R, Di Martino E (eds) Integrated pest control in citrus groves. Proceedings of the Expert's Meeting, Acireale, 26–29 March, 1985. Published for the Commission of the European Communities by A.A. Balkema, Rotterdam Boston, pp 557 559

García M, Becerra V, Reising C (1999) *Harmonia axyridis* Pallas (Coleoptera, Coccinelidae) Estudio biológico. Rev Fac Cien Agrar Univ Nac Cuyo 31(1):85–91

Gordon RD (1985) The Coccinellidae (Coleoptera) of America north of Mexico. J NY Entomol Soc 93: 1–912

Grill CP, Moore AJ, Brodie ED (1997) The genetics of phenotypic plasticity in a colonizing population of the ladybird beetle, *Harmonia axyridis*. Heredity 78:261–269

Hatherly IS, Hart AJ, Tullett AG, Bale JS (2005) Use of thermal data as a screen for the establishment potential of non-native biological control agents in the UK. BioControl 50:687–698

Hodek I, Honěk A (1996) Ecology of Coccinellidae. Series Entomologia, Vol 54. Kluwer Academic Publishers, Dordrecht/Boston/London

Iablokoff-Khnzorian SM (1982) Les coccinelles Coléoptères—Coccinellidae: Tribu Coccinellini des régions Paléarctique et Orientale. Société Nouvelle des Éditions Boubée, Paris

Iperti G, Bertand E (2001) Hibernation of *Harmonia axyridis* (Coleoptera: Coccinellidae) in South-Eastern France. Acta Soc Entomol Bohem 65:207–210

Kabiri F (2006) *Harmonia*: a problem or a distraction? Powerpoint presentation at REBECA Macrobials Workshop: WICC Wageningen April 5–6 2006. http://www.rebeca-net.de/downloads/Risk%20assessment%20macrobials%20Kabiri.pdf. Cited 27 Jul 2007

Kangas E (1945) Ein vertreter der sibirischen Coccinelidenfauna (Col.) in Finnland. Suomen Hyönteisteellinen. Ann Entomol Fenn (Aikarauskirja) 11:63–67

Katsoyannos P, Kontodimas DC, Stathas GJ, Tsartsalis CT (1997) Establishment of *Harmonia axyridis* on citrus and some data on its phenology in Greece. Phytopar 25:183–191

Klausnitzer B (2004) *Harmonia axyridis* (Pallas, 1773) in Basel-Stadt (Coleoptera, Coccinellidae). Mitt Entomol Ges Basel 54:115–122

Koch RL (2003) The multicolored Asian lady beetle, *Harmonia axyridis*: a review of its biology, uses in biological control and non-target impacts. J Insect Sci 3:1–16

Koch RL, Venette RC, Hutchison WD (2006) Invasions by *Harmonia axyridis* (Pallas) (Coleoptera: Coccinellidae) in the Western Hemisphere: implications for South America. Neotrop Entomol 35: 421–434

Komai T, Chino M, Hosino Y (1950) Contributions to the evolutionary genetics of the lady-beetle, *Harmonia*. I. Geographic and temporal variations in the relative frequencies of the elytral pattern types and in the frequency of the elytral ridge. Genetics 35:589–601

Komai T (1956) Genetics of ladybeetles. Genetics 8:155–188

Koval YV, Zderchuk IS (1973) Physiological and biochemical factors favouring acclimatisation of the Fareastern *Harmonia* Coccinellid, *Leis axyridis* Pall. in Soviet Bukovina. Trudy Vses Nauchno-issl skogo Inst Zash Rast 36:33–42

Krafsur ES, Kring TJ, Miller JC, Nariboli P, Obrycki JJ, Ruberson JR, Schaefer PW (1997) Gene flow in the exotic colonizing ladybeetle *Harmonia axyridis* in North America. Biol Control 8:207–214

Kuang RP, Xiao NN (1995) Growth patterns of the ladybird (*Harmonia axyridis* Pallas) relative to variations in population genetics. Insect Sci Applic 16(1):35–38

Kuznetsov VN (1997) Lady beetles of the Russian far east. Memoir No. 1. Center for Systematic Entomology. Gainsville, Sandhill Crane Press, Inc

LaMana ML, Miller JC (1998) Temperature-dependent development in an Oregon population of *Harmonia axyridis* (Coleoptera: Coccinellidae). Environ Entomol 27(4):1001–1005

Lanzoni A, Accinelli G, Bazzocchi GG, Burgio G (2004) Biological traits and life table of the exotic *Harmonia axyridis* compared with *Hippodamia variegata*, and *Adalia bipunctata* (Col., Coccinellidae). J Appl Entomol 128:298–306

Lo X-c, Liu Y-k (1976) Studies on overwintering aggregations of *Leis axyridis* (Pallas) in Northeastern and North China: an observation on the overwintering habitats. Acta Entomol Sin 19(1):115–116

Machado A (2006) El sarantontón asiático *Harmonia axyridis* (Pallas, 1773) presente en Canarias (Coleoptera: Coccinellidae). Vieraea 24:71–72

Majerus MEN, Roy HE, Brown P (2007) The harlequin ladybird survey. website University of Cambridge, Department of Genetics. http://www.harlequin-survey.org/default.htm# Cited 30 Jul 2007

Majerus MEN, Strawson V, Roy HE (2006) The potential impacts of the arrival of the Harlequin ladybird, *Harmonia axyridis* (Pallas) (Coleoptera: Coccinellidae), in Britain. Ecol Entomol 31:207–215

Mader L (1926–1937) Evidenz der paläarktischen Coccinelliden und ihrer Aberrationen in Wort und Bild. 1. Epilachnini, Coccinellini, Halyziini, Synonychini. Published as supplement to: Zeitschrift des Vereines für Naturbeobachter und Sammler (1926–1934), Entomologischer Anzeiger (1935–1936) and entomologisches Nachrichtenblatt, Troppau, 1937

Michaud JP (2002) Invasion of the Florida citrus ecosystem by *Harmonia axyridis* (Coleoptera: Coccinellidae) and asymmetric competition with a native species, *Cycloneda sanguinea*. Environ Entomol 31:827–835

Muggleton J (1978) Selection against the melanic morphs of *Adalia bipunctata* (two-spot ladybird): a review and some new data. Heredity 40:269–280

Müller MJ (1996) Handbuch ausgewählter Klimastationen der Erde. Universität Trier, Forschungsstelle Bodenerosion. Publikation Gerold Richter, Trier (Germany)

Nalepa CA, Kidd KA, Ahlstrom KR (1996) Biology of *Harmonia axyridis* (Coleoptera: Coccinellidae) in winter aggregations. Ann Entomol Soc Am 89:681–685

Nalepa CA, Kennedy GG, Brownie C (2005) Role of visual contrast in the alighting behavior of *Harmonia axyridis* (Coleoptera: Coccinellidae) at overwintering sites. Environ Entomol 34(2):425–431

Ongagna P, Giuge L, Iperti G, Ferran A (1993) Cycle de développement d'*Harmonia axyridis* (Col. Coccinellidae) dans son aire d'introduction: le sud-est de la France. Entomophaga 38:125–128

Orlandini G, Martelucci R (1997) Melone: lotta biologica all'*Aphis gossypii*. Colt Prot 6:33–36

Osawa N (2001) The effect of hibernation on the seasonal variations in adult body size and sex ratio of the polymorphic ladybird beetle *Harmonia axyridis*: the role of thermal melanism. Acta Soc Zool Bohem 65:269–278

Osawa N, Nishida T (1992) Seasonal variation in elytral colour polymorphism in *Harmonia axyridis* (the ladybird beetle): The role of non-random mating. Heredity 69(4):297–307

Poorani J (2004) An annotated checklist of the Coccinellidae (Coleoptera) of the Indian subregion. http://www.angelfire.com/bug2/j_poorani/checklist.pdf Cited 22 Dec 2004

Rabitsch W, Schuh R (2006) First record of the multicoloured Asian ladybird *Harmonia axyridis* (Pallas, 1773) in Austria. Beitr Entomofaun 7:161–164

Saini ED (2004) Presencia de *Harmonia axyridis* (Pallas) (Coleoptera: Coccinellidae) en la provincia de Buenos Aires. Aspectos biológicos y morfológicos. RIA 33:151–160

Sakurai H, Takeda S, Kawai T (1988) Diapause regulation in the lady beetle, *Harmonia axyridis*. In: Niemczyk E, Dixon AFG (eds) Ecology and effectiveness of aphidophaga. SPB Academic Publishing, The Hague, The Netherlands

Samways MJ (2003) Critical response from Professor Michael J. Samways. J Biogeogr 30:817

Sasaji H (1982) A revision of the Formosan Coccinellidae (III) Subfamily Coccinellinae (Coleoptera). Mem Fac Educ Fukui Univ SII (Nat Sci) 31(1):1–49

Savoiskaja GI (1970) The Far-Eastern ladybird against the apple aphid. Zasch Rast 15(6):12–13

Schanderl H, Ferran A, Gracia V (1985) L'élevage de deux coccinelles *Harmonia axyridis* et *Semiadalia undecimnotata* à l'aide d'oeufs d'*Anagasta kuehniella* tués aux rayons ultraviolets. Entomol Exp Appl 49:235–244

Schneider N, Loomans AJM (2006) Sur la présence au Luxembourg de la coccinelle arlequin *Harmonia axyridis* (Pallas, 1773) (Insecta, Coleoptera, Coccinellidae). Bull Soc Nat Lux 106:71–74

Sidlyarevich VI, Voronin KE (1973) Trials on using *Leis axyridis* under glass. Belorusskii Institut Zashchity Rastenii, Belorussian SSR. Zash Rast 6:24

Sinadskii YV, Kozarzhevskaya EF (1980) Biological protection of flowering and decorative plants. Zash Rast 10:21–22

Snyder WE, Clevenger GM, Eigenbrode SD (2004) Intraguild predation and successful invasion by introduced ladybird beetles. Oecologia 140:559–565

Soares O, Coderre D, Schanderl H (2001) Fitness of two phenotypes of *Harmonia axyridis* (Coleoptera: Coccinellidae). Eur J Entomol 98:287–293

Soares O, Coderre D, Schanderl H (2003) Effect of temperature and intraspecific allometry on predation by two phenotypes of *Harmonia axyridis* Pallas (Coleoptera: Coccinellidae). Env Entomol 32:939–944

Sutherst RW (2003) Prediction of species geographical ranges. J Biogeogr 30:805–816

Sutherst RW, Maywald GF (2005) A climate model of the red imported fire ant, *Solenopsis invicta* Buren (Hymenoptera: Formicidae): implications for invasion of new regions, particularly Oceania. Environ Entomol 34(2):317–335

Sutherst RW, Maywald GF, Russell BL (2000) Estimating vulnerability under global change: modular modelling of pests. Agric Ecosyst Environ 82(1–3):303–319

Sutherst RW, Maywald GF, Bottomley W, Bourne A (2004) CLIMEX v.2, CD and User's Guide. Hearne Scientific Software, Melbourne

Tedders WL, Schaefer PW (1994) Release and establishment of *Harmonia axyridis* (Coleoptera: Coccinellidae) in the southeastern United States. Ent News 105:228–243

van Lenteren JC, Babendreier D, Bigler F, Burgio G, Hokkanen HMT, Kuske S, Loomans AJM, Menzler-Hokkanen I, van Rijn PCJ, Thomas MB, Tommasini MG, Zeng Q-Q (2003) Environmental risk assessment of exotic natural enemies used in inundative biological control. BioControl 48:3–38

Voronin KE (1968) The acclimatisation of *Leis axyridis* Pall., a predator of aphids from the Far East, in the Carpathian foothills. In: Zimin LS (ed) The biological method of plant protection. Trudy Vses Nauchno-issl skogo Inst Zash Rast, 31:234–243

Watanabe M (2002) Cold tolerance and *myo*-inositol accumulation in hibernating adults of a lady beetle, *Harmonia axyridis* (Coleoptera: Coccinellidae). Eur J Entomol 99(1):5–9

Watson JR (1929) Report of the Department of Entomology. Rep Florida Agric Expt Sta 1927–1928:42–49

Yuan RC, Zhang FM, Wen GZ (1994) Study on control of soybean aphid by *Harmonia* (*Leis*) *axyridis*. Ji Lin Agric Sci 44:30–32, 57

Harmonia axyridis: What will stop the invader?

António Onofre Soares · Isabel Borges · Paulo A. V. Borges · Geneviève Labrie · Éric Lucas

Abstract In recent years *Harmonia axyridis* (Pallas, 1773) (Coleoptera: Coccinellidae) has become a very popular insect among biological control practitioners and scientists, not only for its potential to be an efficient biological control agent but also because it is considered invasive. Individuals of this species were deliberately introduced into several countries for biological control of different arthropods pests. However the predator itself became an invasive species, affecting the dynamics and composition of several guilds through direct or indirect interactions with established species, including intraguild predation. In this paper we discuss the reasons why the species has a high invasiveness and what are the limits to invasion by this species. It is not clear if the invasiveness of the beetle is linked to its biological, ecological and behavioural abilities, or to other factors such as invasibility and interactions between the invaders, the noninvaders, and the habitat, which may in part explain the reasons of its success and help us to answer the question "what will stop the invader?" We also discuss the reason for the absence of the predator in the Azores islands. Despite the intentional introduction of *H. axyridis* in the Azores and the high number of individuals released, there are no records of this species in the wild, despite recent extensive sampling effort. In this paper we discuss the reasons for the apparent failure or the delay in establishment of the predator. One factor which may hamper the establishment of *H. axyridis* in some of the Azores islands is the absence of winter

A. O. Soares (✉) · I. Borges
Departmento de Biologia, CIRN, Universidade dos Açores, Rua da Mãe de Deus, 13-A, 9501-801 Ponta Delgada, Azores, Portugal
e-mail: onofre@uac.pt

P. A. V. Borges
Departmento de Ciências Agrárias, CITAA, Universidade dos Açores, 9700-851 Angra do Heroísmo, Azores, Portugal

G. Labrie · É. Lucas
Département des Sciences Biologiques, Université du Québec à Montréal, C.P. 8888 Succursale Centre-ville, Montreal, QC, Canada H3C 3P8

Biological Control to Invasion: Ladybird Harmonia axyridis. H.E. Roy, E. Wajnberg (eds.),
doi: 10.1007/978-1-4020-6939-0_9, © IOBC 2007

environmental conditions, mainly the temperature which is seldom lower than 12°C, essential for the induction of diapause. The lack of success in the establishment could be also related to functional diversity saturation, that is species saturation and competitive exclusion of *H. axyridis* by other previously established species may be operating.

Keywords *Harmonia axyridis* · Invader · Invasion · Invasibility · Invasiveness · Europe · Azores

Biological invasions by insect species, why?

The concept of biological invasion is, generally, used to refer to the arrival or introduction, establishment, geographical expansion and integration of a species into a region where it has never been before (Williamson 1996; Shigesada and Kawasaki 1997). Biological invasion includes several processes, such as the importation, introduction and establishment but also the integration of the novel species into the new habitat, after ecological and evolutionary interaction with local species (Vermeij 1996). Invasion is considered a natural phenomenon over evolutionary time scales, with biotic interchange between large regions occurring frequently (Vermeij 2005). However, recent human-assisted invasions are considered to be one of the most important mechanisms of global environmental change and are considered the second leading threat for worldwide biodiversity, after habitat destruction (Vitousek et al. 1996, 1997; Pimentel et al. 2000; Simberloff 2004).

Most invasions occur due to commerce, agriculture and travelling (Williamson 1996; Liebhold et al. 2006). In the USA, the majority of plants and vertebrates were intentionally introduced while the introductions of invertebrate animals were mostly accidental (Pimentel et al. 2000; see also Liebhold et al. 2006). Among the 50,000 species introduced in the United States of America, over 4,500 were arthropods, including insect species, aquatic invertebrates and crabs and 1,300 of them have successfully established (Mooney and Drake 1989; Pimentel et al. 2000, 2005). The economic cost of biodiversity losses, damages and control for the USA was estimated to be $120 billion per year (Pimentel et al. 2005). The figure may continue to rise in the future (Levine and D'Antonio 2003). The introduction and release of exotic biological control agents generally used for classical and inundative biological control programmes are also of concern for invasion biologists and conservationists. The mass production and release of generalist organisms that are flexible (and so easily adapt to their new habitat), increase the probability of impact on non-target organisms. This may lead to displacement and extinction of indigenous species (Simberloff 1996; Simberloff and Stiling 1996; Williamson 1996; Evans 2004).

Predictions and ecological theory about successful introductions of insect species are still imprecise. However, inconclusive evidence suggests that non-indigenous species are limited to those sites under anthropogenic influence, located mainly in marginal places, and the rate of expansion of those species to native sites has still to be tested (Borges et al. 2006a). Simberloff (1989) referred to the importance of habitat modification for the success and failure of insect invasion. He stated that conclusions drawn about pristine habitats may not apply to modified habitats. For this reason the author presented several examples and counter-examples on the importance of opportunity for colonization, suitable habitat, biotic resistance hypothesis and biological traits favouring successful introductions on different modified habitats. The success of an invasion may include deterministic elements (for details see Crawley 1989) but also stochastic elements relating to good luck and good timing (Crawley 1989). However, the importance of chance and timing may differ from case to case and from process to process (Crawley 1989). Does this fact explain why *Harmonia axyridis* Pallas (Coleoptera: Coccinellidae) took so long to become an invader in some regions? The forces contributing to the outcome cannot be specified and measured, and the final result can be viewed as a random draw from a specified distribution (Simberloff 1989). Moreover, crucial empirical data is lacking concerning records of failed invasions (Holt et al. 2005), which is an impediment to our knowledge of the factors contributing to the lack of species invasions.

Marco et al. (2002) presented and defined a set of factors directly involved in conceptual framework to formulate the biological invasion problem into a modelling context. These authors referred to two main factors which may hamper and promote the invasion process: the biological traits that enable the species to invade a new habitat (invasiveness), and the new habitat characteristics that determine its susceptibility to the establishment and spread of the invader (invasibility). A third important aspect linking invasiveness and invasibility is the interaction among the invaders, the noninvaders, and the habitat. There are a small number of studies specifically addressing the reasons why *H. axyridis* has succeeded or failed as an invader in different regions. However taking the conceptual framework of Marco et al. (2002) and the large number of studies and reviews on the biology (Koch 2003), ecology, and behaviour of the predator we can discuss some of the reasons for the successes and failures of the introduction of *H. axyridis* in non-native regions and begin to consider what will stop the invader.

Harmonia axyridis: where and when?

Harmonia axyridis is native to many regions of Asia, such as Formosa, Manchuria, China, Korea, Japan, southern Siberia, Ryukyu Islands and the Bonin Islands (Iablokoff-Khnzorian 1982). Several introductions of the species were made in North America between 1916 and 1985, namely in California in 1916, 1964, and 1965 (Gordon 1985), in Nova Scotia (Canada), Connecticut, Delaware, Georgia, Maine, Maryland, Mississippi, Ohio, Pennsylvania, and Washington (USA) from 1978 to 1982 (Gordon 1985). Specimens were also released in Connecticut in 1985 during a study of the potential of *H. axyridis* to control *Matsucoccus resinosae* Bean and Godwin (Homoptera: Margarodidae) (McClure 1987). This species preys on more than 30 different aphid species and so was considered as a suitable candidate for efficient biological control against many pests including: aphids (see Lucas et al. 2007a; Tedders and Schaefer 1994; Hodek and Honěk 1996), Tetranychidae (Lucas et al. 1997, 2002), Psyllidae (Fye 1981; Michaud 2001, 2002, 2004), Coccoidea (McClure 1986), Chrysomelidae (Snyder and Clevenger 2004), Curculionidae (Kalaskar and Evans 2001; Stuart et al. 2002) and Lepidoptera (Koch et al. 2003, 2005; Musser and Shelton 2003a). Its large body size, high voracity and predation efficiency (Schanderl et al. 1985; Osawa 2000; Soares et al. 2001; Labrie et al. 2006) as well as its good colonization efficiency (Osawa 2000; With et al. 2002) were considered good characteristics to provide successful biological control of many pests. This coccinellid seems to be an effective biological control agent; upon review of 27 studies of *H. axyridis* as a biocontrol agent (Lucas et al. 2007a), 17 studies showed effective biological control of 16 different pest species by this coccinellid. However, in the 24 studies on the impact of *H. axyridis* on competitors, 15 demonstrated a negative impact by exploitative competition or intraguild predation (Lucas et al. 2007a).

Harmonia axyridis: a super invader?

Harmonia axyridis is expected to become ubiquitous in most temperate countries. Why such a great invasive success?

Harmonia axyridis formula; champion of the invaders?

The Champion hypothesis postulates that *H. axyridis* has exceptional capacities allowing it to be successful in any new environment. This means that the biological attributes of the beetle are particularly well designed to permit invasion into new areas. Several morphological, physiological and behavioural traits of the beetle support this hypothesis.

Morphological traits such as the relatively large body size of the beetle compared to other aphidophagous species (Hodek 1973; Michaud 2002), and the presence of spines on the back of third and fourth larval instars could provide a protection from intraguild or higher-order predation (IGP) (Lucas unpublished data). The ovoid shape of adults as well as bright colour may also confer a protection from predators and competitors (De Clercq et al. 2003). In addition, *H. axyridis* is a highly polymorphic species (Soares et al. 2001) and this could also help the species to be an efficient invader. Many modifications, such as elytral patterns, body shape and body size may be adaptive through imparting increased tolerance to adverse conditions (Dobzhansky 1933; Soares et al. 2003a). The relative frequency of phenotypes seems to be related to geographical and seasonal factors (Tan 1946, 1949; Komai 1956; Iablokoff-Khnzorian 1982; Osawa and Nishida 1992), suggesting that some phenotypes may be favorably selected in different parts of the ecosystem or at different times. Thus the genetic polymorphism in *H. axyridis* seems to be the strategy adopted for facing different habitats at different times. Phenotypical macro-geographic variation in Asian populations of *H. axyridis* was related mainly to climatic factors, because pale and melanic phenotypes have different physiological responses to climatic conditions, particularity to relative humidity and temperature (Komai 1956). Environmental temperature and elytra colour in ladybird beetles may affect body temperature (Brakefield and Willmer 1985; Stewart and Dixon 1989; De Jong et al. 1996) and, thus biological activities. The relative frequency of elytral pattern phenotypes also varies with the host plants (micro-geographic variation) suggesting that they could be correlated with the different aphid populations found on these plants (Komai and Hosino 1951). The key point may be that the suitability of different prey species differs for different phenotypes. Recent laboratory study suggests that different phenotypes may thrive on different prey species in the wild (Soares et al. 2005). All these results, driven from field observations and laboratory experiments, suggest that habitats with more complex structures (presenting diverse abiotic field conditions and food sources) increase the possibility of successful establishment by different phenotypes of *H. axyridis*. Nevertheless, correlation between genetic polymorphism, adaptability and invasiveness may not necessarily be very close in some regions. For instance in North America, despite the dominance of the f. *succinea* phenotype across the region, *H. axyridis* was able to succeed as an invader all over that geographic area.

Physiological characteristics, such as development, fecundity and low susceptibility to pathogens could allow its successful invasion in a new environment. A key factor in the invasion process is juvenile growth, as safe conditions in these vulnerable stages can ensure high population growth in the new environment (Marco et al. 2002). A shorter development time of younger larval instars of *H. axyridis* compared to the indigenous

Coleomegilla maculata lengi Timberlake (Coleoptera: Coccinellidae) in Canada was observed in North America (Labrie et al. 2006). Size, strength of the integument and distastefulness of the pupae make this stage less vulnerable to predation (Félix and Soares 2004; Labrie unpublished data). According to several studies, fecundity of *H. axyridis* is higher than other species, with between 703 and 3,800 eggs laid by a single female in laboratory studies (Iablokoff-Khnozorian 1982; Stathas et al. 2001; Mignault et al. 2006). For example, fecundity of *H. axyridis* (2,008 eggs per female) reared on soybean aphids *Aphis glycines* Matsumura (Homoptera: Aphididae) was significantly higher than the invasive *Propylea quatuordecimpunctata* L. (Coleoptera: Coccinellidae) (593 eggs per female) or the indigenous *C. maculata* (390) in Québec, Canada (Mignault et al. 2006). Physiological ability to withstand unfavourable environmental conditions is another key factor. No winter survival was observed in USA (McClure 1987) and Canada (Labrie et al. 2007) when the coccinellid was held outside during winter season. However, survival was observed inside households (Labrie et al. 2007), where this species forms swarms of thousands in the autumn (Kidd and Nalepa 1995; LaMana and Miller 1996; Nalepa et al. 1996; Hagley 1999; Huelsman et al. 2001; Kovach 2004). Households provided a cold-free space for this species (and possibly an enemy-free space), which has allowed geographical extension to the northern part of North America and contributed the invasive success of *H. axyridis*. Indeed the success of *H. axyridis* may be partly linked to the enemy release hypothesis (the potential invader is successful because of the inefficacy of natural enemies in the new area) (Gratton and Welter 1999; Keane and Crawley 2002; Stastny et al. 2005; Halpern and Underwood 2006). An invasive species in a new region is hypothesised to experience a decrease in regulation by natural enemies, resulting in an increase in distribution and abundance (Keane and Crawley 2002). Laboratory and field data report low levels of successful parasitism of *H. axyridis* adults by *Dinocampus coccinellae* Schrank (Hymenoptera: Braconidae) in North America (Hoogendoorn and Heimpel 2002; Firlej et al. 2005) or in Italy (Burgio et al. 2007). Even if *H. axyridis* is sympatric with *D. coccinellae* in Asia (Maeta 1969), the degree of coevolution between *H. axyridis* and *D. coccinellae* remains unknown because the parasitoid native area is unclear (America or Europe) (Balduf 1926). Other entomopathogens, such as *Beauveria bassiana* (Balsamo) Vuillemin (Ascomycota: Hypocreales) (Cottrell and Shapiro-Ilan 2003; Roy et al. 2007), or nematodes (Shapiro-Ilan and Cottrell 2005) did not affect survival of *H. axyridis* in North America.

Behavioural traits of *H. axyridis* have certainly contributed to its invasive success. A superior ability to acquire resources can strongly affect the invaders survival and fecundity, and may confer a large competitive advantage in the invaded range (Petren and Case 1996). An increase in foraging efficiency could allow the invasive species to cope with competition. *Harmonia axyridis* larvae and adults have been shown to have a higher predation and foraging efficiency than indigenous species (Yasuda and Ohnuma 1999; Yasuda et al. 2001; Lanzoni et al. 2004; Labrie et al. 2006). Higher conversion efficiency of prey into biomass was also observed for this species (Labrie et al. 2006), which could increase its exploitative competition ability and population density.

In the same way, the response (or lack thereof) of a non-indigenous species to novel predators and competitors in the new community will likely affect the overall invasion success and potential for impact by the invader. It has been demonstrated that *H. axyridis* is more aggressive than other invasive and indigenous species in many laboratory experiments (Hironori and Katsuhiro 1997; Cottrell and Yeargen 1998; Yasuda and Ohnuma 1999; Snyder et al. 2004; Yasuda et al. 2001; Cottrell 2004; Soares and Serpa 2007; Pell et al. 2007; Ware and Majerus 2007). The incidence of IGP, in laboratory studies, was 65%

for the interactions between all stages of *H. axyridis* (intraguild predator) and the indigenous *C. maculata* (intraguild prey), and 75% for *H. axyridis* and the invasive *P. quatuordecimpunctata* (Labrie et al. unpublished data). This indicates that *H. axyridis* is highly aggressive and has the potential to compete with other species.

The *H. axyridis* formula; one species among the others?

These impressive traits may not be the only ones contributing to the success of *H. axyridis*. Would the coccinellid have such ubiquity on 2007 in absence of human intervention? Obviously no. This species has been extensively and repeatedly released in several areas (see Table 1) and its success may be strongly linked to human activity. Can we postulate that other aphidophagous coccinellids may have a similar response to new environment? From the literature, we can compare several aphidophagous coccinellids with *H. axyridis*. For example, despite shorter developmental time of second and fourth larval instars of *H. axyridis* compared to *C. maculata*, the overall developmental time was similar between the two species (Labrie et al. 2006). Furthermore, other studies demonstrated a slower developmental time of *H. axyridis* in comparison to other species (*Adalia bipunctata* L., *Hippodamia variegata* Goeze, Lanzoni et al. 2004; *P. quatuordecimpunctata*, Mignault et al. 2006; *Olla v-nigrum* Mulsant, Michaud and Olsen 2004).

While, as previously stated, *H. axyridis* showed high fecundity, other studies demonstrated similar or lower fecundity than indigenous species (Bazzochi et al. 2004; Lanzoni et al. 2004; Michaud and Olsen 2004; Soares et al. 2004). For example, in Italy, the mean fecundity of *H. axyridis* (783.8 eggs per female) was only slightly higher than an indigenous species, *A. bipunctata* (720.2 eggs per female) (Bazzochi et al. 2004). Lower fecundity of *H. axyridis* (560.5 eggs per female) compared to *H. variegata* (841.7 eggs per female) was observed in laboratory by Lanzoni et al. (2004). Soares et al. (2004) demonstrated an even lower fecundity than these experiments suggest, with a mean fecundity of 314 eggs per female for the *aulica* phenotype of *H. axyridis* in laboratory. Experiments of Michaud and Olsen (2004) on psyllids demonstrated similar fecundity of *H. axyridis* (242.8 eggs per female) to *O. v-nigrum* (224.5 eggs per female). Fecundity of this invasive species seems thus similar to some other aphidophagous species.

Harmonia axyridis is a highly efficient intraguild predator (Hironori and Katsuhiro 1997; Cottrell and Yeargen 1998; Yasuda and Ohnuma 1999; Snyder et al. 2004; Yasuda et al. 2001; Cottrell 2004; Soares and Serpa 2007; Pell et al. 2007; Ware and Majerus 2007). However, eggs and first and second instar larvae are not well protected from other competitors. Recent laboratory investigations demonstrated that 61–85% of *H. axyridis* eggs were consumed by *C. maculata* and *P. quatuordecimpunctata* (Labrie et al. unpublished data). In these experiments, first and second larval instars were also very vulnerable to predation, with 66% and 31% being consumed by *C. maculata* and *P. quatuordecimpunctata*, respectively. This species is also attacked by hemipteran species, such as *Podisus maculiventris* (Say) (Hemiptera: Pentatomidae) in the laboratory (De Clercq et al. 2003) or repeatedly in the field (Labrie personal observation).

Wise et al. (2001) reported 11 new Coccinellidae species in Manitoba (Canada) including of which *H. axyridis* and *C. maculata lengi*. In Québec (Canada), four aphidophagous coccinellids are dominant in agricultural crops, three invasive species (*H. axyridis*, *Coccinella septempunctata* L. and *P. quatuordecimpunctata*) and one indigenous (*C. maculata*) and coexist widely (Lucas et al. 2007b). Could the majority of aphidophagous species be potential invasive species? To our knowledge, no aphidophagous predators exploit a single

Table 1 Continental areas and islands where the introduction or invasion by the multicoloured Asian ladybeetle has occurred

Countries and Islands	Status	Year	References
Continental areas			
Asia			
China	Indigenous		Iablokoff-Khnzorian (1982)
Korea	Indigenous		Iablokoff-Khnzorian (1982)
Manchuria	Indigenous		Iablokoff-Khnzorian (1982)
Siberia	Indigenous		Iablokoff-Khnzorian (1982)
Ukraine	Invasion	1964	Brown et al. (2007); Katsoyannos et al. (1997)
Europe			
Austria	Invasion	2006	Brown et al. (2007); Rabitsch and Schuh (2006)
Belarus	Introduction	1968	Sidlyarevich and Voronin (1973)
Belgium	Introduction	1997	Adriaens et al. (2003)
Czech Republic	Introduction	2003, 2006	Brown et al. (2007); Nedved and Kovar (unpublished data)
Denmark	Invasion	2006	Brown et al. (2007)
France	Introduction	1982	Schanderl et al. (1985)
Germany	Introduction	1997	Iperti and Bertrand (2001)
Greece	Introduction	1994	Katosyannos et al. (1997)
Holland	Introduction	1996	Iperti and Bertrand (2001)
Italy	Introduction	1990's	Iperti and Bertrand (2001)
Liechtenstein	Invasion	2007	Brown et al. (2007); Loomans (unpublished data)
Luxembourg	Invasion	2004	Brown et al. (2007); Schneider and Loomans (2006)
Norway	Invasion	2006	Brown et al. (2007); Staverloekk et al. (2007)
Portugal	Introduction	1984	Garcia (1986)
Spain	Introduction	1995	Brown et al. (2007); SIFA (2004)
Sweden	Invasion	2007	Brown et al. (2007); Hägg (unpublished data)
Switzerland	Introduction	1996	Brown et al. (2007); Andermatt (1996)
America			
Argentina	Invasion	2001	Saini (2004)
Brazil	Invasion	2002	De Almeida and da Silva (2002)
Canada	Invasion	1994	Coderre et al. (1995)
USA	Introduction	1916–1980	Gordon (1985)
Africa			
South Africa		2007	Stals and Prinsloo (2007)

Table 1 continued

Countries and Islands	Status	Year	References
Islands			
Azores	Introduction	1988–1995	Garcia (1986)
			Schanderl et al. (1991)
			Schanderl and Almeida (1992)
Bonin	Indigenous		Iablokoff-Khnzorian (1982)
Canary	Introduction	2003	Carrillo (2006)
Formosa	Indigenous		Iablokoff-Khnzorian (1982)
Great Britain			
England	Invasion	2004	Majerus et al. (2006)
Wales	Invasion	2006	Brown et al. (2007)
Hawaii	Introduction	1916	Schanderl (1987)
Japan	Indigenous		Iablokoff-Khnzorian (1982)
Jersey, Channel Islands	Invasion	2007	Brown et al. (2007)
Møn	Invasion	2006	Jan Pedersen (personal communication)
Ryukyu	Indigenous		Iablokoff-Khnzorian (1982)

The "year" on the table means the date of the introduction or the first record, after invasion

aphid species, but instead feed on an array of different prey species (euryphagous), on different host plants and in different environments (eurytopic) (Iablokoff-Khnozorian 1982; Hodek and Honěk 1996). They are also good competitors (Elliott et al. 1996; Ellis et al. 1999; Alyokhin and Sewell 2004; Evans 2004) which could allow all these species to invade new environments. Moreover, one may invoke the neutral theory of biodiversity (Bell 2005; Hubbell 2001, 2005, 2006) which suggest that all species are identical and compete equally, once all species can develop on a site. Composition of the community will thus be determined by stochastic dispersion of individuals.

Concluding, we would say that the *H. axyridis* formula is linked to an ability to invade and establish in new environments through growth, development, reproduction and survival under diverse abiotic and biotic conditions. We would also add that several other aphidophagous coccinellids are already invasive or have the potential to invade, and it is not so clear that *H. axyridis* is more adapted or "superior" than these other species in the invasive process. Finally, the *H. axyridis* phenomenon appears to be because of the bio-ecological attributes of this beetle and the exceptional mobilization of the scientific community around this conspicuous and "apparent" beetle.

Where will it end?

Harmonia axyridis has demonstrated its ability to establish in new different areas. Can we predict how this situation will evolve? Which new areas will follow? What would stop or limit *H. axyridis*? Considering Southern countries, *H. axyridis* is present in Korea. But it has not been cited in equatorial or in arid zones. The beetle enters into aestivation when the temperature is too hot, this phase of dormancy is representative of an acclimation to climatic exigencies, but could hamper the optimal functioning of the beetle if such conditions are common.

Considering Northern countries, *H. axyridis* is found in south Siberia (Russia) and also in Canada (Iablokoff-Khnzorian 1982; Coderre et al. 1995) indicating that this beetle has the capacity to deal with cold climates. In United States, McClure (1987) showed however that survival was greatly reduced during winter and Labrie et al. (2007) demonstrated that 0% of the individuals survived outside during the Canadian winter. However survival was assured by the behaviour of overwintering inside human houses. This behaviour allowed survival of about 25%. According to these studies, we may conclude that the northern limitation is more a matter of availability of suitable overwintering location, such as human houses rather than climatic conditions.

Considering the favourable temperate conditions of continental European countries, colonization of European areas is predicted to continue, since no climatic, biophysical traits seem to be detrimental to the beetle. However, considering islands, the isolation could delay colonization despite Great Britain was recently colonized (Brown et al. 2007).

Azores, the first failure of *H. axyridis*?

Until recently, the importation and introduction of exotic species as biocontrol agents was one of the strategies to control exotic pests on crops. Research on how to increase environmental resistance to invasion has not received sufficient attention (Ehler 1998). Theory says that oceanic islands are especially vulnerable to biological invasions due to: (i) many empty available niches leading to a non-saturation of communities, i.e., there are plenty resources available for new species to enter the communities; (ii) taxonomic disharmony, i.e., many families or genera are lacking in island biotas which facilitates the invasion by species with different evolutionary characteristics; (iii) small population size—as species accumulate within an island, competition between indigenous species could led to lower average population sizes, which facilitates the invasion by a new highly competitive exotic species; (iv) facilitation processes, that is, species are nurseries for other species and consequently invasion rate should rise in islands with low functional diversity (Williamson 1981; Gillespie and Roderick 2002; Whittaker and Fernández-Palacios 2007). As a consequence of (i) and (ii) usually poor island communities have low resistance to invasion by non-native species, although this seems to vary with the scale of study (Borges et al. 2006a). In fact, native insect communities that achieved ecological equilibrium diversity in high altitude unfragmented and undisturbed native forests are more resistant to invasion by exotic species (Borges et al. 2006a). For instance, most exotic arthropod species in the Azores are more abundant in human-made disturbed habitats than in native forest (Borges et al. in press a). One other example of the impact of human-caused introduction on the structure and function of ecosystems was the invasion of the Christmas Island by the yellow crazy ant *Anoplolepis gracilipes* (Fr. Smith) (Hymenoptera: Formicidae). The effects of this species on the habitat of that island were evident on the dynamics of the forest floor and species richness of the canopy. Moreover, long-term effects on forest structure and composition were evident in the reduced seedling recruitment, reduced species richness of seedlings, slow litter breakdown and low density of litter invertebrates (O'Dowd et al. 2003). The mutualistic interaction between the invader and scale-insects led to outbreaks of these Homoptera and may explain the decrease of the proportion of growing shoots and dieback in the canopies of large trees (O'Dowd et al. 2003). In the Azores low altitude canopy communities are dominated by introduced ants that replaced the dominant indigenous spider fauna still occurring at pristine high altitude forests (see Borges et al. in press a, unpublished data). As shown by Borges et al. (2006a) invasions of alien arthropod species are an actual and future environmental threat in the Azores, creating a pattern of biotic homogenization that is of great contemporary concern. In the case of the Azorean archipelago, almost 60% of the beetle fauna is non-native (Borges et al. 2005), and the coccinellids community is also mainly composed of exotic species. In fact, at least 14 out of 23 species were considered exotic (Borges et al. 2005).

Since the 1980's, several individuals of *H. axyridis* were introduced into five islands of the Azores (Sta. Maria, Pico, S. Jorge, Faial and Terceira) to control citrus and corn aphid pests (Garcia 1986; Schanderl et al. 1991; Schanderl and Almeida 1992). In 1991, 1,500 *H. axyridis* adults were released in a corn field of Pico Island to study the dispersion and the distribution on plants (Schanderl et al. 1991). From 1988 to 1990, between May and June, 152,996 third larval stages of the coccinellid were released into several citrus orchards on the small island of Sta. Maria (97 km^2) (Schanderl and Almeida 1992). Every single year, and after each release, short-term surveys were performed in Sta. Maria to record the presence and follow the dispersion of the exotic species. During these surveys, it was possible to verify that the species was able to successfully disperse and some individuals were found far away from the release sites; on the wall of the houses or feeding on citrus, apple trees and bean plants (Schanderl and Almeida 1992). Those results suggested the possibility of a successful introduction.

However, recent publications on the terrestrial fauna and flora biodiversity from the Azores did not report the presence of *H. axyridis* in the archipelago (Borges et al. 2005). An investigation of the occurrence patterns of several species of Coccinellidae in the well studied island of Terceira (Borges et al. in press a) shows that nine species were sampled in eight different habitats, none of them were *H. axyridis*. Remarkably, eight out of nine species were sampled mainly in orchards (89% of the species and 95% of the specimens), habitats where *H. axyridis* may be found (Michaud 2002). Thus, the Azorean coccinellid community is composed mainly of recently introduced species well adapted to human-made habitats. These results may suggest that in fact *H. axyridis* did not become established in the Azores. How can we explain this apparent failure of the establishment of this invasive predator in the Azores? Firstly we must not exclude the hypothesis predicting that *H. axyridis* is already established in the Azores and that small populations were not detected because of a lack of long term-surveys specifically addressed to check for its establishment and dispersal over the islands, mainly in Pico and Santa Maria. It is known that approximately 90% of all introduced species fail to establish viable populations in new environments (Williamson 1996). Could the invasion of *H. axyridis* be limited by the absence of suitable habitats? Probably not since the beetle is eurytopic, exploiting agricultural systems, forests, marshes, or urban areas. Could the invasion be limited by the absence of suitable resources? Probably not. *H. axyridis* is euryphagous, exploiting mites (Lucas et al. 1997, 2002), psyllids (Michaud 2001, 2002, 2004) and other insect species. Furthermore, in Canada for example, at the end of the autumn, when all indigenous species have reached their overwintering sites, numerous individuals of the beetle are found in agricultural fields. At this time, there are virtually no preys for the beetles in the field. We noticed that larvae achieved their preimaginal development by cannibalising other individuals. Independently of the results reported before, a question remains unanswered; will invasion and establishment of *H. axyridis* into the Azores be inevitable?

In North America, *H. axyridis* was released without success in many climatically and ecologically different regions over 70 years, but then it became established. In Europe, some releases also failed (e.g. Ukraine, Greece, Italy, and France) before eventually invading Europe. In the case of the Azores, have the particular biotypes released been unsuitable for establishment? Lombaert et al. (2007) state that populations maintained in controlled laboratory conditions should become specialised on a narrow range of environments and hence show lower phenotypic plasticity. These authors found some evidence of differences between invasive and commercialised populations for traits ecologically relevant for the invasive potential of populations. The individuals released in the Azores came from a mass production of INRA-Antibes, which may have limited the introduction to a single biotype of the predator. However, the results of Lombaert et al. (2007) only poorly support the idea of increased phenotypic plasticity being a cornerstone of invasive success of at least some European populations. Thus using a limited strain of the predator may not explain the success or failure in the invasion process.

The Azores presents a temperate oceanic climate, i.e., strongly influenced by the proximity of the ocean and by its topography, which results in high levels of relative atmospheric humidity. The type of climate also restricts temperature fluctuations throughout the year (Azevedo 2005). The milder temperatures occurring over the spring and the summer seasons in the Azores are very close to the thermal optimum for foraging and breeding activities of *H. axyridis* (Schanderl et al. 1985) and thus, the Azorean temperatures seem not to be a limiting factor of its invasiveness. Indeed Poutsma et al. (2007), using the CLIMEX programme (v2), predicted that establishment and spread seem likely in many temperate regions across the world including a large part of temperate and Mediterranean Europe, North and South America, Australia and New Zealand. During the winter, however, the average minimum temperature in the Azores (Azevedo 2005) is very close to the minimum thermal requirement (Schanderl et al. 1985) which could hamper diapause of the adults, and may result in keeping them active during this season.

Harmonia axyridis is a species which is found, preferentially, in arboreal habitats (Iablokoff-Khnzorian 1982). In Belgium, this predator has became dominant on this type of habitat (Adriaens et al. 2007). The vegetation of the Azores archipelago is, however, dominated by the grasslands used for cattle feeding and a wide variety of plant communities, including coastal vegetation, wetland vegetation (lake shore and seashore communities and a variety of bogs), several types of paddies, and different types of native shrub and forest (Borges et al. in press a, b). In general, the vegetation of those plant communities is composed by therophyte, hemycryptophyte and camaephytes non-indigenous life form of plants, found in walking trails, roads, crop margins, pastures, field crops, native shrub and costal areas (Silva and Smith 2006), which may be less suitable for the predator. Arboreal habitats are also present in the Azores but, in general, they are small, fragmented and composed by different combinations of the natives plants *Myrica faya* Ait, *Laurus azorica* (Seub.) Franco, *Persea indica* (L.) Spreng. and the non-indigenous *Pittosporum undulatum* Vent., *Acacia melanoxylon* R. Br., *Cryptomeria japonica* Regius (Silva and Smith 2006). However, aphid species are mostly found on the less suitable annual non-native plants habitats than on the preferential arboreal habitats (Borges et al. 2006b). These factors may limit the establishment of *H. axyridis* in the Azores. On the other hand, alternative and suitable habitats such as field crops and orchards, vineyards, hedgerows and gardens on urban environments may be searched by the invader (Iablokoff-Khnzorian 1982; Adriaens et al. 2007). For instance, corn fields are important for feeding and reproductive activity of *H. axyridis* in many invaded areas (Musser and Shelton 2003b; Nault and Kennedy 2003; Lundgren et al. 2004; Musser et al. 2004; Park and Obrycki 2004; Firlej et al. 2005). Corn fields are common on the Azores and, thus, have the potential to contribute to the establishment of *H. axyridis*.

One factor which may hamper the establishment of *H. axyridis* in the Azores is the absence of winter environmental conditions for the overwintering strategy. In order to face food scarcity and unfavourable weather, most temperate coccinellid do have an obligatory diapause (Dixon 2000; Hodek and Honěk 1996). *Harmonia axyridis* has a true and obligatory diapause (Sakurai et al. 1992), after a period of feeding and reproductive activity, somewhere from April to October (Ongagna et al. 1993; LaMana and Miller 1996; Katsoyannos et al. 1997; Bazzochi et al. 2004). Temperature, photoperiod and scarcity of food are key factors for the induction of the coccinellid diapause. Photoperiod may not be a limiting factor for diapause of *H. axyridis* because geographical position of the Azores (37°–40° N latitude), is very similar to other regions where *H. axyridis* diapause. Would it be possible that *H. axyridis* requires large thermal fluctuations not present in these islands in order to have a satisfactory diapause? In Japan and south-eastern of France, *H. axyridis* interrupts its reproduction when temperature is low (below 12°C) and the daylight short (Ongagna et al. 1993; Watanabe 2002). The optimum temperature for overwintering is between 0 and −5°C (Watanabe 2002). Winter annual temperature occurring in the Azores may not allow a sufficiently long period for diapause by individuals. In the small mountain regions of the Azores, average minimum temperature usually falls below 12°C but for short time, i.e. 2–3 months. A temperature of 0°C is possible only on the Pico Mountain (2,351 m) of Pico Island (Azevedo 2005). In the habitats (crops and orchards) where *H. axyridis* could feed and reproduce, that is, on the lowlands of the islands, the temperature is seldom lower than 12°C. This fact may hamper the induction of diapause. We cannot exclude the possibility of *H. axyridis* feeding during winter throught the presence of alternative food sources like pollen, conspicuous patches of aphids and coccids standing on plants throughout the year on the evergreen plants in the lowlands and even cannibalizing larvae. We may also hypothesise that there could be a possible progressive selection against a tendency for obligatory onset of diapause under the environmental condition of the Azores. Indeed some laboratory studies on the incidence of diapause of *C. septempunctata* over successive generations show a tendency for obligatory onset of diapause under long days conditions (Hodek and Honěk 1996). This tendency was never demonstrated under laboratory conditions for *H. axyridis*.

One important dimension of the invasiveness is the presence of species occupying the same ecological niche. The invaders may be involved in intraguild predation events or competition, either through exploitative or interference competition. The presence of other aphidophagous coccinellids is very common on the islands as *Scymnus*, *Nephus* and *Rhyzobius* species (Soares et al. 2003b; Borges et al. 2005) with which *H. axyridis* may interact. These genera present small individuals, apparently having small competitive abilities. Recent results suggest that *H. axyridis* could endanger the populations of one of the large aphidophagous coccinellids of the islands; *Coccinella undecimpunctata* L. The study considered the magnitude, direction and symmetry of intraguild predation between the developmental stages of the *H. axyridis* and *C. undecimpunctata* and showed that *H. axyridis* was more often the predator than *C. undecimpunctata* and also that eggs were the most vulnerable developmental stage (Félix and Soares 2004). Moreover, laboratory experiments demonstrated that the presence of *H. axyridis* adults significantly affected the reproductive numerical response of *C. undecimpunctata* even though the resource was not in short supply. The results suggested that interference competition with heterospecifics may adversely affect the reproductive capacity of *C. undecimpunctata* (Soares and Serpa 2007).

Recent studies suggest that the invasion success of many species might depend more heavily on their ability to respond to natural selection than on broad physiological tolerance or plasticity (Lee 2002). One possibility for the lack of success in the establishment of *H. axyridis* in the Azores could be related to functional diversity saturation. As mentioned above, most of the coccinellid assemblages in the Azores occurs in orchards and other man-made ecosystems. This could imply species saturation and competitive exclusion of *H. axyridis* by other previously established species. However, this has to be proved through robust experiments and intensive field surveys in several islands of the archipelago (see Borges et al. 2006a, in press a).

It could also be that *H. axyridis* has a lag phase in its establishment. Maybe the establishment success and persistence of *H. axyridis* will be dependent on repeated introductions to avoid the genetic drift effect and increase the genetic diversity of the source population.

Conclusion

The environmental risks associated with the invasion of insects introduced for biological control, like *H. axyridis*, were seldom taken into account. The effectiveness of generalist predators in biological control may be diminished if increased availability of alternative prey causes individual predators to decrease their consumption of the target species. In fact, intraguild predation is common among generalist predators and is an important issue in food web theory. Therefore, increased attention has to be paid to the harmful effects of introduced biological control agents on non-target indigenous species.

Several studies have stressed the biological, ecological and behavioural characteristics that make *H. axyridis* a strong invader, among them, its aggressiveness (as an intraguild predator), polyphagy, fecundity and the fact that it is a less habitat and niche-specific coccinellid than native species. However, the "*H. axyridis* phenomena" may give a false impression of the real potential of the coccinellid to be an invader. This may result from the fact that *H. axyridis* is one of the coccinellid species most extensively studied by the scientific community; it has been released in more locations and more often than any other coccinellid species and by its overwintering behaviour inside human houses, is a great concern for citizens. For the reasons presented before, it is not clear if *H. axyridis* formula is enough to make it "superior" to the other coccinellid species for invasion.

In contrast to the invasiveness of *H. axyridis*, less attention has been paid to the invasibility issue. Invasibility may, in part, explain the reasons for the success (or failure) of establishment of *H. axyridis* and answering the question "what will stop the invader?" In the case of the Azores and despite the deliberated introduction, the occurrence of a temperate climate and crop, garden and arboreal habitats infested with suitable aphids for the predator, it may not be enough to the establishment of the species. Indeed other factors, such as the presence of the aphids on the less suitable annual non-native plants habitats than on the preferential arboreal habitats and the absence of winter annual temperature for the diapause, may explain the apparent failure on its establishment. The lack of success in the establishment of *H. axyridis* in the Azores could be related to functional diversity saturation, that is species saturation and competitive exclusion of *H. axyridis* by other previously established species may be operating. However there is an urgent need to determine accurately the underlying factors that have, so far, prevented the establishment of *H. axyridis* within the sensitive ecosystems of the Azores, focusing on standardized sampling of several habitats in islands subjected to historical releases of specimens (e.g., Sta. Maria, Pico, S. Jorge, Faial, and Terceira).

Acknowledgements The authors would like to thank Peter Brown for his helpful comments on Table 1 and for sending original references about geographical distribution of *H. axyridis* in Europe. Thanks are also due to Ivo Hodek and Luis Silva, for their helpful comments on the following topics, overwintering of ladybird beetles and vegetation cover on the Azores, respectively.

References

Adriaens T, Branquart E, Maes D (2003) The multicolored Asian ladybird *Harmonia axyridis* Pallas (Coleoptera: Coccinellidae), a threat for native aphid predators in Belgium? Belg J Zool 133:195–196

Adriaens T, San Martin y Gomez G, Maes D (2007) Invasion history, habitat preferences and phenology of the invasive ladybird *Harmonia axyridis* in Belgium. BioControl (this issue). doi:10.1007/s10526-007-9137-6

Alyokhin A, Sewell G (2004) Changes in a lady beetle community following the establishment of three alien species. Biol Invasions 6:463–471

Andermatt M (1996) Marienkäferlarven als blattlaus-bekämpfungsmittel? Andermatt Biocontrol 3:10–11

Azevedo EB (2005) "Projecto CLIMAAT—Clima e Meteorologia dos Arquipélagos Atlânticos"—Interreg IIIB, MAC 2.3/A3—Açores, Madeira e Canárias. http://www.climaat.angra.uac.pt. Accessed 19 Sept 2007

Balduf WV (1926) The bionomics of *Dinocampus coccinellae* Schrank. Ann Entomol Soc Am 19:465–489

Bazzochi GG, Lanzoni G, Accinelli G, Burgio G (2004) Overwintering, phenology and fecundity of *Harmonia axyridis* in comparison with native coccinellid species in Italy. BioControl 49:245–260

Bell G (2005) The co-distribution of species in relation to the neutral theory of community ecology. Ecology 86:757–770

Borges I, Soares AO, Hemptinne J-L (2006b) Abundance and spatial distributions of aphids and scales select for different life histories in their ladybird beetle predators. J Appl Entomol 130:461–464

Borges PAV, Cunha R, Gabriel R, Martins AF, Silva L, Vieira V, Dinis F, Lourenço P, Pinto N (2005) Description of the terrestrial Azorean biodiversity. In: Borges PAV, Cunha R, Gabriel R, Martins AMF, Silva L, Vieira V (eds) A list of the terrestrial fauna (Mollusca and Arthropoda) and flora (Bryophyta, Pteridophyta and Spermatophyta) from the Azores. Direcção Regional de Ambiente and Universidade dos Açores, Horta, Angra do Heroísmo and Ponta Delgada

Borges PAV, Lobo JM, Azevedo EB, Gaspar C, Melo C, Nunes VL (2006a) Invasibility and species richness of island endemic arthropods: a general model of endemic vs. exotic species. J Biogeogr 33:169–187

Borges PAV, Ugland KI, Dinis FO, Gaspar C (in press a) Insect and spider rarity in an oceanic island (Terceira, Azores): true rare and pseudo-rare species. In: Fattorini S (ed) Insect ecology and conservation. Research Signpost

Borges PAV, Amorim IR, Cunha R, Gabriel R, Martins AF, Silva L, Costa A, Vieira V. (in press b) Azores—biology. In: Gillespie R, Clagu D (eds) Encyclopedia of Islands. University of California Press, California

Brakefield PM, Willmer PG (1985) The basis of thermal-melanism in the ladybird *Adalia bipunctata*: differences in reflectance and thermal properties between the morphs. Heredity 54:9–14

Brown PMJ, Adriaens T, Bathon H, Cuppen J, Goldarazena A, Hägg T, Kenis M, Klausnitzer BEM, Kovar I, Loomans AJM, Majerus MEN, Nedved O, Pedersen J, Rabitsch W, Roy HE, Ternois V, Zakharov IA, Roy DB (2007) *Harmonia axyridis* in Europe: spread and distribution of a non-native coccinellid. BioControl (this issue). doi:10.1007/s10526-007-9132-y

Burgio G, Lanzoni A, Accinelli G, Maini S (2007) Estimation of mortality by entomophages on exotic *Harmonia axyridis* versus native *Adalia bipunctata* in semi-field conditions in northern Italy. BioControl (this issue). doi:10.1007/s10526-007-9133-x

Carrillo AM (2006) El sarantontón asiático *Harmonia axyridis* (Pallas, 1773) presente en Canárias (Coleoptera: Coccinellidae). Vieraea 34:71–72

Coderre D, Lucas E, Gagne I (1995) The occurrence of *Harmonia axyridis* (Pallas) (Coleoptera, Coccinellidae) in Canada. Can Entomol 127:609–611

Cottrell TE (2004) Suitability of exotic and native lady beetle eggs (Coleoptera: Coccinellidae) for development of lady beetle larvae. Biol Control 31:362–371

Cottrell TE, Shapiro-Ilan DI (2003) Susceptibility of a native and an exotic lady beetle (Coleoptera: Coccinellidae) to *Beauveria bassiana*. J Invertebr Pathol 84:137–144

Cottrell TE, Yeargan KV (1998) Intraguild predation between an introduced lady beetle *Harmonia axyridis* (Coleoptera: Coccinellidae), and a native lady beetle, *Coleomegilla maculata* (Coleoptera: Coccinellidae). J Kans Entomol Soc 71:159–163

Crawley JC (1989) Chance and timing in biological invasions. In: Drake JA, Mooney HA, di Castri F, Groves RH, Kruger FJ, Rejmánek M, Williamson M (eds) Biological invasion: a global perspective. Scope 37. Wiley, Chichester

De Almeida LM, da Silva VB (2002) Primeiro registro de *Harmonia axyridis* (Pallas) (Coleoptera, Coccinellidae): um coccinelideo originario da região Paleartica. Revta Bras Zool 19:941–944

De Clercq P, Peeters I, Vergauwe G, Thas O (2003) Interaction between *Podisus maculiventris* and *Harmonia axyridis*, two predators used in augmentative biological control in greenhouse crops. BioControl 48:39–55

De Jong PW, Gussekloo SWS, Bakefield PM (1996) Differences in thermal balance, body temperature and activity between non-melanic and melanic two-spot ladybird beetles (*Adalia bipunctata*) under controlled conditions. J Exp Biol 199:2655–2666
Dixon AFG (2000) Insect predator–prey dynamics: ladybird beetles & biological control. Cambridge University Press, Cambridge
Dobzhansky ThG (1933) Geographical variations in lady-beetles. Am Nat 709:97–126
Ehler LE (1998) Invasion biology and biological control. Biol Control 13:127–133
Elliott N, Kieckhefer R, Kauffman W (1996) Effects of an invading coccinellid on native coccinellids in an agricultural landscape. Oecologia 105:537–544
Ellis DR, Prokrym DR, Adams RG (1999) Exotic lady beetle survey in northeastern United States: *Hippodamia variegata* and *Propylea quatuordecimpunctata* (Coleoptera: Coccinellidae). Entomol News 110:73–84
Evans EW (2004) Habitat displacement of North American ladybirds by an introduced species. Ecology 85:637–647
Félix S, Soares AO (2004) Intraguild predation between the aphidophagous ladybird beetles *Harmonia axyridis* and *Coccinella undecimpunctata* (Coleoptera: Coccinellidae): the role of body weight. Eur J Entomol 101:237–242
Firlej A, Boivin G, Lucas É, Coderre D (2005) First report of *Harmonia axyridis* Pallas being attacked by *Dinocampus coccinellae* Schrank in Canada. Biol Invasions 7:553–556
Fye RE (1981) Rearing and release of coccinellids for potential control of pear Psylla. Agric Res Serv (Western Region) 20:1–9
Garcia V (1986) Approaches to integrated control of some citrus pests in the Azores and Algarve (Portugal). In: Cavalloro R, Di Martino E (eds) Integrated pest control in citrus groves. Proc. CEC Experts Meeting, Acireale
Gillespie RG, Roderick GK (2002) Arthropods on Islands: colonization, speciation, and conservation. Annu Rev Entomol 47:595–632
Gordon RD (1985) The Coccinellidae (Coleoptera) of America north of Mexico. J N Y Entomol Soc 93:1–912
Gratton C, Welter SC (1999) Does “enemy-free space” exist? Experimental host shifts of an herbivorous fly. Ecology 80:773–785
Hagley EAC (1999) Predatory insects in fruit orchards in Southern Ontario. Publication 208 Agriculture and Agri-Food Canada and Ontario Ministry of Agriculture, Food and Rural Affairs, 32 pp
Halpern SL, Underwood N (2006) Approaches for testing herbivore effects on plant population dynamics. J Appl Ecol 43:922–929
Hironori Y, Katsuhiro S (1997) Cannibalism and interspecific predation in two predatory ladybirds in relation to prey abundance in the field. Entomophaga 42:153–163
Hodek I (1973) Biology of Coccinellidae. Dr. W. Junk NV Publishers, The Hague, Prague
Hodek I, Honěk A (1996) Ecology of Coccinellidae. Kluwer Academic Publishers, Dordrecht, 464 pp
Holt RD, Brakefield M, Gomulkiewicz R (2005) Theories of niche conservation and evolution-could exotic species be potential pest? In: Sax DF, Stachowicz JJ, Gaines SD (eds) Species invasions—insights into ecology, evolution and biogeography. Sinauer Associates, Inc. Publishers, Massachusetts
Hoogendoorn M, Heimpel GE (2002) Indirect interactions between an introduced and a native ladybird beetle species mediated by a shared parasitoid. Biol Control 25:224–230
Hubbell SP (2001) The unified neutral theory of biodiversity and biogeography. Princeton monographs in population biology. Princeton University Press, Princeton
Hubbell SP (2005) Neutral theory in community ecology and the hypothesis of functional equivalence. Funct Ecol 19:166–172
Hubbell SP (2006) Neutral theory and the evolution of ecological equivalence. Ecology 87:1387–1398
Huelsman MF, Jasinski J, Young C, Kovach J (2001) The multicolored Asian lady beetle (*Harmonia axyridis*) as a nuisance pest in households throughout Ohio. http://ipm.osu.edu/lady/icup.htm. Accessed 5 Sept 2007
Iablokoff-Khnzorian SM (1982) Les Coccinelles; Coléoptères-Coccinellidae. Société Nouvelle des Éditions Boubée, Paris, 568 pp
Iperti G, Bertrand E (2001) Hibernation of *Harmonia axyridis* (Coleoptera: Coccinellidae) in South-Eastern France. Acta Soc Zool Bohem 65:207–210
Kalaskar A, Evans EW (2001) Larval responses of aphidophagous lady beetles (Coleoptera: Coccinellidae) to weevil larvae versus aphid as prey. Ann Entomol Soc Am 94:76–81
Katsoyannos P, Kontodimas DC, Stathas GJ, Tsartsalis CT (1997) Establishment of *Harmonia axyridis* on citrus and some data on its phenology in Greece. Phytoparasitica 25:183–191
Keane RM, Crawley MJ (2002) Exotic plant invasions and the enemy release hypothesis. TREE 17:164–170
Kidd KA, Nalepa CA (1995) Distribution of *Harmonia axyridis* (Pallas) (Coleoptera: Coccinellidae) in North Carolina and Virginia. Proc Entomol Soc Wash 97:729–731

Koch RL (2003) The multicoloured Asian lady beetle, *Harmonia axyridis*: a review of its biology, uses in biological control and non-target impacts. J Insect Sci 3:1–16

Koch RL, Hutchison WD, Venette RC, Heimpel GE (2003) Susceptibility of immature monarch butterfly, *Danaus plexippus* (Lepidoptera: Nymphalidae: Danainae), to predation by *Harmonia axyridis* (Coleoptera: Coccinellidae). Biol Control 28:265–270

Koch RL, Venette RC, Hutchison WD (2005) Influence of alternate prey on predation of monarch butterfly (Lepidoptera: Nymphalidae) larvae by the multicolored Asian lady beetle (Coleoptera: Coccinellidae). Environ Entomol 34:410–416

Komai T (1956) Genetics of ladybeetles. Adv Genet 8:155–189

Komai T, Hosino Y (1951) Contributions to the evolutionary genetics of the lady-beetle, *Harmonia*. II. Microgeographic variations. Genetics 36:382–390

Kovach J (2004) Impact of multicolored Asian lady beetles as a pest of fruit and people. Am Entomol 50:159–161

Labrie G, Coderre D, Lucas E (2007) Overwintering strategy of the multicolored Asian ladybeetle (Coleoptera: Coccinellidae): a cold-free space as a factor of invasive success. Ann Entomol Soc Am (in press)

Labrie G, Lucas E, Coderre D (2006) Can developmental and behavioral characteristics of the multicolored Asian lady beetle *Harmonia axyridis* explain its invasive success? Biol Invasions 8:743–754

LaMana ML, Miller JC (1996) Field observations on *Harmonia axyridis* Pallas (Coleoptera: Coccinellidae) in Oregon. Biol Control 6:232–237

Lanzoni A, Accinelli G, Bazzocchi GG, Burgio G (2004) Biological traits and life table of the exotic *Harmonia axyridis* compared with *Hippodamia variegata* and *Adalia bipunctata* (Col. Coccinellidae). J Appl Entomol 128:298–306

Lee CE (2002) Evolutionary genetics of invasive species. TREE 17:386–391

Liebhold AM, Work TT, Mccullough DG, Cavey JF (2006) Airline baggage as a pathway for alien insect species invading the United States. Am Entomol 52:48–54

Levine JM, D'Antonio CM (2003) Forecasting biological invasions with increasing international trade. Conserv Biol 17:322–326

Lombaert E, Malausa T, Devred R, Estoup A (2007) Phenotypic variation in invasive and biocontrol populations of the harlequin ladybird, *Harmonia axyridis*. BioControl (this issue). doi:10.1007/s10526-007-9131-z

Lucas E, Coderre D, Vincent C (1997) Voracity and feeding preferences of two aphidophagous coccinellids on *Aphis citricola* and *Tetranychus urticae*. Entomol Exp Appl 85:151–159

Lucas E, Gagné I, Coderre D (2002) Impact of the arrival of *Harmonia axyridis* on adults of *Coccinella septempunctata* and *Coleomegilla maculata* (Coleoptera: Coccinellidae). Eur J Entomol 99:457–463

Lucas E, Labrie G, Vincent C, Kovach J (2007a) The multicolored Asian ladybeetle, *Harmonia axyridis*—beneficial or nuisance organism? In: Vincent C, Goettel M, Lazarovitz G (eds) Biological control: a global perspective. CABI Publishing, UK

Lucas E, Vincent C, Labrie G, Chouinard G, Fournier F, Pelletier F, Bostanian NJ, Coderre D, Mignault M-P, Lafontaine P (2007b) The multicolored Asian ladybeetle *Harmonia axyridis* in Quebec agro-ecosystems ten year after its arrival. Eur J Entomol 104:737–743

Lundgren JG, Razzak AA, Wiedenmann RN (2004) Population responses and food consumption by predators *Coleomegilla maculata* and *Harmonia axyridis* (Coleoptera: Coccinellidae) during anthesis in an Illinois cornfield. Environ Entomol 33:958–963

Maeta Y (1969) Biological studies on the natural enemies of some Coccinellid beetles. I. On *Perilitus coccinellae* (Schrank). Kontyu 37:147–166

Majerus M, Strawson V, Roy H (2006) The potential impacts of the arrival of the harlequin ladybird, *Harmonia axyridis* (Pallas) (Coleoptera: Coccinellidae), in Britain. Ecol Entomol 31:207–215

Marco DA, Páez SA, Cannas SA (2002) Species invasiveness in biological invasions: a modelling approach. Biol Invasions 4:193–205

McClure MS (1986) Role of predators in regulation of endemic populations of *Matsucoccus matsumurae* (Kuwana) (Homoptera: Margarodidae) in Japan. Environ Entomol 15:976–983

McClure MS (1987) Potential of the Asian predator, *Harmonia axyridis* Pallas (Coleoptera: Coccinellidae), to control *Matsucoccus resinosae* Bean and Godwin (Homoptera: Margarodidae) in the United States. Environ Entomol 16:224–230

Michaud JP (2001) Numerical response of *Olla v-nigrum* (Coleoptera: Coccinellidae) to infestations of Asian citrus psyllid, (Hemiptera: Psyllidae) in Florida. Fla Entomol 84:608–612

Michaud JP (2002) Biological control of Asian citrus psyllid, *Diaphorina citri* (Hemiptera: Psyllidae) in Florida: a preliminary report. Entomol News 113:216–222

Michaud JP (2004) Natural mortality of Asian citrus psyllid (Homoptera: Psyllidae) in central Florida. Biol Control 29:260–269

Michaud JP, Olsen LE (2004) Suitability of Asian citrus psyllid, *Diaphorina citri*, as prey for ladybeetles. BioControl 49:417–431
Mignault M-P, Roy M, Brodeur J (2006) Soybean aphid predators in Québec and the suitability of *Aphis glycines* as prey for three Coccinellidae. BioControl 51:89–106
Mooney HA, Drake JA (1989) Biological invasions: a SCOPE program overview. In: Drake JA, Mooney HA, di Castri F, Groves RH, Kruger FJ, Rejmánek M, Williamson M (eds) Biological invasion: a global perspective. Scope 37. Wiley, Chichester
Musser FR, Shelton AM (2003a) Predation of *Ostrinia nubilalis* (Lepidoptera: Crambidae) eggs in sweet corn by generalist predators and the impact of alternative foods. Environ Entomol 32:1131–1138
Musser FR, Shelton AM (2003b) Factors altering the temporal and within-plant distribution of coccinellids in corn and their impact on potential intraguild predation. Environ Entomol 32:575–583
Musser FR, Nyrop JP, Shelton AM (2004) Survey of predators and sampling method comparison in sweet corn. J Econ Entomol 97:136–144
Nalepa CA, Kidd KA, Ahlstrom KR (1996) Biology of *Harmonia axyridis* (Coleoptera, Coccinellidae) in winter aggregations. Ann Entomol Soc Am 89:681–685
Nault BA, Kennedy G (2003) Establishment of multicolored Asian lady beetle in Eastern North Carolina: seasonal abundance and crop exploitation within an agricultural landscape. BioControl 48:363–378
O'Dowd DJ, Green PT, Lake PS (2003) Invasional "meltdown" on an oceanic island. Ecol Lett 6:812–817
Ongagna P, Giuge L, Iperti G, Ferran A (1993) Cycle de développement d'*Harmonia axyridis* (Col. Coccinellidae) dans son aire d'introduction : le sud-est de la France. Entomophaga 38:125–128
Osawa N (2000) Population field studies on the aphidophagous ladybird beetle *Harmonia axyridis* (Coleoptera: Coccinellidae): resource tracking and population characteristics. Popul Ecol 42:115–127
Osawa N, Nishida T (1992) Seasonal variation in elytral colour polymorphism in *Harmonia axyridis* (the ladybird beetle): the role of non-random mating. Heredity 69:297–307
Park Y-L, Obrycki JJ (2004) Spatio-temporal distribution of corn leaf Aphids (Homoptera: Aphididae) and ladybeetles (Coleoptera: Coccinellidae) in Iowa cornfields. Biol Control 31:210–217
Pell JK, Baverstock J, Roy HE, Ware RL, Majerus MEN (2007) Intraguild predation involving *Harmonia axyridis*: a review of current knowledge and future perspectives. BioControl (this issue). doi: 10.1007/s10526-007-9125-x
Petren K, Case TJ (1996) An experimental demonstration of exploitation competition in an ongoing invasion. Ecology 77:118–132
Pimentel D, Lach L, Zuniga R, Morrison D (2000) Environmental and economic cost of nonindigenous species in the United States. Bioscience 50:53–65
Pimentel D, Zuniga R, Morrison D (2005) Update on the environmental and economic costs associated with alien-invasive species in the United States. Ecol Econ 52:273–288
Poutsma J, Loomans AJM, Aukema B, Heijerman T (2007) Predicting the potential geographical distribution of the harlequin ladybird, *Harmonia axyridis*, using the CLIMEX model. BioControl (this issue). doi:10.1007/s10526-007-9140-y
Rabitsch W, Schuh R (2006) First record of the multicoloured Asian ladybird *Harmonia axyridis* (Pallas, 1773) in Austria. Beiträge Zur Entomofaunistik 7:161–164
Roy HE, Brown PMJ, Rothery P, Ware RL, Majerus MEN (2007) Interactions between the fungal pathogen *Beauveria bassiana* and three species of coccinellid: *Harmonia axyridis*, *Cocinella septempunctata* and *Adalia bipunctata*. BioControl (this issue). doi:10.1007/s10526-007-9122-0
Saini ED (2004) Presencia de *Harmonia axyridis* (Pallas) (Coleoptera: Coccinellidae) en la provincia de Buenos Aires. Aspectos Biologicos y Morfologicos Ria 33:151–160
Sakurai H, Kawai T, Takeda S (1992) Physiological changes related to diapause of the lady beetle, *Harmonia axyridis* (Coleoptera: Coccinellidae). Appl Entomol Zool 4:479–487
Schanderl H (1987) Détermination des conditions optimales d'élevage de la coccinelle *Harmonia axyridis* Pallas (Col., Coccinellidae), et possibilité d'une production continue à l'aide d'une proie de substitution, les œufs *d'Ephestia kuehniella* Zeller (Lep., Pyralidae). Dissertation, Université de Droit, d'Economie et des Sciences, d'Aix-Marseille III
Schanderl H, Almeida JM (1992) Introdução de *Harmonia axyridis* Pallas (Col., Coccinellidae) na ilha de S. Maria. Açoreana 7:401–406
Schanderl H, Ferran A, Larroque MM (1985) Les besoins trophiques et thermiques des larves de la coccinelle *Harmonia axyridis* Pallas. Agronomie 5:417–421
Schanderl H, Ferran A, Coderre D, Ventura A, Soares AO, Almeida JMP, Taveira J (1991) Capacidade de dispersão de *Harmonia axyridis* Pallas (Col., Coccinellidae) após uma largada inundativa para controlo de afídeos do Milho *Rhopalosiphum padi* L. e *Sitobion avenae* F. (Hom., Aphididae). Relatórios e Comunicações do Departamento de Biologia 20:59–64

Schneider N, Loomans AJM (2006) Sur la présence au Luxembourg de la coccinelle arlequin *Harmonia axyridis* (Pallas, 1773) (Insecta, Coleoptera, Coccinellidae). Bulletin de la Société des Naturalistes Luxembourgeois 106:71–74

Shapiro-Ilan DI, Cottrell TE (2005) Susceptibility of lady beetles (Coleoptera: Coccinellidae) to entomopathogenic nematodes. J Invertebr Pathol 89:150–156

Shigesada N, Kawasaki K (1997) Biological invasions: theory and practice. Oxford University Press, Oxford

Sidlyarevich VI, Voronin KE (1973) Trials on using *Leis axyridis* under glass. Zashchita Rastenii 6:24

SIFA (Servicio de información fitosanitaria de Almería) (2004) Organismos de control biológico

Silva L, Smith CW (2006) A quantitative approach to the study of non-indigenous plants: an example from the Azores archipelago. Biodivers Conserv 15:1661–1679

Simberloff D (1996) Risks of species introduced for biological control. Biol Conserv 78:185–192

Simberloff D (1989) Which insect introductions succeed and which fail? In: Drake JA, Mooney HA, di Castri F, Groves RH, Kruger FJ, Rejmánek M, Williamson M (eds) Biological invasion: a global perspective. Scope 37. Wiley, Chichester

Simberloff D (2004) Community ecology: is it time to move on? Am Nat 163:787–799

Simberloff D, Stiling P (1996) How risky is biological control? Ecology 77:1965–1974

Snyder WE, Clevenger GM (2004) Negative dietary effects of Colorado potato beetle eggs for the larvae of native and introduced ladybird beetles. Biol Control 31:353–361

Snyder WE, Clevenger GM, Eigenbrode SD (2004) Intraguild predation and successful invasion by introduced ladybird beetles. Oecologia 140:559–565

Soares AO, Serpa A (2007) Interference competition between ladybird beetle adults (Coleoptera: Coccinellidae): effects on the growth and reproductive capacity. Popul Ecol 49:37–43

Soares AO, Coderre D, Schanderl H (2001) Influence of phenotype on fitness parameters of *Harmonia axyridis* Pallas (Coleoptera: Coccinellidae). Eur J Entomol 98:287–293

Soares AO, Coderre D, Schanderl H (2003a) Effect of temperature and intraspecific allometry on predation by two phenotypes of *Harmonia axyridis* Pallas (Coleoptera: Coccinellidae). Environ Entomol 32:939–944

Soares AO, Elias RB, Resendes R, Figueiredo H (2003b) Contribution to the knowledge of the Coccinellidae (Coleoptera) fauna from the Azores islands. Arquipélago Life Mar Sci 20A:47–53

Soares AO, Coderre D, Schanderl H (2004) Dietary self-selection behaviour by the adults of the aphidophagous ladybeetle *Harmonia axyridis* (Coleoptera: Coccinellidae). J Anim Ecol 73:478–486

Soares AO, Coderre D, Schanderl H (2005) Influence of prey quality on the fitness of two phenotypes of the adults of *Harmonia axyridis*. Entomol Exp Appl 114:227–232

Stals R, Prinsloo G (2007) Discovery of an alien invasive, predatory insect in South Africa. The multicoloured Asian ladybird beetle, *Harmonia axyridis* (Pallas) (Coleoptera: Coccinellidae). S Afr J Sci 103:123–126

Stastny M, Schaffner URS, Elle E (2005) Do vigour of introduced populations and escape from specialist herbivores contribute to invasiveness? J Ecol 93:27–37

Stathas GJ, Eliopoulos PA, Kontodimas DC, Giannopapas J (2001) Parameters of reproductive activity in females of *Harmonia axyridis* (Coleoptera: Coccinellidae). Eur J Entomol 98:547–549

Staverloekk A, Saethre MG, Haagvar E (2007) A review of the biology of the invasive harlequin ladybird Harmonia axyridis (Pallas, 1773) (Coleoptera, Coccinellidae). Nor J Entomol (in press)

Stewart LA, Dixon AFG (1989) Why big species of ladybird beetles are not melanic? Funct Ecol 3:165–177

Stuart RJ, Michaud JP, Olsen LE, McCoy CW (2002) Lady beetles as potential predators of the root weevil *Diaprepes abbreviatus* (Coleoptera: Curculionidae) in Florida citrus. Fla Entomol 85:409–416

Tan CC (1946) Mosaic dominance in the inheritance of color patterns in the lady-bird beetle, *Harmonia axyridis*. Genetics 31:195–210

Tan CC (1949) Seasonal variations of color patterns in *Harmonia axyridis*. Proceedings of the 8th international congress genetics, pp 669–670

Tedders WL, Schaefer PW (1994) Release and establishment of *Harmonia axyridis* (Coleoptera, Coccinellidae) in the southeastern United-States. Entomol News 105:228–243

Watanabe M (2002) Cold tolerance and myo-inositol accumulation in overwintering adults of a lady beetle, *Harmonia axyridis* (Coleoptera: Coccinellidae). Eur J Entomol 99:5–9

Vermeij GJ (1996) An agenda for invasion biology. Biol Conserv 78:3–9

Vermeij GJ (2005) Invasion as expectation-A historical fact of life. In: Sax DF, Stachowicz JJ, Gaines SD (eds) Species invasions—insights into ecology, evolution and biogeography. Sinauer Associates, Inc. Publishers, Massachusetts

Vitousek PM, Di Antonio CM, Loope LL, Westbrooks R (1996) Biological invasions as a global environment change. Am Sci 84:468–487

Vitousek PM, Di Antonio CM, Lloyd L Loope LL, Rejmánek M, Westbrooks R (1997) Introduced species: a significant component of human-caused global change. N Z J Ecol 21:1–16

Ware RL, Majerus MEN (2007) Intraguild predation of immature stages of British and Japanese coccinellids by the invasive ladybird *Harmonia axyridis*. BioControl (this issue). doi:10.1007/s10526-007-9135-8

Whittaker RJ, Fernández-Palacios JM (2007) Island biogeography-ecology, evolution and conservation. Oxford University Press, Oxford

Williamson M (1981) Island populations. Oxford University Press, Oxford

Williamson M (1996) Biological invasions. Chapman & Hall, London

Wise IL, Turnock WJ, Roughley RE (2001) New records of Coccinellid species for the province of Manitoba. Proc Entomol Soc Man 57:5–10

With KA, Pavuk DM, Worchuck JL, Oates RK, Fisher JL (2002) Threshold effects of landscape structure on biological control in agroecosystems. Ecol Appl 12:52–65

Yasuda H, Ohnuma N (1999) Effect of cannibalism and predation on the larval performance of two ladybird beetles. Entomol Exp Appl 93:63–67

Yasuda H, Kikuchi T, Kindlmann P, Sato S (2001) Relationships between attack and escape rates, cannibalism, and intraguild predation in larvae of two predatory ladybirds. J Insect Behav 14:373–384

Intraguild predation involving *Harmonia axyridis*: a review of current knowledge and future perspectives

Judith K. Pell · Jason Baverstock · Helen E. Roy · Remy L. Ware · Michael E. N. Majerus

Abstract As an effective generalist predator of aphids and other hemipteran pests *H. axyridis* has been a successful biological control agent. However, the very functional traits that have contributed to its success in this regard also implicate it as an intraguild predator that poses a significant risk not only to the diversity of other natural enemies of Hemiptera (and their associated ecosystem services), but to biodiversity more widely. In this paper we will specifically review the existing data on intraguild predation involving *H. axyridis*, and consider the strength and symmetry of such interactions both within its native guild and within exotic guilds where it has established as an invasive alien. We will use these studies to interpret the observed population declines in predator diversity in the field, predict species at risk in regions not yet invaded and consider implications for resulting ecosystem services. We will also indicate gaps in our knowledge that require further study in order to identify opportunities for mitigation.

Keywords Coccinellidae · *Harmonia axyridis* · Aphidophagous guild · Intraguild predation · Predators · Parasitoids · Pathogens · Ecosystem services

Introduction

Terrestrial ecosystems support a diversity of species that are directly and indirectly linked to each other within food webs that span multiple trophic levels. Natural enemy species (in the context of this paper we will collectively call these 'predators' as they all have a trophic relationship with their hosts/prey) contribute to the population regulation of species in both

J. K. Pell (✉) · J. Baverstock
Department of Plant and Invertebrate Ecology, Rothamsted Research, Harpenden, Hertfordshire AL5 2JQ, UK
e-mail: judith.pell@bbsrc.ac.uk

H. E. Roy
NERC Centre for Ecology and Hydrology, Monks Wood, Huntingdon PE28 2LS, UK

R. L. Ware · M. E. N. Majerus
Department of Genetics, University of Cambridge, Downing Street, Cambridge CB2 3EH, UK

Biological Control to Invasion: Ladybird Harmonia axyridis. H.E. Roy, E. Wajnberg (eds.),
doi: 10.1007/978-1-4020-6939-0_10,

the same and lower trophic levels (top-down pressure) and in this way influence the structure of the community as a whole. In the case of aphids these natural enemies (aphidophages) include specialist and generalist predators, parasitoids and pathogens (Völkl et al. 2007). Together they represent a 'guild', i.e., a community of species that share the same host/prey resource (Polis et al. 1989; Rosenheim et al. 1995). As aphids are often pests in managed ecosystems these natural enemies provide a valuable pest management ecosystem service that can be manipulated within biological control strategies (e.g., Barbosa 1998; Gurr et al. 2004; Losey and Vaughan 2006; Powell and Pell 2007; Pell 2008).

Different 'predator' taxa play different ecological roles within the guild (Rosenheim et al. 2004a). 'Intermediate predators' function from the third trophic level suppressing herbivores, whereas 'top predators' mainly operate from the fourth trophic level suppressing 'intermediate predators' and, consequently, releasing herbivore populations from control (Polis 1994; Rosenheim 1998). Intraguild predation occurs when one of two species competing for the same host/prey also consumes its competitor, and can be a strong force structuring communities. Omnivorous intraguild predation can be asymmetric when one of the two species (the intraguild predator) always preys on the other (the intraguild prey), or symmetric when both species prey on each other to a greater or lesser extent (Polis et al. 1989; Polis and Holt 1992; Rosenheim et al. 1995). The intraguild predator benefits not only from the nutritive value of the meal, but also from the removal of a competitor. Coincidental intraguild predation occurs when a parasitoid or pathogen is consumed while still developing within its herbivore host and, in this case, the herbivore and the intraguild prey are directly linked (Polis et al. 1989). Intraguild predators function as both 'intermediate' and 'top predators' by feeding on both herbivores and 'intermediate predators'.

The impact of intraguild predation on community structure and diversity can be extremely variable, complex and difficult to predict (Rosenheim et al. 2004a; Snyder and Evans 2006; Straub et al. 2008). Many studies have assessed the prevalence of intraguild predation in aphidophagous guilds (Polis et al. 1989; Rosenheim et al. 1995; Holt and Polis 1997; Müller and Brodeur 2002; Lucas 2005; Pell 2008) and reported it to be a widespread phenomenon with implications for both predator diversity within the guild and the pest management ecosystem services that the guild delivers.

As an effective generalist predator of aphids and other hemipteran pests *Harmonia axyridis* (Pallas) (Coleoptera: Coccinellidae) has been widely and repeatedly augmented or introduced for biological control. As such, it has contributed significantly to pest suppression in a wide variety of managed ecosystems including pecan, apple, sweet corn, alfalfa, cotton, tobacco, wheat and soybean (Tedders and Schaefer 1994; Buntin and Bouton 1997; Brown and Miller 1998; Colunga-Garcia and Gage 1998; Wells and McPherson 1999; Wells et al. 2001; Musser and Shelton 2003; Koch 2003; Majerus et al. 2006; Roy et al. 2006). However, the very functional traits that have made *H. axyridis* an effective biological control agent, also implicate it as an intraguild predator that poses significant risk to the diversity of other natural enemies of Hemiptera and their ecosystem services. Although natural enemy diversity in its native range remains relatively stable in the presence of *H. axyridis* (Kuznetsov 1988, 1997), wherever it has established after introduction as an exotic species, it has been associated with declines in native natural enemies, particularly native coccinellids (Majerus et al. 2006; Roy et al. 2006; Snyder and Evans 2006). These declines have been attributed to direct interspecific competition for resources with less competitive/fecund natural enemies (Michaud 2002) but are also likely to be strongly influenced by its role as an intraguild predator.

There are a number of functional traits that determine the nature, symmetry and outcome of intraguild predation including: relative size (incidence of mortality is often inversely correlated with size) (Majerus 1994; Evans 2000), aggressive strategies and mandibular structure (Yasuda et al. 2001), degree of feeding and habitat specificity, mobility (sessile stages are particularly vulnerable), defence strategies and abundance of extraguild prey (Polis et al. 1989; Lucas et al. 1998; Roy et al. 2006; Straub et al. 2008). In this paper we will specifically review the existing, largely experimental, data on these traits for *H. axyridis*, and discuss their strength and symmetry both within its native guild and within exotic guilds where it has established as an invasive alien. We will use these studies to interpret the observed population declines in natural enemy diversity in the field, predict species at risk in regions not yet invaded and consider implications for resulting ecosystem services. We will also identify gaps in our knowledge that require further study in order to identify opportunities for mitigation.

Current knowledge

Intraguild predation between *H. axyridis* and coccinellids of other species

Eggs

The relative size and mobility of the intraguild predator and prey are known to influence the outcome of intraguild predation, both showing an inverse correlation with the incidence of mortality (Huey and Pianka 1981; Sengonca and Frings 1985; Rosenheim et al. 1995; Lucas et al. 1998). In coccinellids this equates to the immature stages being more vulnerable than adults, and eggs being particularly threatened (Sato and Dixon 2004; Cottrell 2007).

Harmonia axyridis adults and larvae find eggs of many insect species acceptable as prey; laboratory stocks are often maintained on eggs of the Mediterranean flour moth, *Ephestia kuehniella* Zeller (Lepidoptera: Pyralidae) (Berkvens et al. 2007). Cottrell (2004) reports that *H. axyridis* adults attacked eggs of *Coleomegilla maculata* (De Geer) (Coleoptera: Coccinellidae) and *Olla v-nigrum* (Mulsant) (Coleoptera: Coccinellidae) more frequently than adults of these species attacked *H. axyridis* eggs, and attest that intraguild predation of eggs is a serious threat for these species. However, cannibalism and predation of coccinellid eggs are most often associated with larval stages (Dixon 2000; Cottrell 2007) and, for some species, may be affected by the relative abundance of extraguild prey (Sato et al. 2003); at low aphid or coccid densities, coccinellid larvae are more likely to engage in cannibalism or intraguild predation. In addition, the tendency for intraguild predation by larvae varies between species both as intraguild predators and intraguild prey. Cannibalism and intraguild predation by *H. axyridis* are commonly observed, both in the laboratory and the field, even when aphids or coccids are abundant (M.E.N. Majerus and R.L. Ware personal observation). A recent study concluded that *H. axyridis* larvae were more likely to engage in intraguild predation of the eggs of *C. maculata*, *Cycloneda munda* (Say) (Coleoptera: Coccinellidae), *Hippodamia convergens* Guérin-Méneville (Coleoptera: Coccinellidae) and *O. v-nigrum* than for *H. axyridis* eggs to be the intraguild prey of the larvae of these species (Cottrell 2007). This is in agreement with previous work of Cottrell (2004) demonstrating that neither *C. maculata* nor *O. v-nigrum* larvae can complete development on *H. axyridis* eggs whereas *H. axyridis* can complete development on a diet of either *C. maculata* or *O. v-nigrum* eggs.

Defensive adaptations (chemical or physical) are often effective in reducing the susceptibility of sessile life stages, such as eggs, prepupae and pupae to intraguild predation and cannibalism. The chemical defences of coccinellid eggs have been well studied (Agarwala and Dixon 1992; Hemptinne and Dixon 2000). Indeed, while intrinsic toxicity or unpalatability is governed by de novo synthesis of species-specific alkaloids (Pasteels et al. 1973; King and Meinwald 1996), Hemptinne and Dixon (2000) demonstrated the existence of extrinsic chemicals on the surface of some species' eggs that act as deterrents against intraguild predation. Ware et al. (2007) have recently discussed the role of surface deterrents on eggs of the European species *Calvia 14-guttata* (L.) (Coleoptera: Coccinellidae) as a defence against intraguild predation by *H. axyridis*. A similar phenomenon has been speculated for the eggs of the Japanese species *Eocaria muiri* Timberlake (Coleoptera: Coccinellidae) (Ware et al. 2008). The defensive chemistry of *H. axyridis* eggs themselves has been implicated as pivotal to the observed resistance of *H. axyridis* eggs to predation by other aphidophages that appear to find them unpalatable (Alam et al. 2002; Sato and Dixon 2004). Interestingly, similar alkanes to those found on the surface of *H. axyridis* eggs are also present in larval tracks and act as oviposition deterring semiochemicals for other ladybirds, thus reducing egg cannibalism and intraguild predation (Magro et al. 2007).

Larvae and pupae

Ware and Majerus (2007) have comprehensively examined intraguild predation of immature stages of British and Japanese coccinellids by *H. axyridis*. A total of 12 species of coccinellid (Coleoptera: Coccinellidae) were included in this laboratory study: eight derived from British populations (*Coccinella 7-punctata* L., *Adalia 2-punctata* (L.), *Adalia 10-punctata* (L.), *Propylea 14-punctata* (L.), *C. 14-guttata*, *Anatis ocellata* (L.), *Harmonia 4-punctata* (Pontoppidan), *Coccinella 5-punctata* L.) and four from Japanese populations (*Coccinella 7-punctata brucki* Mulsant, *Cheilomenes 6-maculatus* (Fabricius), *Propylea japonica* Thunberg and *E. muiri*). In general, *H. axyridis* was an intraguild predator of all species, with the exception of *A. ocellata*. The important conclusions of this study were that there was no obvious difference between the susceptibilities of Japanese and British coccinellids to intraguild predation by *H. axyridis*, and intraguild predation by *H. axyridis* represents a serious threat for many species in Britain (Ware and Majerus 2007).

The relatively large size of *H. axyridis* throughout its life cycle undoubtedly contributes to its success as an intraguild predator following the 'size matters' hypothesis. In predatory interactions between coccinellid larvae it is generally the larger that eats the smaller, assuming both are mobile (Majerus 1994). Notably, in comparative tests between fourth instar larvae of *H. 4-punctata*, *H. axyridis* and *A. ocellata*, which have similar behavioural and physical defences, the level of intraguild predation was directly correlated to size. Larvae of the largest species, *A. ocellata*, won the majority of encounters with *H. axyridis*, while the smallest species, *H. 4-punctata*, lost most such encounters (Ware and Majerus 2007).

There are few reports of other larval coccinellids successfully attacking *H. axyridis*, and most evidence suggests that the immature stages of *H. axyridis* are resistant to reciprocal attacks. In a laboratory study assessing the interactions between *H. axyridis* and *C. 7-punctata*, Yasuda et al. (2001) attributed the greater success of the former to its higher attack rates and greater escape ability. Recent research is beginning to confirm the importance of chemical defence of *H. axyridis* larvae as a means of preventing counter-attacks (Ware and Majerus 2007), and larvae are known to produce similar defensive alkaloids to those present within eggs and released by adults (Pasteels et al. 1973; King and

Meinwald 1996). Further studies have also implicated the role of superior physical defences (spines rather than hairs) in larval stages of *H. axyridis* compared to other species (Ware and Majerus 2007).

In the field, coccinellid larvae tend to disperse from a plant when prey abundance is low (Sato 2001) and this reduces the incidence of cannibalism and intraguild predation by larger larvae and adults (Sato et al. 2003). However, emigration is a precarious strategy for an immature coccinellid, so there is a trade-off between emigrating or remaining on a plant with a low prey density and risking cannibalism or intraguild predation. *Harmonia axyridis* commonly co-occurs with *C. 7-punctata brucki* and *P. japonica* in their native Japanese range (Yasuda and Shinya 1997; Sato 2001). In a Japanese field study of these three coccinellid species co-occurring on shrubs both *C. 7-punctata brucki* and *H. axyridis* larvae emigrated in response to low prey density whereas *P. japonica* larvae did not (Sato 2001). The prevalence of intraguild predation and cannibalism of *C. 7-punctata brucki* and *H. axyridis* larvae was low whereas no *P. japonica* larvae completed their development on the shrubs and this was assumed to be as a consequence of intraguild predation. In further studies it was confirmed that the early emigration of *C. 7-punctata brucki* larvae enabled them to escape from intraguild predation by *H. axyridis* larvae (Sato et al. 2003) and that the late emigration of *P. japonica* larvae accounted for the high incidence of intraguild predation by *H. axyridis* larvae (Sato et al. 2003).

Propylea japonica is also a smaller species than either *H. axyridis* or *C. 7-punctata brucki* and so this further supports the hypothesis that 'size matters' in intraguild predation. Ware and Majerus (2007) also report *P. japonica* larvae as highly palatable intraguild prey with little physical defence from attack by *H. axyridis*. It is interesting to note from field studies in Japan that *C. 7-punctata brucki* are active earlier in the spring than *H. axyridis* (Takahashi and Naito 1984; Sato 2001). Therefore, *H. axyridis* are more likely to experience a scarcity of aphids, which are abundant in spring and early summer, than *C. 7-punctata brucki*. Consequently, *H. axyridis* is likely to be more reliant on intraguild prey to complete their development than is *C. 7-punctata brucki* (Sato et al. 2003).

Adults

Adult coccinellids are generally less susceptible to predation than immature stages due to their protective elytra and aposematic colour patterns (Majerus 1994). However, they are exposed to a particularly vulnerable period just after eclosion, when their elytra are still soft. Ware and Majerus (2007) report observations of attacks made on eclosing or newly eclosed adults by larvae, when no other food was available. Fourth instar *H. axyridis* larvae were observed to attack and consume new adults of *A. 10-punctata*, *A. ocellata*, *C. 14-guttata*, *C. 7-punctata brucki* and *E. muiri*, after not having attacked them as pupae (Ware and Majerus 2007). However, the consumption of newly eclosed *H. axyridis* adults was rare, and only performed by conspecific larvae and larvae of the congeneric species *H. 4-punctata*. This supports the speculation that the defensive chemistry of *H. axyridis* adults may make them unpalatable to other coccinellids (Hough-Goldstein et al. 1996).

Intraguild predation between *H. axyridis* and non-coccinellid predators

Coccinellids are common within aphidophagous guilds and most studies considering the role of *H. axyridis* as an intraguild predator have focussed on interactions within the

Coccinellidae. However, *H. axyridis* interacts with many other predatory insects at the community level. Lacewings (Neuroptera) are both abundant as aphid predators in natural systems, and as components of biological control strategies. In a laboratory study, Phoofolo and Obrycki (1998) demonstrated that there was no difference in the development time or survival of *H. axyridis* (and also *C. maculata*) fed on a diet of pea aphids, *Acyrthosiphon pisum* (Harris) (Hemiptera: Aphididae), or eggs of the lacewing, *Chrysoperla carnea* (Stephens) (Neuroptera: Chrysopidae). In contrast, *C. carnea* was unable to develop successfully when fed on *H. axyridis* eggs.

A further study examined the interactions between adult *H. axyridis*, *C. carnea* and the gall midge *Aphidoletes aphidimyza* (Rondani) (Diptera: Cecidomyiidae) in the presence of the soybean aphid, *Aphis glycines* Matsumara (Hemiptera: Aphididae), in microcosms and field cages (Gardiner and Landis 2007). *Harmonia axyridis* engaged in intraguild predation with both *C. carnea* and *A. aphidimyza* and was predicted to contribute to declines in both species in the field. Phoofolo and Obrycki (1998) and Gardiner and Landis (2007) both suggested the potential for *H. axyridis* to be an asymmetrical intraguild predator of *C. carnea*. However, *C. carnea* will consume *H. axyridis* eggs (Phoofolo and Obrycki 1998) and recent observations by Fremlin (2007) indicate that *C. carnea* will also attack *H. axyridis* pupae. Therefore, the interaction can be considered to be (weakly) symmetrical. Further research is required to expand our knowledge on interactions between neuropterans and *H. axyridis*.

The spined soldier bug, *Podisus maculiventris* Say (Heteroptera: Pentatomidae) is native to North America and has a broad prey range including over 100 species of insect, primarily soft-bodied, slow-moving larvae of Coleoptera and Lepidoptera, but also aphids (McPherson 1980; Herrick and Reitz 2004). Half of the species on which it preys are important crop pests (Herrick and Reitz 2004). *Podisus maculiventris* is therefore loosely linked with many trophic guilds, including that of aphids (McPherson 1980; Hough-Goldstein et al. 1996), and four species of predaceous coccinellid are listed as prey to this predatory bug (McPherson 1980). In laboratory studies, Hough-Goldstein et al. (1996) demonstrated that although *P. maculiventris* did not consume adult *H. axyridis* they did consume *H. axyridis* larvae in 'no choice' treatments. In this study, *H. axyridis* larvae were described as 'aggressive' and often noted to escape predation, whereas adult *H. axyridis* were rejected, presumably as unpalatable, by *P. maculiventris*. Indeed, *P. maculiventris* took four times longer to capture *H. axyridis* larvae compared to *Spodoptera frugiperda* Smith (Lepidoptera: Noctuidae) larvae (Hough-Goldstein et al. 1996).

A further study demonstrated that interactions between *P. maculiventris* and *H. axyridis* in the presence or absence of extraguild prey (*Spodoptera littoralis* Boisduval (Lepidoptera: Noctuidae) or *Myzus persicae* (Sulzer) (Hemiptera: Aphididae)) were asymmetric in favour of the bug (De Clercq et al. 2003). *Podisus maculiventris* fed on *H. axyridis* eggs and larvae but rarely on adults. As with interactions amongst coccinellids, this interaction was dependent on the life stage of the bug; fourth instar nymphs and adults were more aggressive in their interactions than second instars nymphs. In contrast, *H. axyridis* rarely attacked *P. maculiventris*. Intraguild predation by *P. maculiventris* on *H. axyridis* was reduced in the presence of *S. littoralis* but not *M. persicae*. De Clercq et al. (2003) interpreted this to infer that *H. axyridis* was a less preferred prey in comparison to *S. littoralis* and this was supported by the longer development time of the bug when fed on *H. axyridis* larvae compared to *S. littoralis*. There was a slight difference in the survival to adulthood of pentatomid nymphs fed on *H. axyridis* compared to *S. littoralis* (70% vs. 80–90% respectively) and no nymphs reached adulthood when fed on just aphids (De Clercq et al. 2003).

Intraguild predation between *H. axyridis* and parasitoids

Intraguild predation between predators and parasitoids is asymmetrical and can be described as both coincidental and omnivorous (Polis et al. 1989). In contrast to the literature on intraguild predation between *H. axyridis* and other predators, particularly coccinellids, there is very little information on such interactions between *H. axyridis* and parasitoids. Previous studies have reported the consumption of parasitized aphids by coccinellids (Wheeler et al. 1968; Ferguson and Stiling, 1996) and that the presence, or recent activity, of predatory coccinellids within an aphid colony can reduce the oviposition rate of aphid parasitoids (Taylor et al. 1998).

Takizawa et al. (2000) assessed whether the aphid *Aphis craccivora* Koch (Hemiptera: Aphididae), parasitized by *Aphidius colemani* Viereck (Hymenoptera: Braconidae), were suitable prey for three coccinellid species: *C. 7-punctata*, *P. japonica* and *H. axyridis*. The parasitoid was used at two life stages: 3-day-old larvae within living aphids (coincidental intraguild predation) and sessile aphid 'mummies' containing pupae (asymmetrical omnivorous intraguild predation). Consumption of parasitized aphids containing 3-day-old larvae did not reduce survival or increase development time of any of the coccinellids. In contrast, consumption of aphid 'mummies' increased the development time of all three species and reduced survival to adulthood of *C. 7-punctata* by 70% but did not affect the survival of *H. axyridis* and *P. japonica*. *Coccinella 7-punctata* is considered to be a more aphid-specific predator than *H. axyridis* (Hodek and Honek 1988). It is likely that parasitized 'mummies' are unsuitable prey and, hence, the low survival rates of *C. 7-punctata* fed on parasitized aphids could be attributed to their aphid-specific dietary requirements (Takizawa et al. 2000). In contrast, *H. axyridis* selectively preyed on pea aphids, *A. pisum*, rather than pea aphid 'mummies' parasitized by *Aphidius ervi* Haliday (Hymenoptera: Braconidae) (Snyder and Ives 2003). Although this would not change the impact of coincidental intraguild predation of larval parasitoids by *H. axyridis*, it does demonstrate that interactions are variable depending on the prey concerned. Similarly, in a study at larger spatial scales Snyder et al. (2004a) found that although *H. axyridis* did prey on 'mummies' of the aphid parasitoid *Aphelinus asychis* Walker (Hymenoptera: Aphelinidae) (adult *H. axyridis* showed no discrimination between 'mummies' and aphids although larvae preferred aphids in feeding trials), the overall parasitism level was not affected.

Intraguild predation between *H. axyridis* and pathogens

Entomopathogenic fungi are common pathogens of aphids and can be involved in both coincidental and omnivorous, asymmetrical or symmetrical, intraguild interactions (Roy and Pell 2000; Völkl et al. 2007). The most common fungal pathogens are host-specific species from the Zygomycetes, order Entomophthorales, although other species from the Ascomycetes, order Hypocreales, some with wide host ranges, are used as biopesticides and could, therefore, be involved in intraguild interactions (Powell and Pell 2007; Pell 2008).

During the final stages of infection by entomopathogenic fungi the host dies and the fungus sporulates to produce more infective propagules (conidia) for transmission (Inglis et al. 2001; Pell et al. 2001). As the sporulating aphid cadaver is sessile, it is an easy prey item for generalist predators. In a laboratory study, Roy et al. (2008) assessed the predation of aphid cadavers sporulating with *Pandora neoaphidis* (Remaudière and Hennebert) Humber (Zygomycetes: Entomophthorales) by UK and Japanese-derived *H. axyridis* relative to that of the UK native species *C. 7-punctata*. Unlike *C. 7-punctata*, which showed a strong preference for aphids over sporulating fungal cadavers, *H. axyridis* showed little discrimination between the prey types and would consume whole, sporulating cadavers. The consumption of whole sporulating cadavers could remove a significant quantity of inoculum from the aphid population, which may result in a reduction in further transmission. As *P. neoaphidis* only infects aphids, this represents asymmetrical intraguild predation. The strength of intraguild predation by *H. axyridis* collected in the UK differed from those collected in Japan, with *H. axyridis* (UK) showing less discrimination between prey types compared to *H. axyridis* (Japan) (Roy et al. 2008). Whether feeding on infected cadavers has fitness consequences for *H. axyridis* is unknown.

Other entomopathogenic fungi have wider host ranges and may be infective towards both coccinellids and aphids, making the intraguild interaction potentially symmetrical. For example, isolates of the entomopathogenic fungi *Metarhizium anisopliae* (Metschnikoff) Sorokin (Ascomycete: Hypocreales), *Paecilomyces fumosoroseus* (Wise) Brown and Smith (Ascomycetes: Hypocreales) and *Beauveria bassiana* (Balsamo) Vuillemin (Ascomycetes: Hypocreales) have been recorded infecting both aphids and coccinellids (e.g., Magalhaes et al. 1988; Keller and Zimmerman 1989; Butt et al. 1994; James and Lighthart 1994; Yeo 2000; Pell and Vandenberg 2002; Ormond et al. 2006). In particular, *B. bassiana* is a major overwintering mortality agent of *C. 7-punctata* (Majerus 1994; Ormond et al. 2006). Roy et al. (2007) assessed the susceptibility of *H. axyridis* derived from the UK and Japan to *B. bassiana* relative to that of the UK native species *C. 7-punctata* and *A. 2-punctata*. Only doses of 10^9 conidia ml^{-1} resulted in mortality of *H. axyridis*, in contrast, 80% of *C. 7-punctata* and 70% of *A. 2-punctata* exposed to 10^7 conidia ml^{-1} of *B. bassiana* succumbed to infection. In addition, the results suggested that the UK derived strain may be more resistant to infection than the Japanese derived strain. Interestingly, inoculation with *B. bassiana* at a concentration as low as 10^5 conidia ml^{-1} reduced the fecundity of *H. axyridis* (Roy et al. 2007). Whether any of these coccinellids feed on *B. bassiana*-infected cadavers and the implications of this on the symmetry of intraguild predation is unknown. However, in the study of Pell and Vandenberg (2002) the coccinellid *H. convergens* did consume living *Diuraphis noxia* Kurdjumov (Hemiptera: Aphididae) aphids infected by *P. fumosoroseus* but did not consume sporulating aphid cadavers.

In both studies by Roy et al. (2007, 2008), the interactions involving *H. axyridis* collected from the UK differed from those collected in Japan, with *H. axyridis* from the UK being a stronger intraguild predator of *P. neoaphidis* and more resistant to *B. bassiana* than the strain from Japan. These are the first studies to indicate fundamental differences between *H. axyridis* subpopulations and demonstrate the importance of assessing genetic and ecological variability amongst such subpopulations.

Intraguild predation involving *H. axyridis*: Conclusions to date

There is broad agreement, from the studies described above, that *H. axyridis* is a top predator as it is predominantly a strong asymmetrical intraguild predator of other guild members (Hironori and Katsuhiro 1997; Cottrell and Yeargan 1998; Phoofolo and Obrycki 1998; Yasuda and Ohnuma 1999; Michaud 2002; Sato et al. 2003, 2005; Felix and Soares 2004; Snyder et al. 2004b; Kajita et al., 2006; Pervez and Omkar 2006; Roy et al. 2006, 2007, 2008; Cottrell 2007; Ware and Majerus 2007) and as such can dominate in aggressive intraguild interactions and lead to a decline in guild diversity.

The key functional traits that make *H. axyridis* such an exceptional intraguild predator are its relatively large size, aggressive behaviour, extreme polyphagy and the possession of effective physical and chemical defence strategies. Together these traits provide the mechanism by which *H. axyridis* impacts on guilds and food webs in general (Tedders and Schaefer 1994; Hodek 1996; Yasuda and Shinya 1997; Yasuda and Ohnuma 1999; Kajita et al. 2000; Koch 2003). It therefore seems likely that *H. axyridis* could disrupt aphidophagous/coccidophagous community structure leading to declines in other species in the guild where it establishes as a non-native species. *Harmonia axyridis* has certainly become abundant and widely distributed as a non-native species throughout North America and Europe (Koch 2003; Brown et al. 2007) and field data from the USA has reported associated negative impacts on native coccinellid species in these regions (Elliott et al. 1996; LaMana and Miller 1996; Brown and Miller 1998; Colunga-Garcia and Gage 1998; Lucas et al. 2002; Michaud 2002; Nault and Kennedy 2003). Interestingly, *C. 7-punctata* was also introduced and became established widely in North America prior to the establishment of *H. axyridis* (Gordon 1985). It too displaced native coccinellids and became the dominant predator. However, *H. axyridis* is a more aggressive predator than *C. 7-punctata*; larvae of *H. axyridis* consume those of *C. 7-punctata* but the reverse occurs considerably less frequently (Hironori and Katsuhiro 1997; Yasuda et al. 2004; Ware and Majerus 2007). Since the arrival of *H. axyridis* in the USA, it has displaced *C. 7-punctata* as the 'top predator' demonstrating the considerable dominance of *H. axyridis* as an intraguild predator of native and exotic coccinellids alike (Snyder et al. 2004b).

Similar evidence of declines in native coccinellid species after establishment of *H. axyridis* is accumulating in Europe. For example, evidence suggests that in London numbers of some, but not all, native coccinellids (particularly *A. 2-punctata*) have declined significantly since the arrival of *H. axyridis* in 2004 (Majerus, unpublished data). This is almost certainly due to its superior competitive ability and status as an intraguild predator. From the experimental studies of Ware and Majerus (2007), with consideration of the wider ecologies of *H. axyridis* and British ladybirds, and in the absence of control or mitigation, we anticipate that aphidophagous habitat generalists such as *C. 7-punctata, A. 2-punctata* and *P. 14-punctata* will be most at risk from the establishment of *H. axyridis* in the UK, followed by aphidophagous habitat specialists such as *Myzia oblongoguttata* (L.) (Coleoptera: Coccinellidae), *Myrrha 18-guttata* (L.) (Coleoptera: Coccinellidae) and *Anisosticta 19-punctata* (L.) (Coleoptera: Coccinellidae). The only species that may be relatively unaffected by invasive *H. axyridis* is *A. ocellata*, and indeed *H. axyridis* could actually be at risk from detrimental interactions with this species where they co-inhabit coniferous woodland. It is thought that the aphidophagous specialist *C. 5-punctata* would be particularly threatened if *H. axyridis* invades unstable river shingle, as it is already considered of conservation concern (Ware et al. 2005). Coccidophagous species such as *Exochomus 4-pustulatus* (L.) (Coleoptera: Coccinellidae) and *Chilocorus renipustulatus* (Scriba) (Coleoptera: Coccinellidae) and the *Erica*/*Calluna* heathland specialists *Chilocorus 2-pustulatus* (L.) (Coleoptera: Coccinellidae) and *Coccinella hieroglyphica* L. (Coleoptera: Coccinellidae) are thought to be less at risk, the latter because *H. axyridis* has yet to be recorded from such heathland. Mycophagous, phytophagous and myrmecophilous coccinellids, such as *Thea 22-punctata* (L.) (Coleoptera: Coccinellidae), *Subcoccinella 24-punctata* (L.) (Coleoptera: Coccinellidae) and *Coccinella magnifica* Redtenbacher (Coleoptera: Coccinellidae) respectively, are likely to be the least threatened by the establishment of *H. axyridis* in Britain.

There is clear evidence for declines in diversity of coccinellids in the USA and increasingly in Europe. However, there have been no studies to evaluate impacts on the diversity of other guild members or on biodiversity more widely. These are urgently required, particularly in areas undergoing invasion by *H. axyridis*, if the full implications of its presence are to be evaluated.

Knowledge gaps, future perspectives and implications for biodiversity and ecosystem function

Harmonia axyridis is also a dominant intraguild predator in its native range, as shown by Ware and Majerus (2007) who paired Japanese *H. axyridis* with Japanese native coccinellids in their laboratory interaction studies and found that *H. axyridis* was the successful intraguild predator in the majority of cases. It has also prevented the establishment of the introduced coccinellid *A. 2-punctata* in Japan (Sakuratani et al. 2000; Kajita et al. 2006). However, it appears to co-exist with most other guild members (Kuznetsov 1997) whilst causing declines and competitive exclusion of other coccinellid species in regions where it is an invasive exotic (e.g., Colunga-Garcia and Gage 1998; Michaud 2002). In co-evolved communities that utilise limited resources, species will be under selection to either ensure that they win any competitive interaction or that they avoid competition through resource-use complementarity and niche differentiation (Wilby and Thomas 2002a, b; Pell 2008). It is possible that, although *H. axyridis* is the top predator in its native range, other predator populations can escape intraguild predation through spatial or temporal niche differentiation. Studies to understand the mechanisms for co-existence with *H. axyridis* in its native range are essential if we are to mitigate existing effects in its invasive range.

Temporal niche differentiation

The co-existence of *H. axyridis* and sympatric species in Asia may, in part, be due to phenological differences that lead to temporal niche differentiation. For example, *C. 7-punctata brucki* is active earlier in the spring in Japan than *H. axyridis* and its larvae are therefore larger and more able to survive in intraguild combat with *H. axyridis* (Takahashi and Naito 1984; Sato 2001). Phenological studies of *H. axyridis* in its non-native range are essential if we are to predict the extent of intraguild predation and consequent impacts on biodiversity. Such studies are underway and in the UK, for example, early indications are that *H. axyridis* emerges from winter in synchrony with most native aphidophagous coccinellids, which contributes to the latter's inability to escape intraguild predation. *Harmonia axyridis* is also multivoltine and remains active for longer through the season, allowing numbers of individuals to build up rapidly. All life stages of *H. axyridis* have been recorded in November (early winter) in the UK (Majerus et al. 2006; Brown et al. 2007). In one study from the USA biological control literature, Flowers et al. (2006) considered intraguild interactions amongst three exotic predators attacking the hemlock woolly adelgid, *Adelges tsugae* (Annand) (Hemiptera: Adelgidae) in hemlock stands in West Virginia. The species included were specialists, *Laricobius nigrinus* (Fender) (Coleoptera: Derodontidae) and the coccinellid *Sasajiscymnus tsugae* (Sasaji & McClure) (Coleoptera: Coccinellidae), and the generalist *H. axyridis*. The authors concluded that intraguild predation by *H. axyridis* was not common because they were phenologically separated and active during different seasons; *L. nigrinus* was most active in spring and *H. axyridis* in summer. This also has implications for the wider guild beyond other coccinellids, for which there is currently very little information.

Associations with ants

Other coccinellid species in *H. axyridis*' native range may escape intraguild predation through associations with ants. Although there are limited studies on myrmecophilous interactions in the native range of *H. axyridis*, it is well documented for some coccinellid species in Europe. Many aphids and coccids are myrmecophilous and gain benefits from being associated with ants, including protection from natural enemies, in exchange for food in the form of honeydew (e.g., Hölldobler and Wilson 1990; Jiggins et al. 1993; Majerus et al. 2007). There is considerable evidence to demonstrate that ants display ownership behaviour, whereby they behave more aggressively towards predators in the vicinity of tended colonies than elsewhere (Way 1963). In the case of coccinellids, this aggression may be directed toward both adults and larvae, the former being chased away (e.g., Itioka and Inoue 1996; Sloggett 1998), while the latter may be picked up and dropped off the plant, or killed (e.g., Jiggins et al. 1993; Sloggett and Majerus 2003). As a result of ant aggression, most coccinellids only feed on ant-tended prey when untended prey are scarce (Sloggett and Majerus 2000). Many coccinellids have defences against ant aggression, which may be behavioural, chemical or physical and lead to variation in tolerance to ants amongst different coccinellid species (Majerus et al. 2007). For example, Sloggett and Majerus (2000) showed a hierarchy in six species of coccinellid living in pine woodland in the UK to aggression from *Formica rufa* (L.) (Hymenoptera: Formicidae) from no tolerance (*M. 18-guttata* and *A. ocellata*) to high tolerance (*C. magnifica*). Unfortunately, the level of tolerance of *H. axyridis* to ant aggression in either its native or its invasive range is unknown. Should *H. axyridis* be found to be intolerant of ant presence, it is likely that those aphidophages that have a significant degree of ant tolerance, such as *M. oblongoguttata* and *C. 7-punctata*, will find a refuge from some of the effects of intraguild predation and competition imposed by *H. axyridis* in habitats occupied by ant-tended aphids. Experiments to investigate the level of tolerance of *H. axyridis* to common aphid tending ants, such as *F. rufa*, *Lasius niger* (L.) (Hymenoptera: Formicidae) and *Myrmica ruginodis* Nylander (Hymenoptera: Formicidae), are urgently needed. The small number of coccinellid species that are true myrmecophiles, i.e., have the closest relationships with ants and associated high tolerance, are likely to have the safest refuge from *H. axyridis*, unless *H. axyridis* is also highly tolerant to ants. These species include *C. magnifica* (Sloggett et al. 2002; Majerus et al. 2007) and *Platynaspis luteorubra* (Goeze) (Völkl 1995; Godeau 2000). It is possible that some species that have some tolerance of ants may come under additional selection pressures to improve their tolerance as a result of the presence of *H. axyridis*. In regions now occupied by *H. axyridis*, monitoring the coccinellid presence in the vicinity of nests of *F. rufa* over the next decade would be valuable.

Behavioural interactions

Further work is also required to acquire data on behavioural defence strategies of guild members in the presence of *H. axyridis*, such as dislodgement (dropping behaviour) and escape behaviours which could be extremely important for some species. For example, larvae of the coccinellid *P. 14-punctata* are consumed by *H. axyridis* in laboratory studies conducted in Petri dishes, but it is speculated that the highly mobile larvae of this species may have the capability to successfully escape from *H. axyridis* in field situations (Ware and Majerus 2007). We also know that parasitoids and predators can use volatile and chemical cues to avoid competition; both the coccinellid *C. 7-punctata* and the predatory bug *Anthocoris nemorum* (L.) (Heteroptera: Anthocoridae) can detect and avoid surfaces contaminated by the fungal pathogen *B. bassiana* to which they are susceptible (Meyling and Pell 2006; Ormond 2007). In addition, the aphid parasitoid *A. ervi* detects volatiles from the tracks of *C. 7-punctata* and avoids oviposition in nearby aphid populations (Nakashima et al. 2004). Such behavioural responses can ensure co-existence by reducing the risk of intraguild predation. They may contribute to co-existence of *H. axyridis* in its co-evolved native guild populations and are currently being evaluated for *H. axyridis* in the broader UK guild.

Other intraguild interactions, such as facilitation, have been recorded for some species and shown to mitigate the effects of intraguild predation on guild diversity and may also apply to *H. axyridis*. For example, although *C. 7-punctata* is an intraguild predator of the aphid-specific pathogenic fungus *P. neoaphidis*, it does not consume whole, fungal cadavers and transmission of the fungus is not reduced. Indeed, *C. 7-punctata* enhances transmission of the fungus and passively vectors it between aphid populations, thereby also aiding pathogen dispersal (Roy et al. 1998, 2001; Roy and Pell 2000). Some of these co-evolved behavioural traits may contribute to the maintenance of guild diversity in the native range of *H. axyridis* and may mitigate its effects on some species in the invasive range. In preliminary studies, *H. axyridis* did enhance transmission and dispersal of *P. neoaphidis* in the laboratory (J. Baverstock personal observation).

Natural enemy release and biotype variation

Differences in functional traits of invasive and non-invasive biotypes of *H. axyridis* may have exacerbated their impact in non-native ecosystems. Although direct comparisons between different biotypes of *H. axyridis* and other natural enemies found in its exotic range have only been made for a limited number of functional traits, they have all demonstrated that the UK invasive biotype is a stronger asymmetric intraguild predator than the Japanese biotype. Specifically Roy et al. (2007) demonstrated that *H. axyridis* from a culture derived from Japan were less likely to consume aphid cadavers supporting the beneficial aphid-specific fungal pathogen *P. neoaphidis* than those derived from the UK. They speculated that *H. axyridis* obtained from non-native populations were derived from biological control cultures and, through microevolution or founder effects, have been, perhaps unintentionally, selected for extreme polyphagy and an increased tendency towards intraguild predation compared to *H. axyridis* in its native range (Roderick and Howarth 1999; Hufbauer and Roderick 2005; Roy et al. 2008). It would be extremely interesting to compare the strength and symmetry of intraguild interactions between Japanese, UK and US derived *H. axyridis* for a wider range of guild members, from both native and non-native ranges to determine if they function differently in relation to all guild members with which they compete. In addition, UK-derived *H. axyridis* have greater resistance to some natural enemies, such as the fungus *B. bassiana* and the parasitoid wasp, *Dinocampus coccinellae* Schrank (Hymenoptera: Braconidae) than do Japanese-derived *H. axyridis*, implying that they have also been released from their own population regulators (Koyama and Majerus 2007; Roy et al. 2007). Evolution of resistance to disease, parasitism and insecticides is not uncommon in insects and may be associated with trade offs in fitness (e.g., Foster et al. 2007; Völkl et al. 2007). For example, although UK-derived *H. axyridis* were less suceptible to infection they were also significantly less fecund when infected than *B. bassiana*-infected *H. axyridis* derived from Japan (Roy et al.

2007). Further research is needed to fully evaluate these interactions and determine whether any associated fitness costs could prove to be *H. axyridis*' weakness.

Going beyond managed ecosystems

The focus of research has largely been on interactions between *H. axyridis* and other coccinellids in managed ecosystems and, until recently, mainly in the USA, which has very different agricultural systems to Europe. However, it is clear from the studies described above that *H. axyridis* can have impacts on many other species and that this has implications beyond managed ecosystems to natural habitats. Such impacts require urgent investigation. *Harmonia axyridis* is dispersive and highly polyphagous and therefore likely to impinge on many foodwebs in many different ecosystems, with the potential to disrupt them all. Although native coccinellids with similar prey and habitat requirements are at greatest risk, other non-coccinellid guild members and non-target species beyond the guild, such as Hemiptera of no economic importance (Majerus et al. 2006), lepidopteran eggs and larvae (Koch 2003; Koch et al. 2006) and all the predators, hyperparasites and pathogens in the higher trophic levels (Roy et al. 2007) that may also be negatively affected. This represents a significant threat to biodiversity per se, particularly as some of these species are of considerable conservation concern (Koch et al. 2006).

What about scale?

We must also remember that many of the studies to date have been made on a small scale in the laboratory. This does not detract from their value in identifying key protagonists and 'worst case scenarios': certainly, a species that is not preyed on by *H. axyridis* in a Petri dish experiment is unlikely to be at risk in the field. However, they do not consider the importance of spatial scale and habitat complexity on niche differentiation and the outcome of contests. Testing hypotheses on these subjects requires experiments to move to larger spatial and temporal scales. The work of Snyder et al. (2004a) demonstrated that, although parasitoid 'mummies' of *A. asychis* were susceptible to intraguild predation by *H. axyridis*, overall parasitism rates were unaffected when experiments were done on a larger spatial scale. It is also important to consider more complex communities, i.e. not just pairs of protagonist species but a more representative guild, if we are to determine whether diversity will decline and what implications this will have for ecosystem services.

Potential implications for ecosystem services

Predators are part of the functional biodiversity in managed ecosystems and, while not the only contributor to herbivore population regulation, they are essential for sustainable pest management through the ecosystem services they provide and that we depend upon. While the abundance of particular species can be enhanced through biological control, the question of 'how many species are required to achieve the required service in the crop?' continues to be a matter of debate (e.g., Denoth et al. 2002; Pell 2008; Straub et al. 2008) and must be considered in the context of the arrival of *H. axyridis* and potential declines in guild diversity. The relationship between predator biodiversity and ecosystem function is poorly understood. From the studies described above, and field observations of declines in native aphidophages and coccidophages since the arrival of *H. axyridis*, we must consider whether *H. axyridis* as a top predator, will deliver the same ecosystem service when other guild species are excluded or whether this will reduce aphid regulation (prey release) and the reliability of biological control in the long term.

Straub et al. (2008) reviewed recent literature on the relative pest suppression function of multiple (more than two) compared to single predator species and found that the relationship between predator diversity and pest suppression was context-dependent and could be positive, negative or neutral. Theoretically, if increasing predator diversity encourages intraguild predation, or, as in the case of invasion by *H. axyridis*, an invasive intraguild predator is introduced to the system, then the prediction would be that the equilibrium density of the herbivore would increase and pest control function would be reduced (e.g., Polis et al. 1989; Finke and Denno 2004). Effectively, trophic cascades would be dampened and the herbivore would be released from predation. However, the experimental evidence to support this in a number of communities is mixed. Some studies have documented disruption of pest suppression and others have shown either no change or improved pest suppression in the presence of intraguild predation (Straub et al. 2008). Most recent meta-analysis by Rosenheim and Harmon (2006) and a literature review by Janssen et al. (2006) find no evidence that the presence of intraguild predation disrupts herbivore control. However, there was greater variation than expected between cases, suggesting that pest release depended on system-specific factors, such as the biological traits of the predators and prey, and extrinsic factors, such as the environment, i.e., there was no simple prediction relating intraguild predation to pest release (Straub et al. 2008).

However, Straub et al. (2008) have identified particular traits of relevance from the theoretical literature. The first relates to the life history of the intraguild prey. Coincidental intraguild predation of parasitoids and pathogens is likely to be less disruptive to herbivore suppression than omnivorous intraguild predation of other predators because predation of the herbivore and the parasitoid/pathogen are directly linked and intraguild predation will simultaneously result in predation of the herbivore (Rosenheim et al. 1995). However, this may not hold if several generations of the intraguild prey are considered (Snyder and Ives 2001). Secondly, the relative mobility and size (foraging mode) of the intraguild predator, intraguild prey and the herbivore can all influence the likelihood of prey release. A 'sit and wait' intraguild predator could disrupt suppression of a sedentary herbivore by a mobile intraguild prey species, although this will depend on the size of the intraguild prey species too. If it is large and highly mobile, it will have a large demand for prey, and if removed from the system by intraguild predation could result in release of the herbivore prey (Rosenheim and Corbett 2003). Furthermore, if the intraguild prey is a similar size to the herbivore then adding a larger intraguild predator will enhance herbivore suppression because predation of the herbivore by the intraguild predator would greatly outweigh the small release from predation achieved by the consumption of the intraguild prey (Diehl 1993). The opposite would be the case if the intraguild prey was much larger than the herbivore (Rosenheim and Corbett 2003). There is some experimental support for this second prediction (Rosenheim et al. 2004a, b) but for the most part they remain hypotheses that require experimental evaluation. In addition, they will be influenced by whether the top intraguild predator is co-evolved with the rest of the guild or whether it is an exotic alien species, as is the case with *H. axyridis*. Using the hypotheses above, one could predict from the biological traits of *H. axyridis* that it would release herbivores from control because it consumes intraguild prey that are large and mobile. However, it is itself large, mobile, well defended, multivoltine and highly voracious which could compensate for this, at least in the short term.

Within guilds of predators some species may be functionally redundant if they share traits and their function would be replaced by other species should they be lost (Wilby and Thomas 2002a, b; Ives et al. 2005; Casula et al. 2006). If there are many redundant species within a guild, then a decline in their diversity associated with the arrival of *H. axyridis* may not influence overall function. However, determining whether species are functionally redundant within a guild relies on the ability to divide them into functional groups with similar traits (e.g., prey preference, response to prey density, microhabitat use and phenology). Those with similar traits should compete strongly for resources and are, therefore, likely to be functionally redundant.

Within the aphidophagous guild there have been studies to demonstrate that increasing the number of species in the guild increases prey suppression, suggesting resource-use complementarity (Powell et al. 2006). However, other workers have shown no such effects and suggest significant redundancy of some guild members (Evans 1991; Chang 1996; Straub and Snyder 2006). The former may be more likely because the latter results could be due to negative and positive interactions between species counteracting each other (Snyder and Ives 2003) or because species that appear functionally redundant under some environmental conditions are functionally diverse when environmental conditions change (Naeem and Li 1997). This effect is encompassed in the 'insurance hypothesis' that maintains that different predators perform better or worse in particular environments and that by maintaining them all provides functional compensation and reliable pest suppression despite changing conditions (Loreau et al. 2003). This is essential in managed ecosystems that are fragmented and under constant change (Pell 2008). The multiplicity of responses to change that species from a single functional group are capable of, is critical to landscape scale ecosystem resilience (Elmquist et al. 2003). There are good examples of this in the aphid/predator system (Pell 2008). Key aphid mortality factors, which can be parasitoids on some occasions and syrphids, ground predators or entomopathogenic fungi on other occasions, demonstrate the need for diversity as insurance for function (Krauss and Poehling 1996; Hemmati 1999; Östman et al. 2001; Powell et al. 2003, 2004; Tscharntke et al. 2005; Pell 2008). Declines in guild diversity as a result of introduction of *H. axyridis* could, therefore, reduce the resilience of pest suppression in the long term.

A single prey type, or even species, can provide multiple feeding niches that a diversity of predators can use if there is niche complementarity rather than redundancy and this is achieved by resource partitioning and facilitation amongst predators. If their is complete complementarity then predator diversity should increase pest suppression (Wilby and Thomas 2002a, b; Casula et al. 2006). There is significant evidence for resource partitioning and facilitation within the aphid/predator system (Pell 2008). For example variation between aphid species, or within a species, provides an opportunity for preference amongst predators. The coccinellid *C. 7-punctata* exploited red morphs of the pea aphid more than green morphs whereas the parasitoid *A. ervi* only attacked green morphs, effectively partitioning the resource between them (Losey et al. 1997). Coccinellid species differ in their response to prey density, some being more effective at low densities and others at high densities, which effectively separates them into complementary niches (Evans 2004; Schellhorn and Andow 2005). A further example of complementarity can be seen for the interactions between *C. 7-punctata*, a foliar aphid predator, and the carabid beetle *Harpalus pennsylvanicus* (DeGeer) (Coleoptera: Carabidae), which is restricted to foraging on the soil surface. As the coccinellid forages, it dislodges aphids that are then consumed by the ground predator, enhancing overall aphid suppression (Losey and Denno 1998). Coccinellid predators and parasitoids also facilitate transmission and dispersal of the beneficial aphid pathogen *P. neoaphidis* (Pell et al. 1997; Fuentes-Contreras et al. 1998;

Roy et al. 1998, 2001; Baverstock et al. 2005). Avoidance behaviour of parasitoids and coccinellids in relation to other intraguild predators is common (Nakashima et al. 2004; Meyling and Pell 2006; Ormond 2007). As already described, coccinellid species can also be phenologically separated, thereby partitioning resources temporally and avoiding competition (Takahashi and Naito 1984; Sato 2001; Flowers et al. 2006) or they can have different patch-leaving times (Sato et al. 2003). As there are numerous examples of complementarity amongst aphidophagous species, it follows that the more species there are in the guild, the greater pest suppression will be. This confirms the hypothesis that declines in guild diversity as a result of introduction of *H. axyridis* could reduce effective pest suppression.

Conclusion

Overall, the evidence suggests that predator diversity is necessary for resilient pest suppression and that includes aphid and coccid control (Stiling and Cornelissen 2005; Cardinale et al. 2006; Pell 2008; Straub et al. 2008). Although intraguild predation does not necessarily interfere with the overall ecosystem service it can interfere with pest suppression in some systems, particularly when the intraguild prey are large compared to the extraguild prey and this is the case for *H. axyridis*. Niche complementarity is almost certainly the mechanism by which *H. axyridis* remains in equilibrium with its co-evolved native guild of predators. In its exotic range, the guild of predators are co-evolved with each other but not with *H. axyridis* and as such come into direct competition with *H. axyridis*. This has significant implications for predator diversity, biodiversity per se and also ecosystem services. Pest management strategies aimed at maintaining predator diversity through habitat manipulation and the diversification of our agricultural landscapes could help counter declines in predator diversity associated with the arrival of *H. axyridis* (e.g., Barbosa 1998; Landis et al. 2000; Gurr et al. 2004; Pell 2008). Although implications for biodiversity in natural habitats are of grave concern, it is perhaps through the manipulation of managed land, which, for example, represents 70% of land in the UK, that we have the greatest chance of optimising the environment to favour our native diversity in the presence of *H. axyridis*, thereby also protecting diversity in natural habitats. To achieve this we need robust research data to fill the gaps in knowledge identified in this paper and develop mitigation strategies.

Acknowledgements Judith K. Pell and Jason Baverstock are supported by the Department for Environment, Food and Rural Affairs of the UK (Defra). Rothamsted Research receives grant-aided support from the Biotechnology and Biological Sciences Research Council of the United Kingdom (BBSRC).

References

Agarwala BK, Dixon AFG (1992) Laboratory study of cannibalism and interspecific predation in ladybirds. Ecol Entomol 17:303–309

Alam N, Choi IS, Song KS, Hong J, Lee CO, Jung JH (2002) A new alkaloid from two coccinellid beetles *Harmonia axyridis* and *Aiolocaria haexapilota*. B Kor Chem Soc 23:497–499

Barbosa P (ed) (1998) Conservation biological control. Academic Press, San Diego, CA, 396 pp

Baverstock J, Alderson PG, Pell JK (2005) Influence of the aphid pathogen *Pandora neoaphidis* on the foraging behaviour of the aphid parasitoid *Aphidius ervi*. Ecol Entomol 30:665–672

Berkvens N, Bonte J, Berkvens D, Deforce K, Tirry L, De Clercq P (2007) Pollen as an alternative food for *Harmonia axyridis*. BioControl (this issue). doi:10.1007/s10526-007-9128-7

Brown MW, Miller SS (1998) Coccinellidae (Coleoptera) in apple orchards of eastern West Virginia and the impact of invasion by *Harmonia axyridis*. Entomol News 109:143–151

Brown PMJ, Roy HE, Rothery P, Roy DB, Ware RL, Majerus MEN (2007) *Harmonia axyridis* in Great Britain: analysis of the spread and distribution of a non-native coccinellid. BioControl (this issue). doi:10.1007/s10526-007-9124-y

Buntin GD, Bouton JH (1997) Aphid (Homoptera: Aphididae) management in alfalfa by spring grazing cattle. J Entomol Sci 32:332–342

Butt TM, Ibrahim L, Ball BV, Clark SJ (1994) Pathogenicity of the entomogenous fungi, *Metarhizium anisopliae* and *Beauveria bassiana* against crucifer pests and the honey bee. Biocontrol Sci Techn 4:207–214

Cardinale BJ, Srivastava DS, Duffy JE, Wright JP, Downing AL, Sankaran M, Jouseau C (2006) Effects of biodiversity on the functioning of trophic groups and ecosystems. Nature 443:989–992

Casula P, Wilby A, Thomas MB (2006) Understanding biodiversity effects on prey in multi-enemy systems. Ecol Lett 9:995–1004

Chang GC (1996) Comparison of single versus multiple species of generalist predators for biological control. Environ Entomol 25:207–212

Colunga-Garcia M, Gage SH (1998) Arrival, establishment, and habitat use of the multicolored Asian lady beetle (Coleoptera: Coccinellidae) in a Michigan landscape. Environ Entomol 27:1574–1580

Cottrell TE (2004) Suitability of exotic and native lady beetle eggs (Coleoptera: Coccinellidae) for development of lady beetle larvae. Biol Control 31:362–371

Cottrell TE (2007) Predation by adult and larval lady beetles (Coleoptera: Coccinellidae) on initial contact with lady beetles eggs. Environ Entomol 36:390–401

Cottrell TE, Yeargan KV (1998) Intraguild predation between an introduced lady beetle, *Harmonia axyridis* (Coleoptera: Coccinellidae) and a native lady beetle, *Coleomegilla maculata* (Coleoptera: Coccinellidae). J Kansas Entomol Soc 71:159–163

De Clercq P, Peeters I, Vergauwe G, Thas O (2003) Interaction between *Podisus maculiventris* and *Harmonia axyridis*, two predators used in augmentative biological control in greenhouse crops. BioControl 48:39–55

Denoth M, Frid L, Myers JH (2002) Multiple agents in biological control: improving the odds. Biol Control 24:20–30

Diehl S (1993) Relative consumer sizes and the strength of direct and indirect interactions in omnivorous feeding relationships. Oikos 68:151–157

Dixon AFG (2000) Insect predator-prey dynamics: ladybird beetles and biological control. Cambridge University Press, Cambridge, UK

Elliott N, Kieckhefer R, Kauffman W (1996) Effects of an invading coccinellid on native cocccinellids in an agricultural landscape. Oecologia 105:537–544

Elmquist T, Folke C, Nyström M, Peterson G, Bengtsson J, Walker B, Norberg J (2003) Response diversity, ecosystem change and resilience. Front Ecol Environ 1:488–494

Evans EW (1991) Intra versus interspecific interactions of ladybeetles (Coleoptera; Coccinellidae) attacking aphids. Oecologia 87:401–408

Evans EW (2000) Morphology of invasion: body size patterns associated with establishment of *Coccinella septempunctata* (Coleoptera: Coccinellidae) in western North America. Eur J Entomol 97:469–474

Evans EW (2004) Habitat displacement of North American ladybirds by an introduced species. Ecology 85:637–647

Felix S, Soares AO (2004) Intraguild predation between the aphidophagous ladybird beetles *Harmonia axyridis* and *Coccinella undecimpunctata* (Coleoptera: Coccinellidae): the role of body weight. Eur J Entomol 101:237–242

Ferguson KI, Stiling P (1996) Non-additive effects of multiple natural enemies on aphid populations. Oecologia 108:375–379

Finke DL, Denno RF (2004) Predator diversity dampens trophic cascades. Nature 429:407–410

Flowers RW, Salom SM, Kok LT (2006) Competitive interactions among two specialist predators and a generalist predator of hemlock woolly adelgid, *Adelges tsugae* (Hemiptera: Adelgidae) in south-western Virginia. Agr Forest Entomol 8:253–262

Foster SP, Devine G, Devonshire AL (2007) Insecticide resistance. In: Harrington R, van Emden H (eds) Aphids as crop pests. CABI International, pp 261–285, 717

Fremlin M (2007) Intra-guild predation of harlequin ladybird larvae by lacewing larvae. Bull Amat Entomol Soc 66:110–116

Fuentes-Contreras E, Pell JK, Niemeyer HM (1998) Influence of plant resistance at the third trophic level: interactions between parasitoids and entomopathogenic fungi of cereal aphids. Oecologia 117:426–432

Gardiner MM, Landis DA (2007) Impact of intraguild predation by adult *Harmonia axyridis* (Coleoptera: Coccinellidae) on *Aphis glycines* (Hemiptera: Aphididae) biological control in cage studies. Biol Control 40:386–395

Godeau J-F (2000) Coccinelles amies des fourmis. 2/2/ Groupe de Travail Coccinula. Feuille de Contact 2:10–15

Gordon RD (1985) The Coccinellidae (Coleoptera) of America North of Mexico. J New York Entomol Soc 93:1–912

Gurr GM, Scarratt SL, Wratten SD, Berndt L, Irvin N (2004) Ecological engineering, habitat manipulation and pest management. In: Gurr GM, Wratten SD, Altieri MA (eds) Ecological engineering for pest management: advances in habitat manipulation for arthropods. CSIRO Publishing, pp 1–12

Hemmati F (1999) Aerial dispersal of the entomopathogenic fungus *Erynia neoaphidis*. PhD Thesis, University of Reading 186 pp

Hemptinne JL, Dixon AFG (2000) Defence, oviposition and sex: semiochemical parsimony in two species of ladybird beetles (Coleoptera: Coccinellidae)? A short review. Eur J Entomol 97:443–447

Herrick NJ, Reitz SR (2004) Temporal occurrence of *Podisus maculiventris* (Hemiptera: Heteroptera: Pentatomidae) in North Florida. Fla Entomol 87:587–590

Hironori Y, Katsuhiro S (1997) Cannibalism and interspecific predation in two predatory ladybirds in relation to prey abundance in the field. Entomophaga 42:153–163

Hodek I (1996) Food relationships. In: Hodek I, Honek A (eds) Ecology of Coccinellidae. Kluwer Academic Publishers, Dordrecht pp 143–238

Hodek I, Honek A (1988) Sampling, rearing and handling of aphid predators. In: Minks AK, Harrewijn P (eds) Aphids, their biology, natural enemies and control, Vol. 2B. Amsterdam, Elsevier, pp 311–321

Hölldobler B, Wilson EO (1990) The ants. Springer Verlag, Berlin

Holt RD, Polis GA (1997) A theoretical framework for intraguild predation. Am Nat 149:745–764

Hough-Goldstein J, Cox J, Armstrong A (1996) *Podisus maculiventris* (Hempitera: Pentatomidae) predation on ladybird beetles (Coleoptera: Coccinellidae). Fla Entomol 79:64–68

Huey RB, Pianka ER (1981) Ecological consequences of foraging mode. Ecology 62:991–999

Hufbauer RA, Roderick GK (2005) Microevolution in biological control: mechanisms, patterns and processes. Biol Control 35:227–239

Inglis GD, Goettel MS, Butt TM, Strasser H (2001) Use of hyphomycetous fungi for managing insect pests. In: Butt TM, Jackson C, Magan N (eds) Fungi as biocontrol agents: progress, problems and potential. CABI, Wallingford, pp 23–69

Itioka T, Inoue T (1996) The role of predators and attendant ants in the regulation of a population of the citrus mealybug *Pseudococcus citriculus* in a satsuma orange orchard. Appl Entomol Zool 31:195–202

Ives AR, Cardinale BJ, Snyder WE (2005) A synthesis of subdisciplines: predator-prey interactions and biodiversity and ecosystem functioning. Ecol Lett 8:102–116

James RR, Lighthart B (1994) Susceptibility of the convergent lady beetle (Coleoptera, Coccinellidae) to four entomogenous fungi. Env Entomol 23:190–192

Janssen A, Montserrat M, HilleRisLambers R, de Roos AM, Pallini A, Sabelis MW (2006) Intraguild predation usually does not disrupt biological control. In: Brodeur J, Boivin G (eds) Trophic and guild interactions in biological control. Springer, Dordrecht, The Netherlands, pp 21–44

Jiggins C, Majerus MEN, Gough U (1993) Ant defence of colonies of *Aphis fabae* Scopoli (Hemiptera: Aphididae), against predation by ladybirds. Brit J Entomol Nat Hist 6:129–138

Kajita Y, Takano F, Yasuda H, Agarwala BK (2000) Effects of indigenous ladybird species (Coleoptera: Coccinellidae) on the survival of an species in relation to prey abundance. Appl Entomol Zool 35:473–479

Kajita Y, Yasuda H, Evans EW (2006) Effects of native ladybirds on oviposition of the exotic species, *Adalia bipunctata* (Coleoptera: Coccinellidae), in Japan. Appl Entomol Zool 41:57–61

Keller S, Zimmerman G (1989) Mycopathogens of soil insects. In: Wilding N, Collins NM, Hammond PM, Webber JF (eds) Insect-fungus interactions. Academic Press, London

King A, Meinwald J (1996) Review of the defensive chemistry of coccinellids. Chem Rev 96:1105–1122

Koch RL (2003) The multicoloured Asian lady beetle, *Harmonia axyridis*: a review of its biology, uses in biological control and non-target impacts. J Ins Sci 3:32–47

Koch R, Venette R, Hutchison W (2006) Predicted impact of an exotic generalist predator on monarch butterfly (Lepidoptera: Nymphalidae) populations: a quantitative risk assessment. Biol Invasions 8:1179–1193

Koyama S, Majerus MEN (2007) Interactions between the parasitoid wasp *Dinocampus coccinellae* and two species of coccinellid from Japan and Britain. BioControl (this issue). doi:10.1007/s10526-007-9138-5

Krauss U, Poehling HM (1996) Overwintering, oviposition and population dynamics of hoverflies (Diptera: Syrphidae) in northern Germany in relation to small and large-scale landscape structure. Acta Jutlandica 71:157–169

Kuznetsov VN (1988) The Coccinellidae of the far East in Transcaucasia. Zashchita Rastenii. 5(19) (in Russian)

Kuznetsov VN (1997) Lady beetles of the russian far east, (Memoir Series No. 1). Center for Systematic Entomology, Gainesville, FL

LaMana ML, Miller JC (1996) Field observation on *Harmonia axyridis* (Pallas) (Coleoptera: Coccinellidae) in Oregon. Biol Control 6:232–237

Landis DA, Wratten SD, Gurr GM (2000) Habitat manipulation to conserve natural enemies of arthropod pests in agriculture. Annu Rev Entomol 45:175–201

Loreau M, Mouquet N, Gonzales A (2003) Biodiversity as spatial insurance in heterogeneous landscapes. PNAS USA 100:12765–12770

Losey JE, Denno RF (1998) Positive predator-prey interactions: enhanced predation rates and synergistic suppression of aphid populations. Ecology 79:2143–2152

Losey JE, Vaughan M (2006) The economic value of ecological services provided by insects. Bioscience 56:311–323

Losey JE, Ives AR, Harmon J, Ballantyne F, Brown C (1997) A polymorphism maintained by opposite patterns of parasitism and predation. Nature 388:269–272

Lucas E (2005) Intraguild predation among aphidophagous predators. Eur J Entomol 102:351–364

Lucas E, Coderre D, Brodeur J (1998) Intraguild predation among aphid predators: characterization and influence of extraguild prey density. Ecology 79:1084–1092

Lucas E, Gagne I, Coderre D (2002) Impact of the arrival of adults of *Harmonia axyridis* on adults of *Coccinella septempunctata* and *Coleomegilla maculata* (Coleoptera: Coccinellidae). Eur J Entomol 99:457–463

Magalhaes BP, Lord JC, Wraight SP, Daoust RA, Roberts DW (1988) Pathogenicity of *Beauveria bassiana* and *Zoophthora radicans* to the coccinellid predators *Coleomegilla maculata* and *Eriopis connexa.* J Invertebr Pathol 52:471–473

Magro A, Tene JN, Bastin N, Dixon AFG, Hemptinne JL (2007) Assessment of patch quality by ladybirds: relative response to conspecific and heterospecific larval tracks a consequence of habitat similarity? Chemoecology 17:37–45

Majerus MEN (1994) Ladybirds. HarperCollins, London, 320 pp

Majerus MEN, Strawson V, Roy HE (2006) The potential impacts of the arrival of the harlequin ladybird, *Harmonia axyridis* (Pallas) (Coleoptera: Coccinellidae), in Britain. Ecol Entomol 31:207–215

Majerus MEN, Sloggett JJ, Godeau J-F, Hemptinne J-L (2007) Interactions between ants and aphidophagous and coccidophagous ladybirds. Popul Ecol 49:15–27

McPherson JE (1980) A list of the prey species of *Podisus maculiventris* (Hemiptera: Pentatomidae). Great Lakes Entomol 13:17–24

Meyling NV, Pell JK (2006) Detection and avoidance of an entomopathogenic fungus by a generalist insect predator. Ecol Entomol 31:162–171

Michaud JP (2002) Invasion of the Florida citrus ecosystem by *Harmonia axyridis* (Coleoptera: Coccinellidae) and asymmetric competition with a native species, *Cycloneda sanguinea.* Env Entomol 31:827–835

Müller CB, Brodeur J (2002) Intraguild predation in biological control and conservation biology. Biol Control 25:216–223

Musser FR, Shelton AM (2003) Bt sweet corn and selective insecticides: impacts on pests and predators. J Econ Entomol 96:71–80

Naeem S, Li S (1997) Biodiversity enhances ecosystem reliability. Nature 390:507–509

Nakashima Y, Birkett MA, Pye BJ, Pickett JA, Powell W (2004) The role of semiochemicals in the avoidance of the seven-spot ladybird *Coccinella septempunctata* by the aphid parasitoid, *Aphidius ervi.* J Chem Ecol 30:1103–1116

Nault BA, Kennedy GG (2003) Establishment of multicoloured Asian lady beetle in Eastern North Carolina: seasonal abundance and crop exploitation within an agricultural landscape. BioControl 48:363–378

Ormond E (2007) The overwintering interactions of the seven spot ladybird (*Coccinella septempunctata*) and the entomopathogenic fungus *Beauveria bassiana*. PhD Thesis, Anglia Ruskin University

Ormond EL, Pell JK, Thomas APM, Roy HE (2006) Overwintering ecology of *Coccinella septempunctata*, *Beauveria bassiana* and *Dinocampus coccinellae*. In: Rossing WAH, Eggenschwiler L, Poehling H-M (eds) Working Group "Landscape management for functional biodiversity" at Zürich-Reckenholz (Switzerland), 16–19 May 2006 IOBC/WPRS Bulletin 29, pp 85–88

Östman O, Ekbom B, Bengtsson J (2001) Farming practice and landscape heterogeneity influence biological control. Basic Appl Ecol 2:365–371

Pasteels JM, Deroe C, Tursch B, Braekman JC, Daloze D, Hootele C (1973) Distribution et activités des alcaloïdes défensives des Coccinellidae. J Insect Physiol 19:1771–1784

Pell JK (2008) Ecological approaches to pest management using entomopathogenic fungi; concepts, theory, practice and opportunities. In: Ekesi S, Maniania N (ed) Research Signpost (In press)

Pell JK, Vandenberg JD (2002) Interactions among *Diuraphis noxia*, the fungal pathogen *Paecilomyces fumosoroseus* and the coccinellid *Hippodamia convergens*. Biocontrol Sci Tech 12:217–224

Pell JK, Pluke R, Clark SJ, Kenward MG, Alderson PG (1997) Interactions between two aphid natural enemies, the entomopathogenic fungus, *Erynia neoaphidis* and the predatory beetle, *Coccinella septempunctata*. J Invertebr Pathol 69:261–268

Pell JK, Eilenberg J, Hajek AE, Steinkraus DS (2001) Biology, ecology and pest management potential of entomophthorales. In: Butt TM, Jackson C, Magan N (eds) Fungi as biocontrol agents: progress, problems and potential. CABI International, pp 71–154, 390

Pervez A, Omkar (2006) Ecology and biological control application of multicoloured Asian ladybird, *Harmonia axyridis*. A review. Biocontrol Sci Tech 16:111–128

Phoofolo MW, Obrycki JJ (1998) Potential for intraguild predation and competition among predatory Coccinellidae and Chrysopidae. Entomol Exp Appl 89:47–55

Polis GA (1994) Food web, trophic cascade and community structure. Aust J Ecol 19:121–136

Polis GA, Holt RD (1992) Intraguild predation—the dynamics of complex trophic interactions. TREE 7:151–154

Polis GA, Myers CA, Holt RD (1989) The ecology and evolution of intra-guild predation: potential competitors that eat each other. Annu Rev Entomol 20:297–330

Powell W, Pell JK (2007) Biological control. In: Harrington R, van Emden H (eds) Aphids as crop pests. CABI International, pp 469–513, 717

Powell W, Walters K, A'Hara S, Ashby J, Stevenson H, Northing P (2003) Using field margin diversification in agri-environment schemes to enhance aphid natural enemies. In: Rossing WAH, Poehling HM, Burgio G (eds) Landscape management for functional biodiversity. IOBC/WPRS Bulletin, 26/4, pp 123–128

Powell W, A'Hara S, Harling R, Holland JM, Northing P, Thomas CFG, Walters KFA (2004) Managing biodiversity in field margins to enhance integrated pest control in arable crops ('3D Farming' Project). Project report No 356. HGCA, 226 pp

Powell W, Pell JK, Nakashima Y, Baverstock J, Birkett M (2006) Functional biodiversity of aphid natural enemies: positive intra-guild interactions. In: Rossing WAH, Poehling HM, Burgio G (eds) Landscape management for functional biodiversity. IOBC/ WPRS Bulletin, 29(6), pp 101–104

Roderick GK, Howarth FG (1999) Invasion genetics: natural colonisations, non indigenous species and classical biological control. In: Yano E, Matsuo K, Shiyomi M, Andow D (eds) Biological invasions of pests and beneficial organisms. National Institute of Agro Environmental Sciences, Tsukuba, Japan

Rosenheim JA (1998) Higher-order predators and the regulation of insect herbivore populations. Annu Rev Entomol 43:421–447

Rosenheim JA, Corbett SA (2003) Omnivory and the indeterminancy of predator function: can a knowledge of foraging behaviour help? Ecology 84:2438–2448

Rosenheim JA, Harmon JP (2006) The influence of intraguild predation on the suppression of a shared prey population: an empirical assessment. In: Brodeur J, Boivin G (eds) Trophic and guild interactions in biological control. Springer, Dordrecht, The Netherlands, pp 1–20

Rosenheim JA, Kaya HK, Ehler LE, Marois JJ, Jaffee BA (1995) Intraguild predation among biological control agents: theory and evidence. Biol Control 5:303–335

Rosenheim JA, Glik TE, Goeriz RE, Ramert B (2004a) Linking a predator's foraging behaviour with its effects on herbivore population suppression. Ecology 85:3362–3372

Rosenheim JA, Limburg DD, Colfer RG, Fournier V, Hsu CL, Leonardo TE, Nelson EH (2004b) Herbivore population suppression by an intermediate predator, *Phyoseiulus macropilis*, is insensitive to the presence of an intraguild predator: an advantage of small body size? Oecologia 140:577–585

Roy HE, Pell JK (2000) Interactions between entomopathogenic fungi and other natural enemies: implications for biological control. Biocontrol Sci Tech 10:737–752

Roy HE, Pell JK, Clark SJ, Alderson PG (1998) Implications of predator foraging on aphid pathogen dynamics. J Invertebr Pathol 71:236–247

Roy HE, Pell JK, Alderson PG (2001) Targeted dispersal of the aphid pathogenic fungus *Erynia neoaphidis* by the aphid predator *Coccinella septempunctata*. Biocontrol Sci Tech 11:99–110

Roy HE, Brown P, Majerus MEN (2006) *Harmonia axyridis*: a successful biocontrol agent or an invasive threat? In: Eilenberg J, Hokkanen H (eds) An ecological and societal approach to biological control. Kluwer Academic Publishers, Netherlands

Roy HE, Brown PMJ, Rothery P, Ware RL, Majerus MEN (2007) Interactions between the fungal pathogen *Beauveria bassiana* and three species of coccinellid: *Harmonia axyridis*, *Coccinella septempunctata* and *Adalia bipunctata*. BioControl (this issue). doi:10.1007/s10526-007-9122-0

Roy HE, Baverstock J, Ware RL, Clark SJ, Majerus MEN, Baverstock KE, Pell JK (2008). Intraguild predation of the aphid pathogenic fungus *Pandora neoaphidis* by the invasive coccinellid *Harmonia axyridis*. Ecol Entomol (In press)

Sakuratani Y, Marsumoto Y, Oka M, Kubo T, Fuji A, Uotani M, Teraguchi T (2000) Life history of *Adalia bipunctata* (Coleoptera: Coccinellidae) in Japan. Eur J Entomol 97:555–558

Sato S (2001) Ecology of ladybirds: factors influencing their survival. PhD Thesis, University of East Anglia, UK

Sato S, Dixon AFG (2004) Effect of intraguild predation on the survival and development of three species of aphidophagous ladybirds: consequences for invasive species. Agr Forest Entomol 6:21–24

Sato S, Dixon AFG, Hironori Y (2003) Effect of emigration on cannibalism and intraguild predation in aphidophagous ladybirds. Ecol Entomol 28:628–633

Sato S, Yasuda H, Evans EW (2005) Dropping behaviour of larvae of aphidophagous ladybirds and its effect on incidence of intraguild predation: interactions between the intraguild prey, *Adalia bipunctata* (L.) and *Coccinella septempunctata* (L.), and the intraguild predator, *Harmonia axyridis* Pallas. Ecol Entomol 30:220–224

Schellhorn NA, Andow D (2005) Response of coccinellids to their aphid prey at different spatial scales. Popul Ecol 47:281–288

Sengonca C, Frings B (1985) Interference and competitive behaviour of the aphid predators, *Chrysoperla carnea* and *Coccinella septempunctata* in the laboratory. Entomophaga 30:245–251

Sloggett JJ (1998) Interactions between Coccinellids (Coleoptera) and ants (Hymenoptera: Formicidae), and the evolution of Myrmecophily in *Coccinella magnifica* Redtenbacher. PhD thesis, University of Cambridge

Sloggett JJ, Majerus MEN (2000) Aphid-mediated coexistence of ladybirds (Coleoptera: Coccinellidae) and the wood ant *Formica rufa*: seasonal effects, interspecific variability and the evolution of a coccinellid myrmecophile. Oikos 89:345–359

Sloggett JJ, Majerus MEN (2003) Adaptations of *Coccinella magnifica*, a myrmecophilous coccinellid to aggression by wood ants (*Formica rufa* group). II. Larval behaviour, and ladybird oviposition location. Eur J Entomol 100:337–344

Sloggett JJ, Völkl W, Schulze W, von der Schulenberg JH, Majerus MEN (2002) The ant-associations and diet of the ladybird *Coccinella magnifica* (Coleoptera: Coccinellidae). Eur J Entomol 99:565–569

Snyder WE, Evans EW (2006) Ecological effects of invasive arthropod generalist predators. Annu Rev Ecol Evol S 37:95–122

Snyder WE, Ives AR (2001) Generalist predators disrupt biological control by a specialist parasitoid. Ecology 82:705–716

Snyder WE, Ives AR (2003) Interactions between specialist and generalist natural enemies: parasitoids, predators and pea aphid biocontrol. Ecology 84:91–107

Snyder WE, Ballard SN, Yang S, Clevenger GM, Miller TD, Ahn JJ, Hatten TD, Berryman AA (2004a) Complementary biocontrol of aphids by the ladybird beetle *Harmonia axyridis* and the parasitoid *Aphelinus asychis* on greenhouse roses. Biol Control 30:229–235

Snyder WE, Clevenger GM, Eigenbrode SD (2004b) Intraguild predation and successful invasion by introduced ladybird beetles. Oecologia 140:559–565

Stiling P, Cornelissen T (2005) What makes a successful biological control agent? A meta-analysis of biological control agent performance. Biol Control 34:236–246

Straub CS, Snyder WE (2006) Experimental approaches to understanding the relationship between predator diversity and conservation biological control. In: Brodeur J, Boivin G (eds) Trophic and guild interactions in biological control. Springer, Dordrecht, The Netherlands, pp 221–239

Straub CS, Finke DL, Snyder WE (2008) Are the conservation of natural enemy biodiversity and biological control compatible goals? Biol Control (In press)

Takahashi K, Naito A (1984) Seasonal occurrence of aphids and their predators (Col. Coccinellidae) in alfalfa fields. Bulletin of the National Grassland Research Institute 29:62–66 [In Japanese with English summary]

Takizawa T, Yasuda H, Agarwala BK (2000) Effect of three species of predatory ladybirds on oviposition of aphid parasitoids. Entomol Sci 3:465–469

Taylor AJ, Müller CB, Godfray HCJ (1998) Effect of aphid predators on oviposition behaviour of aphid parasitoids. J Insect Behav 11:297–302

Tedders WL, Schaefer PW (1994) Release and establishment of *Harmonia axyridis* (Coleoptera: Coccinellidae) in the south-eastern United States. Entomol News 105:228–243

Tscharntke T, Klein AM, Kruess A, Steffan-Dewenter I, Thies C (2005) Landscape perspectives on agricultural intensification and biodiversity—ecosystem service management. Ecol Lett 8:857–874

Völkl W (1995) Behavioural and morphological adaptations of the coccinellid, *Platynaspis luteorubra* for exploiting ant-attending resources (Coleoptera: Coccinellidae). J Insect Behav 8:653–670
Völkl W, Mackauer M, Pell JK, Brodeur J (2007) Predators, parasitoids and fungal pathogens. In: Harrington R, van Emden H (eds) Aphids as crop pests. CABI International, pp 187–233
Ware RL, Majerus MEN (2007) Intraguild predation of immature stages of British and Japanese coccinellids by the invasive ladybird *Harmonia axyridis*. BioControl (this issue). doi:10.1007/s10526-007-9135-8
Ware RL, Majerus MEN, Roy HE, Symington F (2005) The harlequin ladybird arrives in Britain: a threat to our native species? Bull Amat Ent Soc 64:175–186
Ware RL, Ramon-Portugal F, Magro A, Ducamp C, Hemptinne JL, Majerus, MEN (2007) Chemical protection of *Calvia quatuordecimguttata* eggs against intraguild predation by the invasive ladybird *Harmonia axyridis*. BioControl (this issue). doi:10.1007/s10526-007-9129-6
Ware RL, Evans N, Malpas L, Michie LJ, O'Farrell K, Majerus MEN (2008) Intraguild predation by the invasive ladybird *Harmonia axyridis*: 1. British and Japanese coccinellid eggs. Neobiota (In press)
Way MJ (1963) Mutualism between ants and honey-dew producing Homoptera. Annu Rev Entomol 8: 307–344
Wells ML, McPherson RM (1999) Population dynamics of three coccinellids in flue-cured tobacco and functional response of *Hippodamia convergens* (Coleoptera: Coccinellidae) feeding on tobacco aphids (Homoptera: Aphididae). Environ Entomol 28:768–773
Wells ML, McPherson RM, Ruberson JR, Herzog GA (2001) Coccinellids in cotton: population response to pesticide application and feeding response to cotton aphids (Homoptera: Aphididae). Environ Entomol 30:785–793
Wheeler AG, Hayes JT, Stephens JL (1968) Insect predators of mummified aphids. Can Entomol 100: 221–222
Wilby A, Thomas MB (2002a) Natural enemy diversity and pest control: patterns of pest emergence with agricultural intensification. Ecol Lett 5:353–360
Wilby A, Thomas MB (2002b) Are the ecological concepts of assembly and function of biodiversity useful frameworks for understanding natural pest control? Agr Forest Entomol 4:237–243
Yasuda H, Ohnuma N (1999) Effect of cannibalism and predation on the larval performance of two ladybird beetles. Ent Exp Appl 93:63–67
Yasuda H, Shinya Y (1997) Cannibalism and interspecific predation in two predatory ladybirds in relation to prey abundance in the field. Entomophaga 42:153–163
Yasuda H, Kikuchi T, Kindlmann P, Sato S (2001) Relationships between attacks and escape rates, cannibalism, and intraguild predation in larvae of two predatory ladybirds. J Ins Behav 14:373–384
Yasuda H, Evans EW, Kajita Y, Urakawa K, Takizawa T (2004) Asymmetric larval interactions between introduced and indigenous ladybirds in North America. Oecologia 141:722–731
Yeo H (2000) Mycoinsecticides for aphid management: a biorational approach. PhD Thesis, University of Nottingham 308 pp

Intraguild predation of immature stages of British and Japanese coccinellids by the invasive ladybird *Harmonia axyridis*

Remy L. Ware · Michael E. N. Majerus

Abstract Declines in native aphidophages in North America have been linked to intraguild predation (IGP) by the invasive coccinellid *Harmonia axyridis* (Pallas) (Coleoptera: Coccinellidae). It is feared that many British species will face a similar fate following the recent establishment of *H. axyridis* in the UK. Meanwhile, *H. axyridis* exists in apparent ecological equilibrium with other members of its guild in Japan. The impact of *H. axyridis* on British coccinellids is uncertain but intraguild predatory interactions do occur, particularly amongst immature stages. This study investigates IGP between immature stages of *H. axyridis* and various British and Japanese coccinellids. The only asymmetric IG predator of *H. axyridis* at first instar was *Anatis ocellata* (Linnaeus). *Harmonia axyridis* engaged in symmetric IGP with *Coccinella septempunctata* Linnaeus, *Calvia quatuordecimguttata* (Linnaeus), *Harmonia quadripunctata* (Pontoppidan) and *Eocaria muiri* Timberlake, but was the asymmetric IG predator of all other species studied. The level of IGP was high between fourth instar larvae, and frequently biased towards *H. axyridis*, except in the case of *A. ocellata*, which again was the only IG predator of *H. axyridis*. In interactions between fourth instar larvae and pre-pupae, IGP was unidirectional towards *H. axyridis* for all species except *A. ocellata*, which acted as both IG predator and IG prey. Pupae were better protected against IGP than pre-pupae but most species were still susceptible to attack by *H. axyridis*, although IGP was symmetric with *A. ocellata*, and *H. quadripunctata* pupae were never attacked. The differences in susceptibility of the various species and developmental stages to IGP by *H. axyridis* are discussed in relation to physical defence structures. We find no evidence that Japanese species have superior defences to British ones and suggest that behavioural strategies may enable co-existence in the native range. We discuss the relevance of IGP by *H. axyridis* to the species it is likely to encounter in Britain.

R. L. Ware (✉) · M. E. N. Majerus
Department of Genetics, University of Cambridge, Downing Street, Cambridge CB2 3EH, UK
e-mail: r.ware@gen.cam.ac.uk

M. E. N. Majerus
e-mail: m.majerus@gen.cam.ac.uk

Biological Control to Invasion: Ladybird Harmonia axyridis. H.E. Roy, E. Wajnberg (eds.),
doi: 10.1007/978-1-4020-6939-0_11, © IOBC 2007

Keywords Behavioural defence · Coccinellidae · Co-evolution · Coleoptera · Harlequin ladybird · *Harmonia axyridis* · Intraguild predation · Invasive alien species · Physical defence

Introduction

The establishment and spread of the invasive coccinellid *Harmonia axyridis* (Pallas) (Coleoptera: Coccinellidae) in Britain (Brown et al. 2007) has provoked concerns for native ecosystems centred on aphids (Majerus et al. 2006). An increasing body of evidence from North America suggests that *H. axyridis* has the potential to negatively impinge on a wide range of species through both resource competition and interspecific predation, including intraguild predation (Koch and Galvan 2007). Such interactions are accepted to be a common cause of declines in native species following the arrival of an invasive alien (Williamson 1996). The observation that *H. axyridis* coexists with many other aphidophages in its natural Asian range (Kuznetsov 1997; R. Ware and M. Majerus, pers. obs.), might suggest that species in its introduced range are at a disadvantage through their lack of co-evolved defensive strategies.

Intraguild predation (IGP) is well reported from aphidophagous guilds (Rosenheim et al. 1995; Lucas 2005). It occurs when two species that share a host or prey, and may therefore compete, also interact trophically (Polis et al. 1989). There is controversy over whether the killing of a potential competitor constitutes IGP if the victim is not consumed. Some authors prefer to classify such instances as 'interspecific killing', while others accept that, with the exception of nutritive gain for the predator, the same benefits and costs are involved here as in IGP sensu stricto and therefore both situations should be included when considering IGP (Lucas 2005). In the context of this paper we are primarily interested in the vulnerabilities of different species as intraguild prey to *H. axyridis*: i.e. whether an individual is killed or not. Thus we adopt Lucas' (2005) revised definition when referring to IGP.

The size and mobility of the IG predator and the IG prey are known to have considerable influence on the outcome of their interaction (Şengonca and Frings 1985; Phoofolo and Obrycki 1998; Lucas et al. 1998; Rosenheim et al. 1995; Felix and Soares 2004). Thus, in coccinellids, eggs, younger larvae, ecdysing larvae, pre-pupae and pupae constitute particularly vulnerable stages (Agarwala and Dixon 1992; Majerus 1994; Dixon 2000; Sato and Dixon 2004). Most coccinellids possess some form of defence against predation, which may be chemical, physical or behavioural, and many show such adaptations throughout their life cycle.

Chemical defence has been particularly well studied at the egg stage, where both internal toxins (Pasteels et al. 1973; Agarwala and Dixon 1992) and external feeding deterrents (Hemptinne et al. 2000; Omkar et al. 2004) have been reported. Indeed, we know that the eggs of at least one Japanese species, *Eocaria muiri* Timberlake, and one British species, *Calvia quatuordecimguttata* (Linnaeus), are particularly well defended from IGP by *H. axyridis* (Ware et al. in press; Ware et al. 2007). Coccinellid larvae, pre-pupae and pupae are also equipped with a chemical armoury and contain similar defensive alkaloids to those present within eggs (Pasteels et al. 1973). In addition, adults, larvae and pre-pupae will 'reflex bleed' when disturbed (Majerus 1994). The exudate contains alkaloids and pyrazines and is known to be an effective deterrent to many predators (Majerus and Kearns 1989; de Jong et al. 1991; Holloway et al. 1991).

When IGP between larvae does occur, it is usually the larger that attacks the smaller (Dixon 2000; Felix and Soares 2004). Thus, being larger than average at each developmental stage may itself be an adaptation for success in IGP. However, there are a variety of other physical defences employed by coccinellid immature stages. Larvae of many species possess hairs or spines which render them unpalatable or difficult to 'handle' or subdue (R. Ware, pers. obs.), while others, such as *Scymnus* spp., are covered in a waxy coating, which provides an effective defence against ants and syrphid larvae (Pope 1979; Völkl and Vohland 1996; Agarwala and Yasuda 2001). It is notable, however, that ecdysing or newly ecdysed larvae of all species are particularly vulnerable to attack, as it takes time for their integument to harden or waxes to develop fully (Lucas et al. 2000; Agarwala and Yasuda 2001). Before the larval skin splits in preparation for pupation, the physical defences of a coccinellid pre-pupa are largely the same as those of a final instar larva. However, some species are reported to use 'gin-traps' at the pupal stage as an effective defence against enemies such as mites (Hinton 1955) and ants (Eisner and Eisner 1992).

Most coccinellids also have some form of behavioural defence against predation. For mobile stages, such as larvae, a first response to a perceived threat would simply be to run away (R. Ware, pers. obs.). Further defensive tactics of aphidophagous larvae include dropping from a host plant if a potential predator approaches (Sato et al. 2005) or migrating to a new site before aphid populations decline (Sato et al. 2003). Predation of vulnerable sessile stages, such as eggs, ecdysing larvae and pupae, may be avoided by species-specific oviposition sites (Schellhorn and Andow 1999) and selection of moulting and pupation sites (Lucas et al. 2000) respectively. Although coccinellid pre-pupae and pupae appear to be extrinsically quiescent, they are not completely immobile and are able to flick up their anterior end rapidly when disturbed, which is thought to be a defence against parasitoid attack (Disney et al. 1994).

Variability in defences of members of an aphidophagous guild to IGP is likely to have a marked influence on the guild's overall structure. Intraguild predation of immature stages is considered a likely scenario between *H. axyridis* and British coccinellids (Majerus et al. 2006; Ware et al. in press) but its precise nature and impact is uncertain. Here we investigate the level and symmetry of IGP between *H. axyridis* and British coccinellids at four different stages of development: first instar larvae, fourth instar larvae, newly formed pre-pupae and 3-day-old pupae. British species were chosen from a range of different habitats on which *H. axyridis* is likely to encroach (Table 1). Four species of Japanese coccinellid were also investigated with a view to comparing the vulnerabilities and defences of species from the native range of *H. axyridis* with those from part of its invaded range.

Materials and methods

General protocol

The coccinellid species used in this study and the areas of origin of their stock cultures are listed in Table 1. All individuals used in experiments were F1 or F2 offspring from field collected samples. Mating pairs of all species were housed in 9 cm Petri dishes and maintained in a constant environment room at 22°C, 35% humidity and 14 hL:10 hD. Adults were supplied with excess pea aphids, *Acyrthosiphon pisum* (Harris), daily, and eggs were collected daily prior to transfer of adults to a clean dish to induce oviposition (Majerus et al. 1989).

Table 1 Details of species (Coleoptera: Coccinellidae) used in this study

Species	Code	Origin	Habitat in UK or Japan
Harmonia axyridis (Pallas 1773)	HA-B	Battersea, London, UK	Generalist
Coccinella septempunctata Linnaeus 1758	C7	Thetford, Norfolk, UK	Herbaceous generalist
Adalia bipunctata (Linnaeus 1758)	A2	Cambridge, Cambridgeshire, UK	Generalist
Adalia decempunctata (Linnaeus 1758)	A10	Thetford, Norfolk, UK	Broad-leaved deciduous woodland specialist
Propylea quatuordecimpunctata (Linnaeus 1758)	P14	Cambridge, Cambridgeshire, UK	Generalist
Calvia quatuordecimguttata (Linnaeus 1758)	C14	Thetford, Norfolk, UK	Broad-leaved deciduous woodland specialist
Anatis ocellata (Linnaeus 1758)	AO	Thetford, Norfolk, UK	Host plant restricted specialist: needled conifers
Harmonia quadripunctata (Pontoppidan 1763)	H4	Thetford, Norfolk, UK	Host plant restricted specialist: needled conifers
Coccinella quinquepunctata Linnaeus 1758	C5	Aberystwyth, Wales, UK	Generalist with environmental constraints: unstable river shingle
Harmonia axyridis (Pallas 1773)	HA-J	Fuchu, Honshu, Japan	Generalist
Coccinella septempunctata brucki Mulsant 1866	CSB	Fuchu, Honshu, Japan	Generalist
Cheilomenes sexmaculatus (Fabricus 1781)	CHS	Fuchu, Honshu, Japan	Generalist
Propylea japonica (Thunberg 1781)	PJ	Fuchu, Honshu, Japan	Herbaceous generalist
Eocaria muiri Timberlake 1943	EM	Fuchu, Honshu, Japan	Broad-leaved deciduous woodland specialist

Habitat classes for British species amended from Majerus (1994, pp. 141–143); information for Japanese species from R. Ware and M. Majerus (pers. obs.)

All experiments were carried out in a constant environment room at 22°C, 35% humidity and 14 hL:10 hD.

First instar larval contests

Newly hatched larvae were given excess aphids for 24 h and those used in first instar contests were age-matched based on the date of hatching. A single *H. axyridis* larva was transferred to one pole of a clean 5 cm Petri dish using a single-bristle paintbrush. A single larva of a non-conspecific species was then added to the opposite pole of the dish. No other food or water was provided thereafter. For all interactions, except those between *H. axyridis* and *Harmonia quadripunctata* (Pontoppidan), dishes were monitored twice daily until both larvae had died, and times of death from the time of set-up were recorded. Larval identification was carried out under a dissecting microscope on the basis of spine morphology (R. Ware, pers. obs.). Due to the difficulty in distinguishing *H. axyridis* and *H. quadripunctata* at first instar, the larva that survived longest in each contest was reared to fourth instar when identity could be established by use of abdominal markings. Causes of death of larvae were classified as: 'unknown' (where no obvious injury was visible but larvae may have appeared somewhat shrivelled due to desiccation), 'attacked, but not eaten after 48 h' (evidence of injury, e.g. bite marks, reflex blood, but corpse otherwise intact) or 'attacked, and eaten within 48 h' (body parts missing, evidence of fluid being sucked). First instar larvae of British *H. axyridis* were matched with first instar larvae of each of the eight British species, while first instar larvae of Japanese *H. axyridis* were matched with those of each of the four Japanese species. Each trial was replicated 20 times.

Fourth instar larval contests

First instar larvae not used in the above experiments were reared on aphids in 9 cm Petri dishes (five larvae per dish) to generate material for fourth instar contests. Larvae used were age-matched based on the date of third ecdysis. A single fourth instar *H. axyridis* larva that had ecdysed that day was transferred to one pole of a clean 9 cm Petri dish using a fine bristled paintbrush. A single fourth instar larva of a non-conspecific species was then added to the opposite pole of the dish. No other food or water was provided thereafter. Dishes were observed regularly over the next 2 days (eight checks in 48 h) and once daily thereafter. Evidence of IGP, time of larval deaths and pupations were noted. The time elapsed from the set-up of the experiment until death of one protagonist was analysed as an indication of the ease with which an IG predator could subdue its prey. All species were identifiable by eye at fourth instar (R. Ware, pers. obs.). Again, British *H. axyridis* larvae were matched with larvae of the British species, and Japanese *H. axyridis* were pitted against larvae of the Japanese species. Twenty replicates were performed for each contest.

IGP of pre-pupae by fourth instar larvae

Fourth instar larvae not used in the above experiments were isolated in 9 cm Petri dishes, provided with excess aphids daily, and allowed to pupate in the dish. Following formation of the pre-pupa, the dish was cleaned with an alcohol swab to remove any aphid debris and waste material, and a fourth instar *H. axyridis* larva, which had undergone third ecdysis that day, was added using a fine bristled paintbrush. No other food or water was provided thereafter. Dishes were observed regularly over the next 2 days (eight checks in 48 h) and once daily thereafter. Pre-pupae were observed for attack/consumption by the larva or eclosion. Larvae were observed for death or pupation. *Harmonia axyridis* larvae were presented with sympatric pre-pupae of all British and Japanese species in this manner, and larvae of all these species were presented with pre-pupae of *H. axyridis*. Cannibalism of *H. axyridis* pre-pupae by non-sibling conspecific larvae was also tested. Twenty replicates were performed for each trial, with the exception of *H. axyridis* larvae versus *Coccinella quinquepunctata* Linnaeus pre-pupae, in which only ten replicates were achieved due to low culture output.

IGP of pupae by fourth instar larvae

Pupal trials followed the same protocol as above, but using 3-day-old pupae rather than pre-pupae. Again, problems with producing enough *C. quinquepunctata* meant reduced replicates for *H. axyridis* larvae versus *C. quinquepunctata* pupae ($N = 13$) and *C. quinquepunctata* larvae versus *H. axyridis* pupae ($N = 8$).

Statistical analysis

All statistical manipulations were performed using Minitab 14, apart from Fisher's exact test and the Fisher-Freeman-Halton test, which were carried out using StatXact 7. British species are considered separately from Japanese species throughout.

Levels and symmetry of IGP

Outcomes of larval fights were classified as: *H. axyridis* acted as the IG predator, *H. axyridis* acted as the IG prey, or no IGP event occurred. For each species-pair comparison, the level of IGP between *H. axyridis* larvae and larvae of another species is determined as the proportion of replicates in which IGP occurred out of the total number of replicates for that species-pair, and an index of symmetry is given by the proportion of replicates in which *H. axyridis* was the IG predator out of the total number of replicates in which IGP occurred (Lucas et al. 1998). Thus, a symmetry index of >0.5 shows that IGP was in favour of *H. axyridis*, while an index of <0.5 indicates IGP was in favour of the other species. For first and fourth instar larvae, symmetry of IGP was analysed using a one-sample χ^2 goodness of fit test with an expected symmetry of 50%. Comparisons of levels and symmetry between species were made using a fully orthogonal analysis of contingency tables using Fisher's exact test (for 2×2 tables) (FEt) or the Fisher-Freeman-Halton test (for $r \times 2$ tables) (FFHt).

The concept of 'symmetry' is not applicable to trials involving larvae versus (pre)pupae as the latter can never be the IG predator. Instead, we considered the relative level of IGP when *H. axyridis* is the larva, compared to when *H. axyridis* is the (pre)pupa, in 2×2 contingency tables of numbers of (pre)pupae 'killed' versus 'not killed' for each species, using Fisher's exact test. Levels of IGP of (pre)pupae of different species by *H. axyridis* were compared using a fully orthogonal analysis of contingency tables with Fisher's exact test or the Fisher-Freeman-Halton test as appropriate. Similar analysis was conducted to compare the levels of IGP of *H. axyridis* (pre)pupae by larvae of different species.

Time until death of IG prey

For fourth instars, pre-pupae and pupae, the time until death of the IG prey (measured from time of experimental set-up) was compared non-parametrically as the Kolmogorov–Smirnov tests revealed a significant difference from normality in some cases ($P < 0.05$). The Kruskal–Wallis test and multiple pairwise Mann–Whitney U tests adjusting for ties were used. Alpha values were adjusted according to the Bonferroni correction for multiple comparisons (referred to in text as αB).

Results

First instar larval contests

Levels and symmetry of IGP

Both larvae scored as attacked but not eaten after 48 h and those scored as attacked and eaten after 48 h were considered to be the subject of IGP. The level and symmetry of IGP between *H. axyridis* and all British and Japanese species at first instar is shown in Table 2. The number of replicates in which the pairing of *H. axyridis* with a non-conspecific competitor resulted in IGP was similar for all British species (FFHt = 10.32, 7 d.f., $P = 0.167$). When IGP does occur between two species, A and B, it is clear that there are three possibilities: IGP may be symmetric (with an equal likelihood of either species acting as predator or prey), asymmetric in favour of species A (species A is the IG predator and

Table 2 The number of replicates in which *H. axyridis* acted as the IG predator or IG prey in pairings with British and Japanese species at first and fourth instar. Levels and indices of symmetry of IGP are also shown

Species	First instar				Fourth instar			
	HA acts as IG predator	HA acts as IG prey	Level of IGP	Symmetry of IGP	HA acts as IG predator	HA acts as IG prey	Level of IGP	Symmetry of IGP
British								
C7	6	4	0.50	0.60	20	0	1.00	1.00
A2	10	3	0.65	0.77	20	0	1.00	1.00
A10	18	0	0.90	1.00	20	0	1.00	1.00
P14	13	0	0.65	1.00	20	0	1.00	1.00
C14	6	6	0.60	0.50	11	3	0.70	0.79
AO	2	13	0.75	0.13	2	17	0.95	0.11
H4	7	9	0.80	0.44	16	4	1.00	0.80
C5	12	1	0.65	0.92	17	0	0.85	1.00
Japanese								
CSB	17	2	0.95	0.89	20	0	1.00	1.00
CHS	13	2	0.75	0.87	20	0	1.00	1.00
PJ	18	0	0.90	1.00	20	0	1.00	1.00
EM	9	3	0.67	0.75	19	1	1.00	0.95

$N = 20$ for each pairing. For species codes, see Table 1

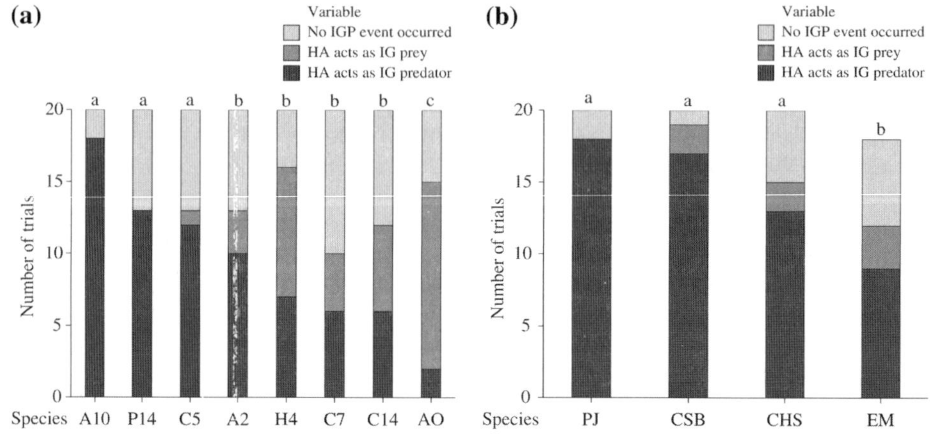

Fig. 1 Symmetry of IGP between *H. axyridis* and (**a**) British coccinellids and (**b**) Japanese coccinellids at first instar. Analysis shows that with species marked a, IGP was asymmetric in favour of *H. axyridis* (referred to as 'class 1' in text); with species marked b, IGP was symmetric ('class 2'); and with species marked c, IGP was asymmetric in favour of the other species ('class 3'). Species are placed along the *x*-axes from left to right in order of decreasing susceptibility to IGP by *H. axyridis*. For species codes, see Table 1

species B is the IG prey) or asymmetric in favour of species B (species B is the IG predator and species A is the IG prey).

In respect of the relationship of *H. axyridis* with British coccinellids at first instar, this study provides examples of all three (Fig. 1a). In contests with *Adalia decempunctata* (Linnaeus), *Propylea quatuordecimpunctata* (Linnaeus) and *C. quinquepunctata*, *H. axyridis* was predominantly the IG predator, hereafter referred to as 'class 1'. *Harmonia axyridis* engaged in symmetric IGP with *Coccinella septempunctata* Linnaeus, *Adalia bipunctata* (Linnaeus), *C. quatuordecimguttata* and *H. quadripunctata*, hereafter referred to as 'class 2'. Only one British species, *Anatis ocellata* (Linnaeus) acted as an asymmetric IG predator of *H. axyridis*, hereafter referred to as 'class 3'. The validity of grouping species into classes 1 and 2 is shown by homogeneity within each class (FFHt = 2.14, 2 d.f., $P = 0.591$ for class 1 and FFHt = 3.518, 3 d.f., $P = 0.348$ for class 2). *Harmonia axyridis* acted as the IG predator significantly more often in class 1 than in class 2 (FEt = 24.22, 1 d.f., $P < 0.001$) and acted as the IG prey significantly less often in class 2 than in class 3 (FEt = 9.04, 1 d.f., $P = 0.003$).

The level of IGP was similar for all Japanese species studied (FFHt = 6.40, 3 d.f., $P = 0.082$). IGP was symmetric with *E. muiri*, but asymmetric in favour of *H. axyridis* in interactions with all other species (Fig. 1b). However, there was no significant difference in symmetry found between Japanese species (FFHt = 4.72, 3 d.f., $P = 0.160$).

Fourth instar larval contests

Levels and symmetry of IGP

The level and symmetry of IGP between *H. axyridis* and all British and Japanese species at fourth instar is shown in Table 2 and Fig. 2. Intraguild predation occurred on 100% of occasions when *H. axyridis* was paired with *C. septempunctata*, *H. quadripunctata*,

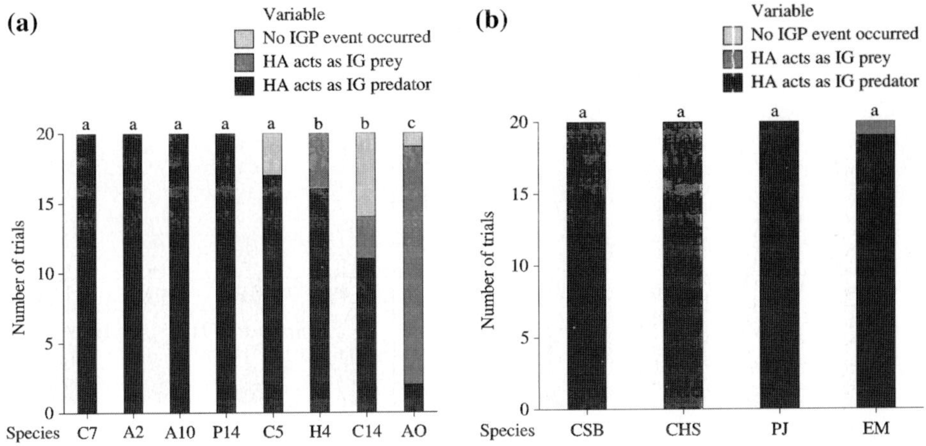

Fig. 2 Symmetry of IGP between *H. axyridis* and (**a**) British coccinellids and (**b**) Japanese coccinellids at fourth instar. Analysis shows that with species marked a, IGP was asymmetric in favour of *H. axyridis* ('class 1'); with species marked b, IGP was also asymmetric in favour of *H. axyridis* but significantly less so than for species marked a ('class 2'); and with species marked c, IGP was asymmetric in favour of the other species ('class 3'). Species are placed along the *x*-axes from left to right in order of decreasing susceptibility to IGP by *H. axyridis*. For species codes, see Table 1

A. bipunctata, *A. decempunctata* and *P. quatuordecimpunctata*, but less frequently with *C. quatuordecimguttata*, *A. ocellata* and *C. quinquepunctata* (FEt = 18.00, 1 d.f., $P < 0.001$), for which the level of IGP was similar (FFHt = 4.27, 3 d.f., $P = 0.130$). *Harmonia axyridis* was universally the IG predator of *C. septempunctata*, *A. bipunctata*, *A. decempunctata*, *P. quatuordecimpunctata* and *C. quinquepunctata* larvae ('class 1'). Intraguild predation was asymmetric in favour of *H. axyridis* in contests with *H. quadripunctata* and *C. quatuordecimguttata* ('class 2'), but asymmetric in favour of *A. ocellata* in interactions with this species ('class 3'). *Harmonia axyridis* acted as the IG predator significantly more often in class 1 than class 2 (FEt = 17.8, 1 d.f., $P < 0.001$), and significantly more often in class 2 than class 3 (FEt = 24.36, 1 d.f., $P < 0.001$). There was no difference in symmetry within a class (IGP was 100% asymmetric in class 1; FEt = 0.095, 1 d.f., $P \sim 1$ for class 2).

All replicates involving fourth instars of *H. axyridis* and Japanese species resulted in an IGP event. Intraguild predation was fully asymmetric towards *H. axyridis* for *Coccinella septempunctata brucki* Mulsant, *Cheilomenes sexmaculatus* (Fabricus) and *Propylea japonica* (Thunberg) and significantly so for *E. muiri* (Fig. 2b). There was no significant difference in symmetry between Japanese species (FFht = 2.84, 3 d.f., $P \sim 1$).

Time until death of IG prey

When H. axyridis *is the IG predator*. In considering attacks made by *H. axyridis* on British species, *A. ocellata* was excluded from analysis as only two individuals were killed. There was significant variation in the time until death of other British larvae (Kruskal–Wallis test: $H = 30.04$, 6 d.f., $P < 0.001$) (Fig. 3a). Multiple pairwise Mann–Whitney U tests (αB (21 tests) = 0.002) revealed the following significant results: *C. septempunctata* larvae were killed more quickly than *H. quadripunctata* and *C. quatuordecimguttata* larvae ($W = 262.0$,

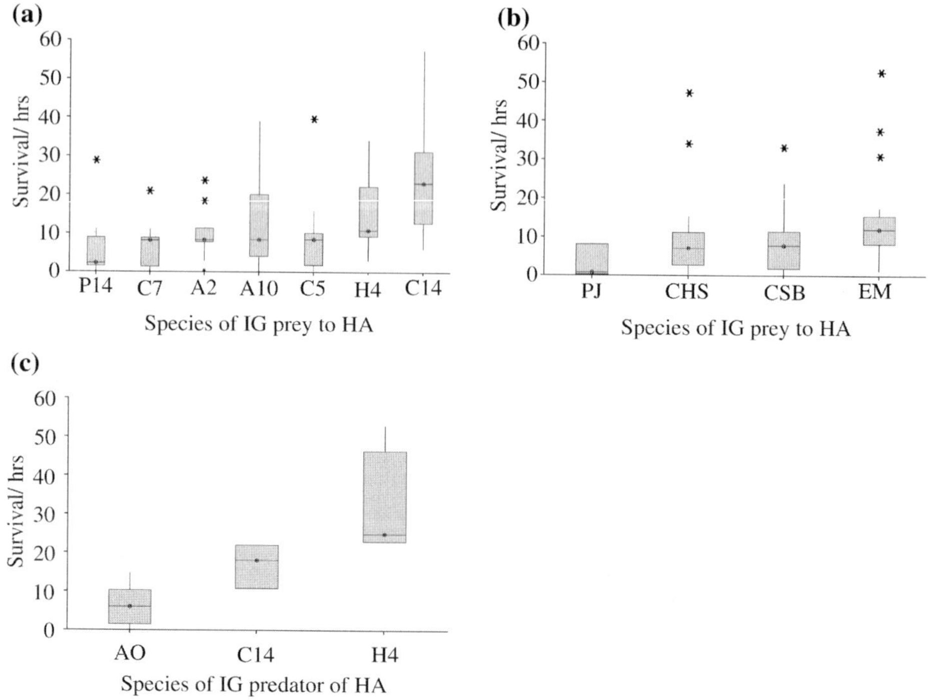

Fig. 3 Time until death of fourth instar larvae of (**a**) British and (**b**) Japanese coccinellids when acting as IG prey to *H. axyridis* fourth instar larvae, and (**c**) time until death of fourth instar *H. axyridis* larvae when acting as IG prey to *C. quatuordecimguttata*, *A. ocellata* and *H. quadripunctata* fourth instar larvae. Species are placed along the *x*-axes from left to right in order of increasing survival time. Medians (circles with stars), interquartile ranges (boxes), ranges (vertical lines) and outliers (stars) shown. For species codes, see Table 1

$P < 0.001$ and $W = 238.0$, $P < 0.001$ respectively); and *P. quatuordecimpunctata* larvae were killed more quickly than *H. quadripunctata* and *C. quatuordecimguttata* larvae ($W = 258.0$, $P < 0.001$ and $W = 228.5$, $P < 0.001$ respectively). All other species-pairs were not significantly different from each other at the corrected α value (all $P > 0.002$), but see Fig. 3a for order of increasing times.

Japanese fourth instar larvae also differed in how quickly they were attacked and killed by *H. axyridis* (Kruskal–Wallis test: $H = 19.96$, 3 d.f., $P < 0.001$) (Fig. 3b). Multiple pairwise Mann–Whitney *U* tests (αB (6 tests) = 0.008) showed that *P. japonica* larvae were killed more quickly than *C. sexmaculatus* larvae ($W = 514.4$, $P = 0.005$) and *E. muiri* larvae ($W = 246.5$, $P < 0.001$). All other species-pairs were not significantly different from each other at the corrected α value (all $P > 0.008$).

When H. axyridis *is the IG prey*. *Harmonia axyridis* larvae were killed significantly more quickly by *A. ocellata* than by *H. quadripunctata*, (Mann–Whitney *U* test: αB (3 tests) = 0.017, $W = 153.0$, $P = 0.003$) but there was no difference in the timing of deaths caused by *A. ocellata* and *C. quatuordecimguttata* (Mann–Whitney *U* test: αB (3 tests) = 0.017, $W = 53.0$, $P = 0.026$) or between *C. quatuordecimguttata* and *H. quadripunctata* (Mann–Whitney *U* test: αB (3 tests) = 0.017, $W = 6.0$, $P = 0.052$) (Fig. 3c).

IGP of pre-pupae by fourth instar larvae

Levels of IGP

The levels of IGP between *H. axyridis* and British coccinellids in interactions between fourth instar larvae and pre-pupae are shown in Table 3 and Fig. 4a. Intraguild predation occurs significantly more often when *H. axyridis* is the larva compared to when it is the pre-pupa in interactions with *C. septempunctata*, *A. bipunctata*, *A. decempunctata*, *P. quatuordecimpunctata*, *C. quatuordecimguttata* and *C. quinquepunctata*. Intraguild predation between *H. axyridis* and *H. quadripunctata* is also significantly unidirectional towards *H. axyridis*. *Anatis ocellata* larvae preyed upon *H. axyridis* pupae at a similar level to the reverse.

There was no difference in the susceptibility of different British species' pre-pupae to IGP by *H. axyridis* (FFHt = 5.60, 7 d.f., $P = 0.585$) and the level of IGP of heterospecific pre-pupae was similar to that of British conspecific pre-pupae (FEt = 2.82, 1 d.f., $P = 0.148$). The susceptibility of *H. axyridis* pre-pupae to IGP by larvae of different species did vary (FFHt = 72.97, 7 d.f., $P < 0.001$). *Anatis ocellata* and *H. quadripunctata* larvae showed higher levels of IGP of *H. axyridis* pre-pupae than the other British species (FEt = 70.65, 1 d.f., $P < 0.001$), who all showed similarly low levels (FFHt = 2.96, 5 d.f., $P = 0.871$). There was no statistically significant difference in the level of predation by *A. ocellata* and *H. quadripunctata* (FEt = 4.67, 1 d.f., $P = 0.065$).

Table 3 The number of British and Japanese (pre)pupae killed by fourth instar *H. axyridis* larvae (and level of IGP), and the number of *H. axyridis* (pre)pupae killed by fourth instar larvae of those species

Species	Fourth instar larvae versus pre-pupae				Fourth instar larvae versus pupae			
	HA as larva		HA as pre-pupa		HA as larva		HA as pupa	
	Number pre-pupae killed	Level of IGP	Number pre-pupae killed	Level of IGP	Number pupae killed	Level of IGP	Number pupae killed	Level of IGP
British								
HA-B	18	0.90	N/A	N/A	0	0	N/A	N/A
C7	20	1.00	2	0.10	4	0.20	0	0.00
A2	18	0.90	2	0.10	3	0.15	0	0.00
A10	20	1.00	0	0.00	10	0.50	0	0.00
P14	20	1.00	1	0.05	18	0.90	0	0.00
C14	19	0.95	2	0.10	15	0.75	0	0.00
AO	20	1.00	18	0.90	7	0.35	11	0.55
H4	19	0.95	12	0.60	0	0.00	2	0.10
C5	10	1.00	1	0.05	4	0.31	1	0.08
Japanese								
HA-J	20	1.00	N/A	N/A	2	0.10	N/A	N/A
CSB	19	0.95	2	0.10	7	0.35	0	0.00
CHS	19	1.00	0	0.00	18	0.90	0	0.00
PJ	20	1.00	0	0.00	19	0.95	0	0.00
EM	20	1.00	1	0.05	16	0.80	0	0.00

$N = 20$ for each pairing (except those involving *C. quinquepunctata*, for which N is indicated in Materials and methods). For species codes, see Table 1

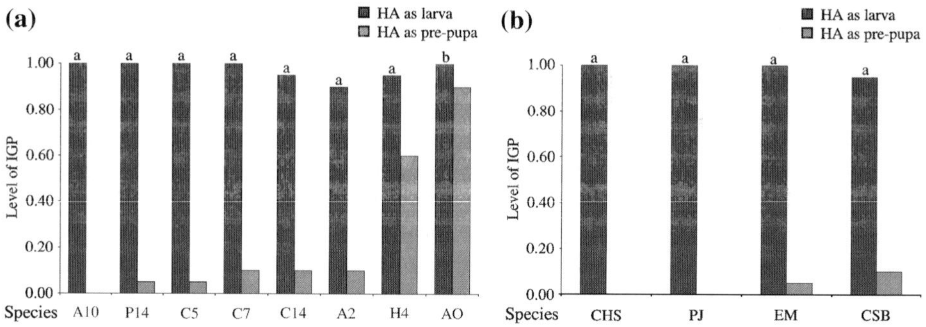

Fig. 4 Levels of IGP between *H. axyridis* and (**a**) British and (**b**) Japanese coccinellids in interactions between fourth instar larvae and pre-pupae. Analysis shows that with species marked a, IGP was significantly unidirectional towards *H. axyridis*; while with species marked b, IGP was bidirectional. Species are placed along the *x*-axes from left to right in order of decreasing susceptibility to IGP by *H. axyridis*. For species codes, see Table 1

The levels of IGP between *H. axyridis* and Japanese coccinellids in interactions between fourth instar larvae and pre-pupae are shown in Table 3 and Fig. 4b. Intraguild predation is significantly unidirectional towards *H. axyridis* for all species, in that *H. axyridis* larvae predate other species' pre-pupae at a much higher level than those species' larvae predate *H. axyridis* pre-pupae.

There was no difference in the susceptibility of different Japanese species' pre-pupae to IGP by *H. axyridis* (FFHt = 2.16, 3 d.f., $P \sim 1$) and the level of IGP of heterospecific pre-pupae was similar to that of Japanese conspecific pre-pupae (FEt = 0.45, 1 d.f., $P \sim 1$). The level of IGP of *H. axyridis* pre-pupae was similarly low for all species of Japanese larvae (FFHt = 3.00, 3 d.f., $P = 0.611$).

Time until death of IG prey

When H. axyridis *is the IG predator*. The time elapsed from a *H. axyridis* larva being placed in a dish containing a British pre-pupa until the larva killed the pre-pupa is shown in Fig. 5a. This time showed significant variation between species (Kruskal–Wallis test: $H = 38.05$, 8 d.f., $P < 0.001$). Multiple pairwise Mann–Whitney U tests (αB (36 tests) = 0.0014) showed that *P. quatuordecimpunctata* pre-pupae were killed more quickly than pre-pupae of *H. axyridis* ($W = 489.0$, $P < 0.001$), *A. bipunctata* ($W = 489.0$, $P < 0.001$), *H. quadripunctata* ($W = 230.0$, $P < 0.001$), *A. ocellata* ($W = 275.0$, $P < 0.001$) and *C. quatuordecimguttata* ($W = 277.5$, $P < 0.001$). All other pairwise comparisons were not statistically significant at the Bonferroni-corrected α value.

The different species of Japanese pre-pupae also differed with respect to how quickly they were killed (Kruskal–Wallis test: $H = 29.49$, 4 d.f., $P < 0.001$) (Fig. 5b). Multiple pairwise Mann–Whitney U tests (αB (10 tests) = 0.005) revealed that *C. sexmaculatus* and *P. japonica* pre-pupae were killed more quickly than *H. axyridis* and *C. septempunctata brucki* pre-pupae (CHS-HA: $W = 504.0$, $P = 0.004$; PJ-HA: $W = 570.0$, $P < 0.001$; CHS-CSB: $W = 480.5$, $P = 0.001$; PJ-CSB: $W = 530.0$, $P < 0.001$). No other species-pair comparisons gave significant values (all $P > 0.005$).

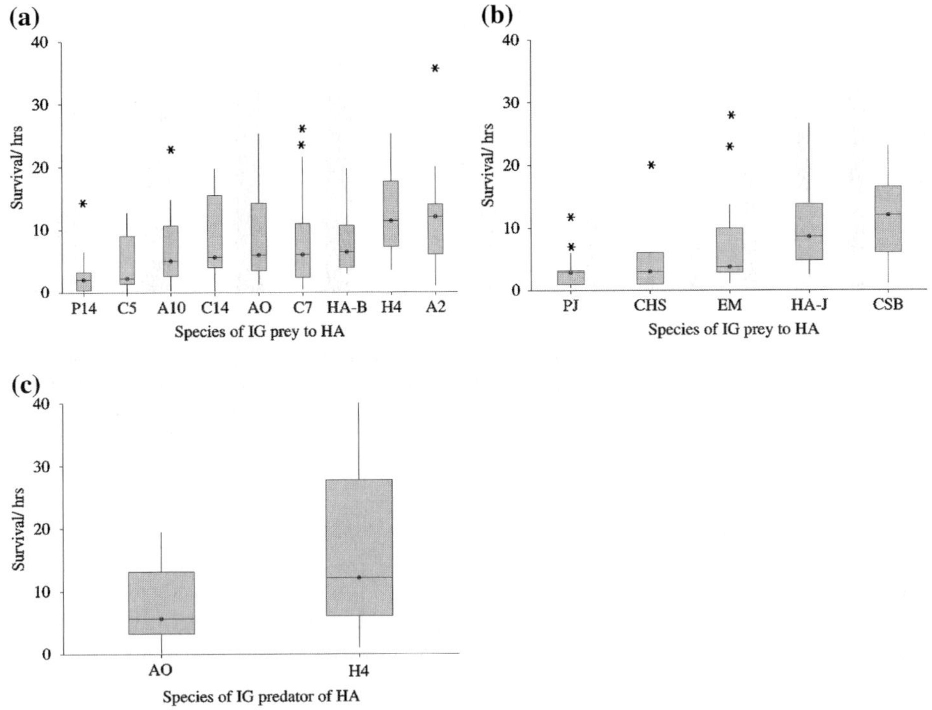

Fig. 5 Time until death of pre-pupae of (**a**) British and (**b**) Japanese coccinellids when acting as IG prey to fourth instar *H. axyridis* larvae, and (**c**) time until death of *H. axyridis* pre-pupae when acting as IG prey to *A. ocellata* and *H. quadripunctata* fourth instar larvae. Species are placed along the *x*-axes from left to right in order of increasing survival time. Medians (circles with stars), interquartile ranges (boxes), ranges (vertical lines) and outliers (stars) shown. For species codes, see Table 1

When H. axyridis *is the IG prey*. The only British species that acted as IG predators of *H. axyridis* pre-pupae at appreciable frequency were *H. quadripunctata* and *A. ocellata* ($N \leq 2$ for all other species). There was no significant difference in how quickly *H. axyridis* pre pupae were killed by larvae of these two species (Mann–Whitney U test: $W = 230.0$, $P = 0.066$) (Fig. 5c).

IGP of pupae by fourth instar larvae

Levels of IGP

The levels of IGP between *H. axyridis* and British coccinellids in interactions between fourth instar larvae and pupae are shown in Table 3 and Fig. 6a. Intraguild predation of pupae of *C. septempunctata*, *A. bipunctata* and *C. quinquepunctata* by *H. axyridis* larvae occurred at low levels and therefore showed no significant difference from the reciprocal interactions, which occurred rarely, or not at all. No *H. axyridis* larvae preyed upon *H. quadripunctata* pupae, and the reciprocal interaction only occurred on two occasions, again resulting in no significant difference in the level of IGP. Intraguild predation of *A. ocellata* pupae by *H. axyridis* occurred at a similar frequency to the reverse interaction.

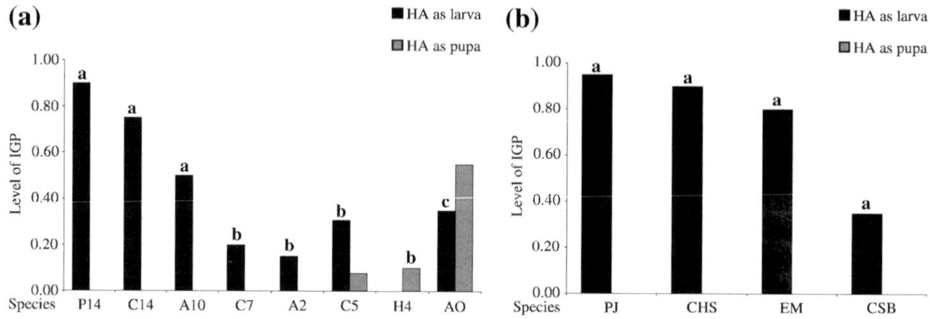

Fig. 6 Levels of IGP between *H. axyridis* and (**a**) British and (**b**) Japanese coccinellids in interactions between fourth instar larvae and pupae. Analysis shows that with species marked a, IGP was significantly unidirectional towards *H. axyridis*; with species marked b, levels of IGP were too small to reveal the directionality of IGP; while with species marked c, IGP was bidirectional. Species are placed along the *x*-axes from left to right in order of decreasing susceptibility to IGP by *H. axyridis*. For species codes, see Table 1

Intraguild predation of pupae by larvae was significantly unidirectional towards *H. axyridis* with respect to *A. decempunctata*, *P. quatuordecimpunctata* and *C. quatuordecimguttata*.

The level of IGP of British conspecific and non-conspecific pupae by *H. axyridis* larvae differed with respect to the species of pupa (FFHt = 74.03, 8 d.f., $P < 0.001$). *Propylea quatuordecimpunctata* and *C. quatuordecimguttata* pupae were preyed upon at similar levels (FEt = 1.49, 1 d.f., $P = 0.408$) and at a higher rate than pupae of *C. septempunctata*, *A. bipunctata*, *A. decempunctata*, *A. ocellata* and *C. quinquepunctata* (FEt = 31.90, 1 d.f., $P < 0.001$), which showed similar levels of IGP by *H. axyridis* (FFHt = 6.88, 4 d.f., $P = 0.143$). Finally, *H. quadripunctata* and conspecific pupae were never killed by *H. axyridis* larvae. The susceptibility of *H. axyridis* pupae to IGP by larvae of different species also varied (FFHt = 72.97, 7 d.f., $P < 0.001$), with *A. ocellata* larvae showing significantly higher levels of IGP of *H. axyridis* pupae than the other British species, including conspecifics (FEt = 40.09, 1 d.f., $P < 0.001$), which all showed similarly low levels (FFHt = 7.23, 7 d.f., $P = 0.062$).

The levels of IGP between *H. axyridis* and Japanese coccinellids in interactions between fourth instar larvae and pupae are shown in Table 3 and Fig. 6b. Intraguild predation of pupae by larvae was significantly unidirectional towards *H. axyridis* with respect to all Japanese species studied, as no non-conspecific larvae killed *H. axyridis* pupae.

The level of IGP of Japanese conspecific and non-conspecific pupae by *H. axyridis* larvae differed with respect to the species of pupa (FFHt = 47.79, 4 d.f., $P < 0.001$). *Cheilomenes sexmaculatus*, *P. japonica* and *E. muiri* pupae fell victim to IGP by *H. axyridis* larvae at a similar level to each other (FFHt = 1.52, 2 d.f., $P = 0.77$) but more often than did conspecific pupae and pupae of *C. septempunctata brucki* (FEt = 45.84, 1 d.f., $P < 0.001$), which were killed with similar frequency (FEt = 3.45, 1 d.f., $P = 0.127$).

Time until death of IG prey

When H. axyridis *is the IG predator.* The time elapsed from a *H. axyridis* larva being placed in a dish containing a British pupa until the larva killed the pupa is shown in

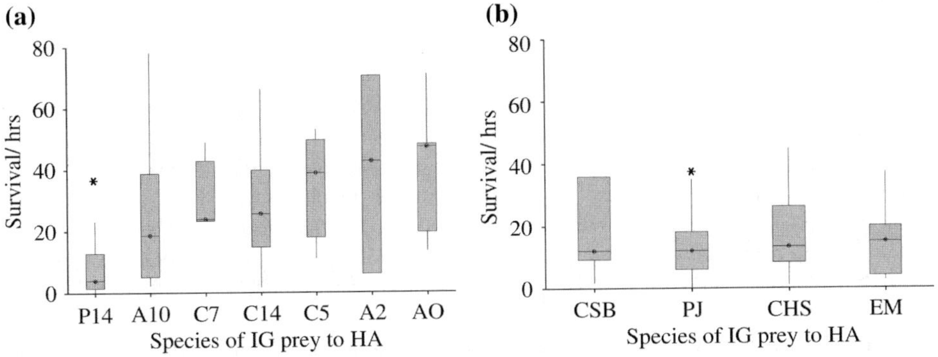

Fig. 7 Time until death of pupae of (**a**) British and (**b**) Japanese coccinellids when acting as IG prey to *H. axyridis*. Species are placed along the *x*-axes from left to right in order of increasing survival time. Medians (circles with stars), interquartile ranges (boxes), ranges (vertical lines) and outliers (stars) shown. For species codes, see Table 1

Fig. 7a. Note that the sample sizes being analysed for *A. bipunctata*, *C. septempunctata* and *C. quinquepunctata* were low, at $N = 3$, $N = 4$ and $N = 4$ respectively. The time until death of pupae showed significant variation with respect to species (Kruskal–Wallis test: $H = 23.06$, 6 d.f., $P = 0.001$). It is clear from Fig. 7a that *P. quatuordecimpunctata* pupae were killed very quickly, and significantly more quickly than pupae of *C. septempunctata*, *A. ocellata* and *C. quatuordecimguttata* (Mann–Whitney U tests: αB (6 tests) $= 0.008$; $W = 174.5$, $P = 0.006$; $W = 178.0$, $P = 0.001$ and $W = 207.5$, $P = 0.004$ respectively) but not significantly more quickly than *A. bipunctata*, *A. decempunctata* and *C. quinquepunctata* pupae (all $P > 0.008$). There was no significant difference in the time until death of pupae of the other species if *P. quatuordecimpunctata* is excluded (Kruskal–Wallis test: $H = 3.64$, 5 d.f., $P = 0.603$).

There was no significant variation in the time until death of non-conspecific Japanese pupae (Kruskal–Wallis: $H = 0.83$, 3 d.f., $P = 0.840$) (Fig. 7b).

When H. axyridis *is the IG prey*. Sample sizes of pupae killed by *H. quadripunctata* and *C. quinquepunctata* were too low to analyse. *Anatis ocellata* larvae killed *H. axyridis* pupae within a median time of 19.5 h.

Discussion

Introduced species often seem to perform better than conspecifics in their native range (Williamson 1996). A prominent hypothesis explaining this success is that introduced species are typically free of, or are less affected by, the natural enemies (competitors, predators, and parasites) that they would normally encounter in their native range (Torchin et al. 2003). Furthermore, while sympatric competitors might be expected to possess co-evolved defence strategies for dealing with detrimental interactions, such as competition and intraguild predation, such co-evolved strategies are unlikely between taxa whose ranges do not overlap. Indeed, the ability of an alien species to exploit native members of a guild, while avoiding exploitation itself, is thought to be important in determining the success of establishment and spread (Sato and Dixon 2004). With respect to *H. axyridis*, it is hypothesised that superior defensive strategies of native aphidophages, compared to those inhabiting the introduced range of the species, may explain why it exists in apparent ecological equilibrium with a range of species in Asia but has had serious detrimental impacts where it has established as an invasive alien.

This study examined the nature of IGP between immatures of *H. axyridis* and a selection of coccinellids from parts of its native and introduced ranges. It is not apparent from the results presented here that Japanese species are any better protected against IGP by *H. axyridis* than are their British counterparts, at least in a laboratory setting. Intraguild predation events between larvae of *H. axyridis* and all other species were common at first and fourth instar. At first instar, the British species *C. septempunctata*, *C. quatuordecimguttata* and *H. quadripunctata*, and the Japanese species *E. muiri*, were found to engage in symmetric IGP with *H. axyridis*, and the only species to act as an asymmetric intraguild predator of *H. axyridis* was the British coccinellid *A. ocellata*. Meanwhile, all other species were the IG prey in such interactions. For many species, IGP seemed more one-sided at fourth instar, and was actually asymmetric towards *H. axyridis* for all species except *A. ocellata*, which again emerged as the only IG predator of *H. axyridis*. In interactions between fourth instar larvae and pre-pupae, IGP was considered unidirectional towards *H. axyridis* for all Japanese species and all British species with the exception of *A. ocellata*, which was both IG predator and IG prey of *H. axyridis*. Pupae were clearly much better defended than pre-pupae, as the level of IGP was noticeably lower. However, the same patterns emerged, with the only bidirectional interaction occurring between *H. axyridis* and *A. ocellata*. Conspecific pupae and pupae of *H. quadripunctata* were never attacked, and *H. quadripunctata* larvae only attacked *H. axyridis* on two occasions.

The susceptibility of different species to IGP is likely to be affected by the nature of any physical and/or chemical defences they possess. Although there was no obvious demarcation between Japanese and British species in this regard, there was some general variation between species. The outcome of IGP between larvae is known to be influenced by the relative size of the protagonists (Lucas et al. 1998; Felix and Soares 2004) with the larger species generally acting as the IG predator and the smaller as the IG prey. Thus, large size may itself be considered as an adaptation against IGP. Fourth instar *H. axyridis* larvae are considerably larger than fourth instar larvae of most other species studied, which may partly explain their dominance as IG predators. *Anatis ocellata* larvae are larger than *H. axyridis* (this species is the largest British coccinellid) and are therefore more likely to act as the IG predator. However, size was clearly not the only factor determining the susceptibility of a larva to IGP, as larvae of *C. septempunctata* and *C. septempunctata brucki* consistently acted as the IG prey in interactions with *H. axyridis*, despite being roughly comparable in size. Such size differences between larvae were less pronounced at first instar (R. Ware, pers. obs.), which may explain both the apparently lower level of IGP at this stage, and the fact that more species were able to avoid predation by, or even kill, *H. axyridis*.

Variation in physical defensive structures is also likely to play a role in governing the outcome of IG interactions. Larvae of *H. axyridis* are extremely well protected by a covering of thick dorsal spines present at all instars. The only other species which possessed similar structures were *A. ocellata*, *H. quadripunctata* and *C. quatuordecimguttata*. Fourth instar larvae of these species were the least commonly attacked by *H. axyridis*, and when *C. quatuordecimguttata* and *H. quadripunctata* larvae were killed, it was after a much longer time period than the other species, indicating they were more difficult to subdue. *Harmonia quadripunctata* larvae are structurally very similar to *H. axyridis*, so the fact that fourth instar *H. axyridis* still emerged as asymmetric IG predators of fourth instar *H. quadripunctata* larvae and pre-pupae may be a function of the smaller size of *H. quadripunctata*, or of greater aggression by *H. axyridis*, or both. *Calvia quatuordecimguttata* larvae were able to fight back against *H. axyridis* at first instar but this ability was not upheld at fourth instar, presumably due to the larger size difference at this stage. The dorsal spines of *A. ocellata* are clearly an effective defence as this species rarely succumbed to IGP by *H. axyridis*. In addition, it was able to overpower the aggressive and well protected fourth instar *H. axyridis* larvae within a relatively short time period. Species which consistently succumbed to IGP by *H. axyridis* at the larval stages were *P. quatuordecimpunctata*, *A. decempunctata*, *C. quinquepunctata*, *A. bipunctata*, *C. septempunctata* and all Japanese coccinellids. All of these species are physically less well protected than *H. axyridis*, *H. quadripunctata*, *C. quatuordecimguttata* and *A. ocellata*, in that their dorsal surface is covered in fine hairs rather than sclerotised spines. Note that the least well defended of these species, *P. quatuordecimpunctata* and *P. japonica*, have almost smooth surfaces, and were extremely susceptible to IGP by *H. axyridis* and the most rapidly subdued of all species studied. Although not studied here, we suspect that the smooth larvae of *Myzia oblongoguttata* (Linnaeus) would be similarly vulnerable, despite the larger size of this species.

The pre-pupae of a particular species are likely to be similarly, if not more, vulnerable to IGP than their fourth instar larvae. Their physical defence structures are essentially the same, since they possess the fourth instar larval skin, but they are practically immobile and so potentially represent an easy meal for an IG predator. Indeed, immobility is often cited as a significant risk factor in the context of IGP (Lucas et al. 1998). This is reflected in the results presented here, with *H. axyridis* pre-pupae being extremely well protected and only attacked to a significant extent by *A. ocellata* and *H. quadripunctata* larvae, while pre-pupae of all other species were readily attacked by *H. axyridis* larvae. Again, *P. quatuordecimpunctata* and *P. japonica* were the most quickly killed. Pre-pupae of *C. quinquepunctata* and *C. sexmaculatus* were also killed relatively quickly. It is interesting that cannibalism of *H. axyridis* pre-pupae occurred at high frequency, indicating that the physical defences of *H. axyridis* pre-pupae are more easily overpowered by larvae of the same species. Once the larval skin has split and the pupa is formed, most coccinellids are notably less susceptible to IGP, as the tough pupal integument affords better protection than the soft larval skin. The results presented here clearly show reduced levels of IGP of pupae compared to pre-pupae for all British species except *P. quatuordecimpunctata* and *C. quatuordecimguttata*, which were still highly vulnerable and killed with relative ease. All Japanese species' pupae were highly vulnerable to IGP by *H. axyridis*. *Harmonia axyridis* pupae are clearly very well defended, even more so than their pre-pupae, as IGP by conspecific larvae, *H. quadripunctata* larvae and *A. ocellata* larvae was considerably reduced.

The results of this work support the perceived status of *H. axyridis* as a top intraguild predator of immatures of other coccinellids (Dixon 2000; Pell et al. 2007, and references therein). However, we found no evidence that native sympatric species have better defences against IGP by *H. axyridis* than species in its introduced range. This was also found to be the case for IGP of eggs (Ware et al. in press). Indeed, the three species that did emerge as being the best defended were all British. It therefore seems that another explanation of the discrepancy in the intraguild interactions of *H. axyridis* between its native and introduced range must be invoked. It is possible that *H. axyridis* is subject to more rigid population control in its Asian range due to the higher incidence of natural enemies such as male-killers (Majerus et al. 1999), parasitoids (LaMana and Miller 1996) and fungi (Riddick and Schaefer 2005). This warrants further investigation. Behavioural adaptations of sympatric species may also provide an explanation, whereby particular life history strategies make encounters with *H. axyridis* less likely. One obvious limitation of this study is that interactions were confined to Petri-dish arenas. Englund (1997) highlights the importance of spatial scale and prey movements in predation experiments. In natural conditions, escape behaviour is a likely possibility and may reduce the risk of IGP. For example, *P. quatuordecimpunctata* and *P. japonica* larvae were extremely vulnerable to IGP by *H. axyridis* in our experiments but the larvae of both are long-legged and mobile and may therefore be able to escape more easily in the wild. Many species are also known to exploit species-specific oviposition sites (Schellhorn and Andow 1999) and *P. japonica* females are reported to reduce oviposition in response to faecal cues from *H. axyridis* (Agarwala et al. 2003). *Coccinella septempunctata brucki* is known to lays its eggs earlier in spring than *H. axyridis* (Takahashi 1989), which means that *H. axyridis* rarely encounters eggs and younger larvae of *C. septempunctata brucki* (Sato and Dixon 2004). Sato et al. (2003, 2005) also report the dropping behaviour and emigration of *C. septempunctata* larvae as defensive tactics against IGP. Particularly vulnerable stages such as newly ecdysed larvae and pupae may be protected from IGP if species-specific moulting and pupation sites are selected (Lucas et al. 2000). The possibility of such behavioural avoidance tactics highlights the need for realistic field studies to be conducted when evaluating the risk and impact of IGP.

Results from the laboratory experiments conducted in this study show that a range of British coccinellids face considerable risk from the recent establishment of invasive *H. axyridis*. These findings must be viewed in conjunction with knowledge of the likely extent of habitat overlap between these species and *H. axyridis* (Ware et al. 2005; Table 1). *Harmonia axyridis* is an extremely generalist species (Roy et al. 2006) and is therefore likely to encroach on the ranges of both other habitat generalists and of more habitat-specific species. The susceptibility of *C. quinquepunctata* to IGP by *H. axyridis* is of particular concern as this species is considered endangered in Britain (Majerus 1994) and could be severely threatened if *H. axyridis* invades unstable river shingles. *Coccinella septempunctata*, *A. bipunctata* and *P. quatuordecimpunctata* are all considered to be fairly generalist species (Majerus and Kearns 1989; Majerus 1991) and therefore likely to interact with *H. axyridis* commonly. *Adalia decempunctata* and *C. quatuordecimguttata* are mainly deciduous tree specialists and so also potentially at risk, although the results presented here show some level of protection for *C. quatuordecimguttata* larvae. *Anatis ocellata* and *H. quadripunctata* are coniferous tree specialists so could encounter *H. axyridis* where it invades coniferous woodland (*H. axyridis* has already been found breeding on pine trees, *Pinus sylvestris*, in Britain, R. Ware and M. Majerus, pers. obs.) but have superior defence against *H. axyridis* than other British species. We have also shown that *H. axyridis* could even be at risk from *A. ocellata*. Overall it is clear that further research into the co-existence of British species with *H. axyridis* is needed, in particular focussing on the different timings of reproduction, in order to assess the full impact of Britain's newest invasive insect.

Acknowledgements The authors wish to acknowledge the help of Mr. Ian Wright, Mr. Sherif Elnagdy, Mr. Sami Saeed and Miss Laura-Jane Michie for help with culture maintenance, and Dr. Francois Balloux for advice on statistical techniques. We are grateful for the continued support shown by Dr. Helen Roy and Mr. Peter Brown at the Centre for Ecology and Hydrology, Monks Wood, UK, and thank Mr. Robert Frost for help with UK collections. We also thank Professor Yoshiaki Obara, Dr. Toshiyuki Satoh and Mr. Satoshi Koyama, from the Tokyo University of Agriculture and Technology, for help with collections and facilities in Japan.

References

Agarwala BK, Dixon AFG (1992) Laboratory study of cannibalism and interspecific predation in ladybirds. Ecol Entomol 17:303–309

Agarwala BK, Yasuda H (2001) Larval interactions in aphidophagous predators: effectiveness of wax cover as defence shield of *Scymnus* larvae against predation from syrphids. Entomol Exp Appl 100:101–107

Agarwala BK, Yasuda H, Kajita Y (2003) Effect of conspecific and heterospecific feces on foraging and oviposition of two predatory ladybirds: role of fecal cues in predator avoidance. J Chem Ecol 29: 357–376

Brown PMJ, Roy HE, Rothery P, Roy DB, Ware RL, Majerus MEN (2007) *Harmonia axyridis* in Great Britain: analysis of the spread and distribution of a non-native coccinellid. BioControl (this issue). doi:10.1007/s10526-007-9124-y

de Jong PW, Holloway GJ, Brakefield PM, de Vos H (1991) Chemical defence in ladybird beetles (Coccinellidae). II. Amount of reflex fluid, the alkaloid adaline and individual variation in defence in 2-spot ladybirds (*Adalia bipunctata*). Chemoecology 2:15–19

Disney RHL, Majerus MEN, Walpole MJ (1994) Phoridae (Diptera) parasitising Coccinellidae (Coleoptera). Entomologist 113:28–42

Dixon AFG (2000) Insect predator-prey dynamics: ladybird beetles and biological control. Cambridge University Press, Cambridge

Eisner T, Eisner M (1992) Operation and defensive role of "gin traps" in a Coccinellid pupa (*Cycloneda sanguinea*). Psyche 99:265–273

Englund G (1997) Importance of spatial scale and prey movements in predator caging experiments. Ecology 78:2316–2325

Felix S, Soares AO (2004) Intraguild predation between the aphidophagous ladybird beetles *Harmonia axyridis* and *Coccinella undecimpunctata* (Coleoptera: Coccinellidae): the role of body weight. Eur J Entomol 101:237–242

Hemptinne JL, Lognay G, Gauthier C, Dixon AFG (2000) Role of surface chemical signals in egg cannibalism and intraguild predation in ladybirds (Coleoptera: Coccinellidae). Chemoecology 10:123–128

Hinton HE (1955) Protective devices of endopterygote pupae. Trans Soc Br Entomol 12:49–93

Holloway GJ, de Jong PW, Brakefield PM, de Vos H (1991) Chemical defence in ladybird beetles (Coccinellidae). I. Distribution of coccinelline and individual variation in defence in 7-spot ladybirds (*Coccinella septempunctata*). Chemoecology 2:7–14

Koch RL, Galvan TL (2007) Bad side of a good beetle: the North American experience with *Harmonia axyridis*. BioControl (this issue). doi:10.1007/s10526-007-9121-1

Kuznetsov VN (1997) Lady beetles of the Russian Far East (Memoir series no. 1). Center for Systematic Entomology, Florida

LaMana ML, Miller JC (1996) Field observations on *Harmonia axyridis* Pallas (Coleoptera: Coccinellidae) in Oregon. Biol Control 6:232–237

Lucas É (2005) Intraguild predation among aphidophagous predators. Eur J Entomol 102:351–364

Lucas É, Coderre D, Brodeur J (1998) Intraguild predation among aphid predators: characterization and influence of extraguild prey density. Ecology 79:1084–1092

Lucas É, Coderre D, Brodeur J (2000) Selection of molting and pupation sites by *Coleomegilla maculata* (Coleoptera: Coccinellidae): avoidance of intraguild predation. Environ Entomol 29:454–459

Majerus MEN (1991) Habitat and host plant preferences of British ladybirds. Entomol Month Mag 127: 167–175

Majerus MEN (1994) Ladybirds. Harper Collins, London

Majerus MEN, Kearns PWE (1989) Ladybirds (Naturalists' handbook series). Richmond Publishing, Slough

Majerus MEN, Kearns PWE, Forge H, Ireland H (1989) Ladybirds as teaching aids: 1 collecting and culturing. J Biol Educ 23:85–95

Majerus TMO, Schulenburg JHGvd, Majerus MEN, Hurst GDD (1999) Molecular identification of a male-killing agent in the ladybird *Harmonia axyridis* (Pallas) (Coleoptera: Coccinellidae). Insect Mol Biol 8:551–555

Majerus MEN, Strawson V, Roy HE (2006) The potential impacts of the arrival of the harlequin ladybird, *Harmonia axyridis* (Pallas) (Coleoptera: Coccinellidae), in Britain. Ecol Entomol 31:207–215

Omkar, PA, Gupta AK (2004) Role of surface chemicals in egg cannibalism and intraguild predation by neonates of two aphidophagous ladybirds, *Propylea dissecta* and *Coccinella transversalis*. J Appl Entomol 128:692–695

Pasteels JM, Deroe C, Tursch B, Braekman JC, Daloze D, Hootele C (1973) Distribution et activités des alcaloïdes défensifs des Coccinellidae. J Insect Physiol 19:1771–1784

Pell JK, Baverstock J, Roy HE, Ware RL, Majerus MEN (2007) Intraguild predation involving *Harmonia axyridis*: a review of current knowledge and future perspectives. BioControl (this issue). doi: 10.1007/s10526-007-9125-x

Phoofolo MW, Obrycki JJ (1998) Potential for intraguild predation and competition among predatory Coccinellidae and Chrysopidae. Entomol Exp Appl 89:47–55

Polis GA, Myers CA, Holt RD (1989) The ecology and evolution of intraguild predation: potential competitors that eat each other. Annu Rev Ecol Syst 20:297–330

Pope RD (1979) Wax production by coccinellid larvae. Syst Entomol 4:171–196

Riddick EW, Schaefer PW (2005) Occurrence, density and distribution of parasitic fungus *Hesperomces virescens* (Laboulbeniales: Laboulbeniaceae) on multicoloured Asian lady beetle (Coleoptera: Coccinellidae). Ann Entomol Soc Am 103:799–804

Rosenheim JA, Kaya HK, Ehler LE, Marois JJ, Jaffee BA (1995) Intraguild predation among biological control agents: theory and evidence. Biol Control 5:303–335

Roy HE, Brown P, Majerus MEN (2006) *Harmonia axyridis*: a successful biocontrol agent or an invasive threat? In: Eilenberg J, Hokkanen H (eds) An ecological and societal approach to biological control. Kluwer Academic Publishers, Netherlands

Sato S, Dixon AFG (2004) Effect of intraguild predation on the survival and development of three species of aphidophagous ladybirds: consequences for invasive species. Agric Forest Entomol 6:21–24

Sato S, Dixon AFG, Yasuda H (2003) Effect of emigration on cannibalism and intraguild predation in aphidophagous ladybirds. Ecol Entomol 28:628–633

Sato S, Yasuda H, Evans EW (2005) Dropping behaviour of larvae of aphidophagous ladybirds and its effects on incidence of intraguild predation: interactions between the intraguild prey, *Adalia bipunctata* and *Coccinella septempunctata* and the intraguild predator, *Harmonia axyridis* Pallas. Ecol Entomol 30:220–224

Schellhorn NA, Andow DA (1999) Cannibalism and interspecific predation: role of oviposition behavior. Ecol Appl 9:418–428

Şengonca Ç, Frings B (1985) Interference and competitive behaviour of the aphid predators, *Chrysoperla carnea* and *Coccinella septempunctata* in the laboratory. Entomophaga 30:245–251

Takahashi K (1989) Intra- and inter-specific predations of lady beetles in spring alfalfa fields. Jpn J Entomol 57:199–203

Torchin ME, Lafferty KD, Dobson AP, Mckenzie VJ, Kuris AM (2003) Introduced species and their missing parasites. Nature 421:628–630

Völkl W, Vohland K (1996) Wax covers in larvae of two *Scymnus* species: do they enhance coccinellid larval survival? Oecologia 107:498–503

Ware RL, Majerus MEN, Roy HE, Symington F (2005) The harlequin ladybird arrives in Britain: a threat to our native species? Bull Amat Ent Soc 64:175–186

Ware RL, Ramon-Portugal F, Magro A, Ducamp C, Hemptinne J-L, Majerus MEN (2007) Chemical protection of *Calvia quatuordecimguttata* eggs against intraguild predation by the invasive ladybird *Harmonia axyridis*. BioControl (this issue). doi:10.1007/s10526-007-9129-6

Ware RL, Evans N, Malpas L, Michie LJ, O'Farrell K, Majerus MEN (in press) Intraguild predation by the invasive ladybird *Harmonia axyridis*: 1: British & Japanese coccinellid eggs. Neobiota

Williamson M (1996) Biological invasions (population and community biology series). Kluwer Academic Publishers, England

Chemical protection of *Calvia quatuordecimguttata* eggs against intraguild predation by the invasive ladybird *Harmonia axyridis*

Remy Lian Ware · Felipe Ramon-Portugal · Alexandra Magro · Christine Ducamp · Jean-Louis Hemptinne · Michael E. N. Majerus

Abstract Intraspecific and interspecific predation of eggs is a well documented phenomenon amongst aphidophagous coccinellids. The invasive species *Harmonia axyridis* (Pallas) (Coleoptera: Coccinellidae) is known to be a top intraguild predator and reported to attack the eggs of many coccinellid species both in a laboratory setting and in the wild. A previous laboratory study highlighted that while many species' eggs were highly palatable to *H. axyridis*, the eggs of *Calvia quatuordecimguttata* (Linnaeus) (Coleoptera: Coccinellidae) appeared to be extremely well protected from attack. Here we present the results of behavioural experiments testing the hypothesis that substances on the egg surface are responsible for this protection, and report preliminary results of GC-MS analysis of these compounds. When the coatings of *C. quatuordecimguttata* eggs were removed using hexane, they became significantly more susceptible to predation by neonate *H. axyridis* larvae. However, their overall palatability was not affected, in that complete consumption was never or rarely observed. This suggests that the surface compounds are a

R. L. Ware (✉) · M. E. N. Majerus
Department of Genetics, University of Cambridge, Downing Street, Cambridge CB2 3EH, UK
e-mail: r.ware@gen.cam.ac.uk

M. E. N. Majerus
e-mail: m.majerus@gen.cam.ac.uk

F. Ramon-Portugal · A. Magro · C. Ducamp · J.-L. Hemptinne
Laboratoire d'Agroécologie, UMR CNRS 5174 «Evolution et Diversité Biologique»,
Ecole Nationale de Formation Agronomique, BP 22687, 31326 Castanet-Tolosan, France

F. Ramon-Portugal
e-mail: felipe.ramonportugal@educagri.fr

A. Magro
e-mail: alexandra.magro@educagri.fr

C. Ducamp
e-mail: christine.ducamp@educagri.fr

J.-L. Hemptinne
e-mail: jean-louis.hemptinne@educagri.fr

Biological Control to Invasion: Ladybird Harmonia axyridis. H.E. Roy, E. Wajnberg (eds.),
doi: 10.1007/978-1-4020-6939-0_12,

true indicator of unpalatability in this species. The effect of hexane-washing on already palatable conspecific eggs was also analysed but had no significant effect on the susceptibility of eggs to cannibalism. We conclude that the eggs of at least one European species are effectively protected by surface deterrents from intraguild predation by *H. axyridis*. This effect might be due to both the diversity and abundance of hydrocarbons present within the egg coating, the presence of alkenes and/or the presence of patches of a red substance on the eggs' surface, which is thought to belong to the acid group. In conjunction with data on the susceptibility of other immature stages of *C. quatuordecimguttata*, this finding may indicate a decreased risk of the species falling victim to invasive *H. axyridis*, despite their coincident habitat ranges.

Keywords *Calvia quatuordecimguttata* · Chemical defence · Chemoecology · Coccinellidae · Cream-spot ladybird · Harlequin ladybird · *Harmonia axyridis* · Intraguild predation

Introduction

The egg stage represents a particularly vulnerable period for many vertebrates and invertebrates. Insect eggs are immobile and, in the absence of any parental protection, constitute an easy meal for a range of oophagous predators (Blum and Hilker 2002), in addition to being a target for various parasitoids and pathogens. Amongst aphidophagous coccinellids, the commonest cause of mortality at the egg stage is predation by conspecific or heterospecific larvae and adults (Banks 1956; Mills 1982; Majerus 1994; Dixon 2000; Cottrell 2004). Cannibalism of eggs by hatching larvae is known to dramatically increase survival of the latter (Pienkowski 1965; Majerus and Majerus 1997; Snyder et al. 2000). Through the process of intraguild predation, defined by Polis et al. (1989) as the killing and eating of a potential competitor, the consumption of an egg of another guild member has obvious benefits in that it can both provide a significant nutritional advantage and remove a future source of competition (Hemptinne et al. 2000a; Lucas 2005). Although non-conspecific eggs generally provide a poorer quality meal than conspecific eggs, intraguild predation is preferable to death from starvation and commonly occurs when local aphid populations decline (Agarwala et al. 1998; Obrycki et al. 1998; Agarwala and Yasuda 2001).

As an adaptation against predation, eggs of many coccinellid species are protected by unpalatable or toxic chemicals. Chemical protection of eggs is well reported from a diverse array of insect species (Blum and Hilker 2002), and in coccinellids usually involves de novo synthesis of species-specific alkaloids, which are present throughout the life cycle (Pasteels et al. 1973; King and Meinwald 1996). Although the production of this defence can be costly (Holloway et al. 1991, 1993), the benefits in terms of deterring a range of predators are significant (Marples 1993; Hemptinne et al. 2000a; Blum and Hilker 2002; Sato and Dixon 2004). The possession of intrinsic protective chemicals is thought only to make evolutionary sense if their existence is 'flagged' in some way by a signal on the egg surface: they are of little value if the eggs are killed before a predator learns their distasteful or toxic properties (Hemptinne et al. 2000b). An alternative explanation may come in the form of kin selection in species laying their eggs in clutches: if a naïve predator samples one unpalatable egg, that egg is killed, but further eggs in the same clutch may be avoided (Agarwala and Dixon 1993).

Various studies report the apparent toxicity or unpalatability of the eggs of *Harmonia axyridis* (Pallas) (Coleoptera: Coccinellidae) to larvae of other species (Burgio et al. 2002;

Cottrell 2004; Sato and Dixon 2004). Other life stages of *H. axyridis* appear equally well defended (Ware and Majerus 2007). This species is native to Asia but invasive in North America, continental Europe and more recently, the UK. The ability of *H. axyridis* to successfully exploit native members of the aphidophagous guild, while avoiding exploitation itself, is thought to be an important factor in its widespread establishment and spread (Sato and Dixon 2004; R. L. Ware et al. unpublished data). Dramatic declines in native coccinellids have been witnessed in North America following the establishment of *H. axyridis*, and have been attributed to both competition and intraguild predation of their immature stages (Hironori and Katsuhiro 1997; Cottrell and Yeargan 1998; Michaud 2002). Intraguild predation of eggs and other immature stages is now thought to pose a significant threat to many European coccinellids.

A previous study investigated the vulnerabilities and relative palatabilities of the eggs of a range of British and Japanese coccinellids to *H. axyridis* (Ware et al., in press). While most species' eggs were readily consumed by first instar larvae, eggs of one British species, *Calvia quatuordecimguttata* (Linnaeus) (Coleoptera: Coccinellidae), were almost always avoided. Other workers have reported the reluctance of other coccinellid species to consume *C. quatuordecimguttata* eggs: Dyson (1996) demonstrated violent rejection behaviour of *Harmonia quadripunctata* (Pontoppidan), *Adalia bipunctata* (Linnaeus) and *Propylea quatuordecimpunctata* (Linnaeus) larvae towards *C. quatuordecimguttata* eggs, and Vanhove (1998) presented similar findings for *Calvia decemguttata* (Linnaeus) and *Adalia decempunctata* (Linnaeus) larvae. However, this repellent effect was not witnessed with conspecific larvae and cannibalism occurs commonly (Dyson 1996; Vanhove 1998). It has been suggested that the agent responsible for deterring non-conspecific predators is on the surface of the egg, as larvae withdraw rapidly upon first contact (Majerus 1994; Dyson 1996; Vanhove 1998; Ware et al., in press). Dyson (1996) reports preliminary results showing the deterrent to be hexane-soluble, as eggs washed in hexane became palatable to all predators studied. Hemptinne et al. (2000b) provide evidence for similar surface deterrents on *A. bipunctata* and *Coccinella septempunctata* Linnaeus eggs. Extracts of eggs washed in hexane were shown to be composed mainly of alkanes, which act as signals to both conspecific and heterospecific larvae.

Here we present the results of preliminary behavioural experiments testing the hypothesis that substances on the egg surface are responsible for protection of *C. quatuordecimguttata* eggs against intraguild predation by *H. axyridis*. We also present the results of chemical analysis of the coating of *C. quatuordecimguttata* eggs, and make comparisons with the substances already identified from *A. bipunctata* and *C. septempunctata* eggs. Finally, we discuss the results in relation to the threat posed to *C. quatuordecimguttata* by the establishment of *H. axyridis* in Europe, and the role of egg surface chemistry in providing signals to conspecific and heterospecific predators.

Materials and methods

Coccinellid rearing

All larvae and eggs used in experiments were derived from field collected pairs or F1 material. *H. axyridis* stocks were derived from samples sent into the Harlequin Ladybird Survey from around the UK. Stocks of *C. quatuordecimguttata* were collected from either Cambridge or London, UK. Mating pairs of all species were housed in 9 cm Petri dishes lined with filter paper and kept in the laboratory at 22°C and 16 hL: 8 hD. Adults were supplied daily with excess pea aphids, *Acyrthosiphon pisum* (Harris), and filter papers were replaced daily. A few clutches of *H. axyridis* eggs were retained to generate the larvae required for the behavioural experiments, while the remaining *H. axyridis* eggs and all *C. quatuordecimguttata* eggs were subjected to the treatments detailed below.

Extraction of egg coating

Eggs were detached from the filter paper on which they were laid using the tip of a closed glass Pasteur pipette, after wetting the filter paper with a drop of de-ionised water to loosen the ovipositional glue. Half of the eggs available per day for each species were placed in a 5 ml test tube and washed in hexane (treatment A), while the other half were left unwashed but otherwise treated similarly (treatment B). Treatment A eggs were washed in 3 ml of n-hexane for liquid chromatography (Merck, Hohenbrumn, Germany) for 2 min, without agitation. The supernatant was carefully removed using a Pasteur pipette and the remaining solvent evaporated under a gentle stream of nitrogen for 3 min. After extraction, the eggs were observed under a dissecting microscope and, whenever broken eggs were detected, the extract was discarded. The extract of *C. quatuordecimguttata* eggs was conserved at 4°C until GC-MS analysis. Intact washed eggs were used for the behavioural experiments.

Gas chromatography-mass spectrometry (GC-MS) analysis

The extracts obtained from washing 42 *C. quatuordecimguttata* eggs were pooled. The solvent was evaporated under a gentle stream of nitrogen and the dry residue dissolved in 8.4 μl of n-hexane (20 μl per 100 eggs). Elucidation of the structure of the compounds in the n-hexane extract was carried out by GC-MS (Finnigan Trace 2000 operated by electron impact at 70 eV) using a helium carrier at 1.2 ml/min, operated under the following temperatures: source: 200°C, interface between GC and MS modules: 250°C, splitless injector: 280°C. Samples of 1 μl were injected using an apolar capillary column (Restek RTX-5MS, 30 × 0.25 mm^2, 0.25 μm film thickness). The temperature programme of the oven was set to: 50°C for 1 min, then from 50 to 140°C at 20°C per min, then from 140 to 300°C at 3°C per min, and finally held at 300°C for 3 min. The mass spectra were scanned from 60 to 450 m/z. On-line data acquisition was carried out with Xcalibur software (Thermo Finnigan, MA, USA). Detection limits are automatically established by the Xcalibur software, which only considers peaks over 0.4% of the highest peak. Compounds were identified by their fragmentation patterns compared with the NIST library and by comparison with previously injected known compounds. The quantification of each compound was carried out using a calibration curve at different concentrations of a mixture of hydrocarbons from C12 to C60 (Sigma-Aldrich, Tanfkirchen, Germany). The non-adecane [$CH_3(CH_2)_{17}CH_3$; Sigma-Aldrich] was used as internal standard. In order to quantify each detected compound, 3 μl of extract were mixed with 3 μl of a non-adecane solution (at 0.0781 mg l^{-1} in n-hexane).

Behavioural experiments

Two neonate sibling larvae of *H. axyridis* were carefully removed from their clutch before dispersal, using a fine bristled paintbrush, and placed individually in clean 4 cm Petri dishes. One larva was provided with a single *C. quatuordecimguttata* egg that had been washed in hexane (treatment A) while the other was provided with a single unwashed *C. quatuordecimguttata* egg (treatment B). Eggs were manipulated using a blunt mounted entomological pin. No other food or water was provided thereafter. The larvae were observed at regular intervals and times of death recorded. No larvae underwent ecdysis during this time. Survival times were calculated from the time of experimental set-up. After 24 h the 'status' of the egg was observed carefully using a 15× hand lens and placed into one of three categories as in Ware et al. (in press): not eaten (N), when the egg was completely untouched; partially eaten (P), when the chorion was broken but a substantial amount of yolk remained; and completely eaten (C), when the whole egg had been consumed or yolk remained only at the poles. The experiment was replicated 20 times. The reaction of *H. axyridis* larvae to washed and unwashed non-sibling conspecific eggs was also monitored in this way, with a view to assessing the effect of the washing process on a normally palatable egg. Survival controls involved placing a neonate larva in a dish with no egg (or other food or water), and noting its time of death.

When the chorion of an egg is broken by a sampling predator, the embryo within it is killed, whether it is subsequently eaten or dies through desiccation. From the perspective of the egg's individual fitness, therefore, there is essentially no difference between category 'C' and category 'P'. Eggs falling into category 'N' will be the only potential survivors. On the other hand, if an egg is palatable, it will presumably be completely consumed, rather than being sampled and then left behind. Thus, the proportion of eggs falling into category 'C' gives an indication of the palatability of the egg to the larva.

Statistical analysis

The effect of hexane washing on the number of eggs falling into each category was analysed for each species using a fully orthogonal approach. First, we tested for heterogeneity in the data using the Fisher-Freeman-Halton test on a 3 × 2 contingency table (2 *df*). If heterogeneity was found, Fisher's exact tests (1 *df*) were performed on the number of eggs falling into category 'C' versus 'P', and the numbers of eggs falling into 'C + P' versus 'N'. This analysis considers the palatability and the susceptibility of the eggs, respectively. Survival times were compared using the Kruskal-Wallis test and multiple pairwise Mann-Whitney *U*-tests adjusting for ties. Alpha values were adjusted according to the Bonferroni correction for multiple comparisons (referred to in text as αB). The Fisher-Freeman-Halton test and Fisher's exact test were carried out in StatXact 7, and Minitab 14 was used to perform Kruskal-Wallis and Mann-Whitney *U*-tests.

Results

Gas chromatography-mass spectrometry (GC-MS) analysis

It is unfortunate that only a relatively small number of *C. quatuordecimguttata* eggs (n = 42) were available for GC-MS analysis, due to difficulty in manipulating the eggs without breaking them and logistical constraints of culturing this species in the time available. Normally the authors would aim to pool the extracts of three lots of 100 eggs each. We therefore report the following results as preliminary data only. Figure 1 shows the mass spectrum for the extract of 42 *C. quatuordecimguttata* eggs. The peaks are

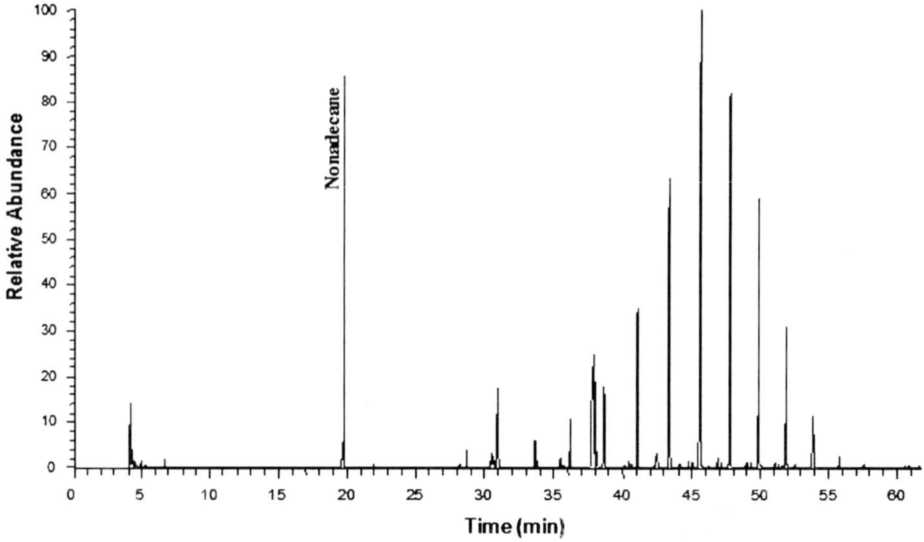

Fig. 1 Mass spectrum of the extract of *Calvia quatuordecimguttata* egg coatings. Non-adecane is the internal standard

identified and quantified in Table 1. Twenty-four peaks were obtained, of which 20 have been positively identified. A total quantity of 17.60 mg of surface hydrocarbons per 50 eggs was obtained, of which 90.28% were linear alkanes, 8.59% were alkenes, and the remaining 1.13% represented the four unidentified peaks. The alkane Hentriacontane was the major compound, and represented 20% of the total quantity (3.50 mg). These results, particularly in respect of the alkenes and double bond positioning, need to be confirmed by compound-specific chemical analysis.

During manipulation of *C. quatuordecimguttata* eggs, the authors observed the presence of patches of a red substance on their surface, visible under a 15× hand lens. This substance was found to be soluble in both water and hexane, which indicates that it could belong to the acid group with low relative molecular weight. Further analysis of its chemical nature is still needed.

Behavioural experiments

The effect of washing *C. quatuordecimguttata* eggs in hexane on its status after 24 h when left with a neonate *H. axyridis* larva is shown in Fig. 2. There is significant heterogeneity in the numbers of eggs falling into each category between treatments A and B ('C' versus 'P' versus 'N': Fisher statistic = 11.42, 2 *df*, $P = 0.002$). Further analysis indicates there is no difference in palatability between treatments ('C' versus 'P': Fisher statistic = 0.483, 1 *df*, $P \sim 1.00$) but that more washed eggs were killed than unwashed eggs ('C + P' versus 'N': Fisher statistic = 11.09, 1 *df*, $P = 0.002$).

The results of the hexane treatment of *H. axyridis* eggs on cannibalism are shown in Fig. 3. There was no significant heterogeneity within this data, showing that washing conspecific eggs in hexane did not affect how many were completely eaten, partially eaten, or not eaten, by larvae after 24 h. ('C' versus 'P' versus 'N': Fisher statistic = 2.94, 2 *df*, $P = 0.216$).

Table 1 Compounds identified from extracts of *C. quatuordecimguttata* egg coatings

Compounds	*C. quatuordecimguttata*		*A. bipunctata*		*C. septempunctata*	
	mg l^{-1}	%	mg l^{-1}	%	mg l^{-1}	%
Heneicosane			0.50	11.90		
Docosane	0.01	0.04	0.10	2.38	0.04	3.64
6-Tricosene	0.03	0.19				
Unidentified	0.13	0.71				
Tricosane	0.20	1.15	0.30	7.14	0.13	11.82
7-Methytricosane			1.20	28.57		
9-Methytricosane			2.10	50.00		
Tretracosane	0.07	0.38			0.11	10.00
Unidentified	0.02	0.12				
6-Pentacosene	0.02	0.14				
Pentacosane	0.15	0.86			0.17	15.45
6-Hexacosene	1.24	7.06				
Hexacosane	0.29	1.65			0.05	4.55
6-Heptacosene	0.04	0.21				
Heptacosane	0.67	3.83			0.44	40.00
6-Octacosene	0.06	0.37				
Octacosane	1.50	8.53			0.04	3.64
Unidentified	0.02	0.11				
6-non-acosene	0.04	0.23				
Non-acosane	2.99	17.01			0.12	10.91
6-Triacontene	0.07	0.40				
Triacontane	3.01	17.08				
Unidentified	0.03	0.19				
Hentriacontane	3.50	19.87				
Dotriacontane	1.95	11.09				
Tritriacontane	1.16	6.60				
Tetratriacontane	0.38	2.19				
Total	17.60	100.00	4.20	100.00	1.10	100.00

Compounds from published data for *A. bipunctata* and *C. septempunctata* egg coatings (Hemptinne et al. 2000b) are given for comparison. Results are expressed in mg per litre for 50 eggs, and as relative % amounts

There was a significant difference in survival between larvae subjected to the five different feeding regimes shown in Fig. 4 (Kruskal-Wallis test: $H = 53.07$, $df = 4$, $P < 0.001$). For larvae that had consumed an untreated or treated *C. quatuordecimguttata* egg before their death, survival was significantly increased relative to the control (Mann-Whitney *U*-tests: αB (10 tests) = 0.005: $W = 214.0$, $P < 0.001$ and $W = 219.0$, $P < 0.001$, respectively). However, there was no difference in survival between larvae that had received untreated or hexane-treated *C. quatuordecimguttata* eggs (Mann-Whitney *U*-test: αB (10 tests) = 0.005: $W = 309.5$, $P = 0.056$) (Fig. 3). Consumption of an untreated or treated conspecific egg significantly increased larval survival compared to the control (Mann-Whitney *U*-test: αB (10 tests) = 0.005: $W = 210.0$, $P < 0.001$ and $W = 231.0$, $P < 0.001$, respectively). There was no difference in survival between larvae that had consumed untreated conspecific eggs or hexane-treated conspecific eggs (Mann-Whitney

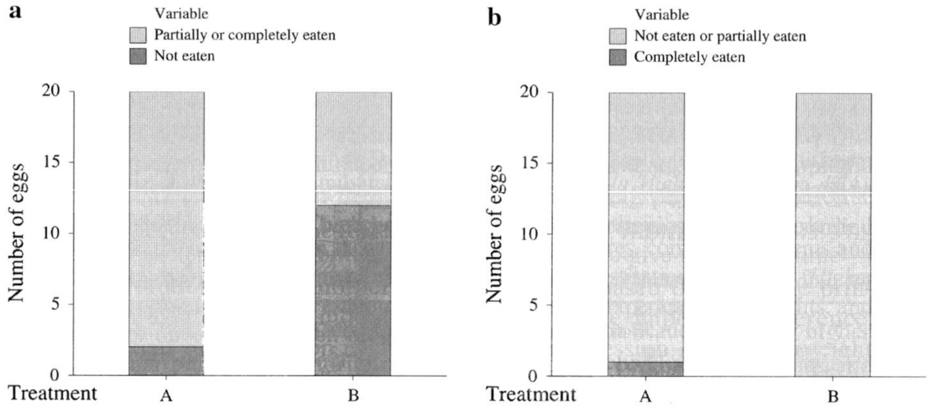

Fig. 2 (**a**) Susceptibility and (**b**) palatability of *Calvia quatuordecimguttata* eggs to intraguild predation by neonate *Harmonia axyridis* larvae after 24 h under treatments A (hexane washed) and B (unwashed). $n = 20$ for each treatment

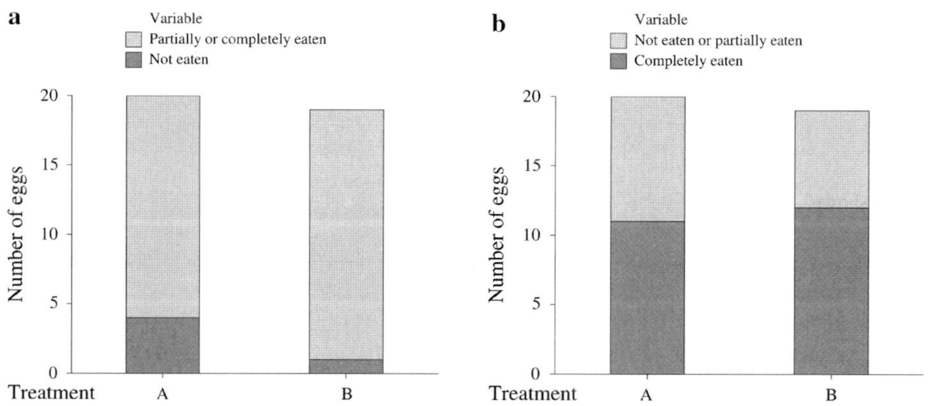

Fig. 3 (**a**) Susceptibility and (**b**) palatability of *Harmonia axyridis* eggs to cannibalism by neonate non-sibling larvae after 24 h under treatments A (hexane washed) and B (unwashed). $n = 20$ for each treatment

U-test: αB (ten tests) = 0.005: $W = 322.5$, $P = 0.596$). There was no statistically significant difference in survival between larvae that had consumed either treated or untreated conspecific eggs and those that had consumed untreated *C. quatuordecimguttata* eggs (Mann-Whitney *U*-test: αB (ten tests) = 0.005: $W = 225.0$, $P = 0.094$ and $W = 223.0$, $P = 0.051$, respectively). However, the consumption of treated *C. quatuordecimguttata* eggs resulted in a lower survival than the consumption of treated or untreated conspecific eggs (Mann-Whitney *U*-test: αB (ten tests) = 0.005: $W = 279.0$, $P = 0.002$ and $W = 256.5$, $P < 0.001$, respectively) (Fig. 4).

Discussion

Hemptinne et al. (2000b) argue that unpalatability or toxicity of coccinellid eggs should be advertised by a chemical signal on the egg's surface. They show that eggs of *A. bipunctata*

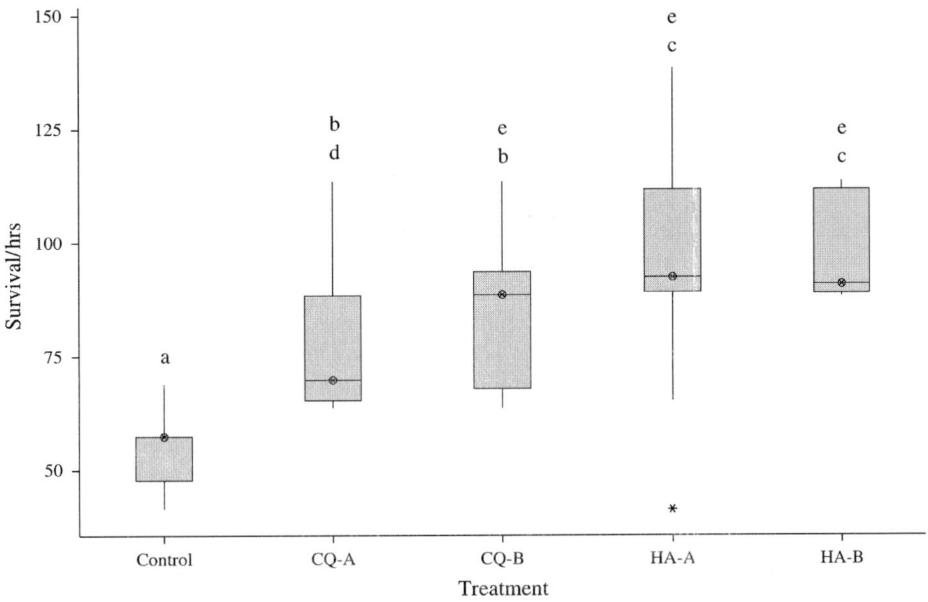

Fig. 4 Survival of neonate *Harmonia axyridis* larvae after having partially or completely consumed one egg of *Calvia quatuordecimguttata* (CQ) or of its own species (HA) under treatments A (hexane washed) and B (unwashed). No statistically significant difference was found between treatments plotted with the same letter. Medians (circles with stars), interquartile ranges (boxes), ranges (vertical lines) and outliers (stars) shown

are coated with hexane-soluble hydrocarbons that deter predation by both conspecific larvae and larvae of *C. septempunctata*, and that these chemicals are similar to those used in larval tracks (Doumbia et al. 1998; Magro et al. 2007) and mate recognition (Hemptinne et al. 1998). Meanwhile, *C. septempunctata* eggs are coated in surface deterrents that prove an effective defence against *A. bipunctata* larvae but not conspecific larvae (Hemptinne et al. 2000b). Other studies suggest that when eaten, the eggs of these two species impose a nutritional cost on larvae of the other species (Agarwala and Dixon 1992; Hemptinne et al. 2000a; Sato and Dixon 2004). This implies that surface chemical deterrents are honest signals of the cost of intraguild predation between these two species.

This study has confirmed that eggs of *C. quatuordecimguttata* are also protected by chemical deterrents on their surface. Preliminary chemical analysis indicates that the egg coating consists mainly of hydrocarbons, and highlights another, as yet unidentified, substance, thought to be a member of the acid group. Hydrocarbons are long-lasting and highly resistant molecules that are found commonly in nature and used by a range of species as cues for communication purposes. Hydrocarbons have been identified in the egg coatings of *A. bipunctata* and *C. septempunctata* by Hemptinne et al. (2000b). There are, however, some striking differences between the compounds found on the eggs of *C. quatuordecimguttatta* and these two species (Table 1). First, the egg coating of *C. quatuordecimguttata* contains four times the amount of hydrocarbons found in *A. bipunctata*, and 16 times the amount isolated from *C. septempunctata* eggs. Second, a more diverse range of compounds were isolated from *C. quatuordecimguttata* compared to the other two species: the extract from *C. quatuordecimguttata* contained 24 different compounds, while those of *A. bipunctata* and *C. septempunctata* contained only five and eight, respectively. If the chemical results are confirmed, the extract of *C. quatuordecimguttatta* is unique in containing alkenes.

Dyson (1996) had already noted the presence of patches of a red substance on the surface of *C. quatuordecimguttata* eggs and speculated that they were a deterring agent. However, we suggest that their solubility in water would make them rather unreliable as a form of chemical defence. Further analysis is in progress in order to fully understand the nature and role of these patches. Nevertheless, the work presented here clearly shows that the removal of surface chemicals from *C. quatuordecimguttata* eggs using a hexane solvent significantly increases their susceptibility to attack by neonate *H. axyridis* larvae, and is therefore in agreement with the results previously found using larvae of other coccinellid species (Dyson 1996). Furthermore, the results show that while the removal of surface molecules increases the proportion of eggs attacked, they remain significantly unpalatable to *H. axyridis* larvae and are rarely completely consumed within 24 h. This suggests that here the surface signals are a reliable indicator of intrinsic toxic or distasteful chemicals, as theorised by Hemptinne et al. (2000b). We predict that coating *H. axyridis* eggs with the surface extract from *C. quatuordecimguttata* will make them unpalatable to *H. axyridis* larvae. Work to test this prediction is in progress.

When *H. axyridis* larvae did consume all or part of a *C. quatuordecimguttata* egg, no toxic effect was shown, in that it did not reduce survival below that of a larva given no food at all. We therefore suggest that the intrinsic chemicals are significantly distasteful to *H. axyridis* but not toxic, at least in the short-term. This should be confirmed by studies into the longer-term effects on development. While the consumption of a *C. quatuordecimguttata* egg can provide some short-term nutritional benefit to a neonate *H. axyridis* larva, it seems less than that afforded by a conspecific egg. This is particularly evident when considering larvae that had consumed treated *C. quatuordecimguttata* eggs, although this may simply be due to the larger sample size of individuals that sampled an egg when the egg had been washed. This result is in accordance with other work citing the increased value of consuming conspecific eggs over heterospecific eggs (Agarwala and Yasuda 2001; Cottrell 2004; Sato and Dixon 2004; Omkar et al. 2004).

It is thought that co-occurring species should be protected from intraguild predation by each other, either by means of chemical defence, or through behavioural adaptations that reduce the likelihood of encounters (Sato and Dixon 2004). For example, in Japan, *H. axyridis* and *C. septempunctata brucki* Mulsant are frequently found together in alfalfa fields (Takahashi 1989). *H. axyridis* is relatively resistant to intraguild predation from *C. septempunctata brucki* due the possession of defensive chemicals and robust physical defence structures (Ware and Majerus 2007). Moreover, as *C. septempunctata brucki* oviposits earlier in the year than *H. axyridis*, the eggs and younger larvae of *C. septempunctata brucki* are rarely encountered by *H. axyridis* larvae (Takahashi 1989; Sato and Dixon 2004). While sympatric species might be expected to possess co-evolved defence strategies against intraguild predation, such co-evolved strategies are unlikely between taxa whose ranges do not overlap. Ware et al. (in press) discuss the risk posed to many British species by the arrival of invasive *H. axyridis*. One exception appears to be *C. quatuordecimguttata*. This species is usually considered to be a deciduous tree specialist (Majerus 1991) and although the habitat range of *H. axyridis* is considerably more generalist (Roy et al. 2006), it is likely that the two species will interact in both Britain and Europe where their ranges overlap. However, this study has provided evidence that *C. quatuordecimguttata* eggs are effectively protected from predation by *H. axyridis*, and Ware and Majerus (2007) have also shown its larvae and pupae to be relatively well defended from attack. It therefore seems that at least one European species is already well-equipped to deal with the encroachment of *H. axyridis.*

Acknowledgements The authors wish to acknowledge the help and support of members of the Cambridge University Department of Genetics Field Station in collecting and maintaining stocks in the UK, and of the Laboratoire d'Agroécologie (UMR CNRS 5174) at École Nationale de Formation Agronomique, Toulouse. We would also like to thank Dr. Francois Balloux from the Department of Genetics, University of Cambridge, for advice on statistical analysis.

References

Agarwala BK, Bhattacharya S, Bardhanroy P (1998) Who eats whose eggs? Intra- versus inter-specific interactions in starving ladybird beetles predaceous on aphids. Ethol Ecol Entomol 10:361–368

Agarwala BK, Dixon AFG (1992) Laboratory study of cannibalism and interspecific predation in ladybirds. Ecol Entomol 17:303–309

Agarwala BK, Dixon AFG (1993) Why do ladybirds lay eggs in clusters? Funct Ecol 7:541–548

Agarwala BK, Yasuda H (2001) Overlapping oviposition and chemical defense of eggs in two co-occurring species of ladybird predators of aphids. J Ethol 19:47–53

Banks CJ (1956) Observations on the behaviour and mortality in Coccinellidae before dispersal from the eggshells. P R Soc Lond A 31:56–61

Blum MS, Hilker M (2002) Chemical protection of insect eggs. In: Hilker M, Meiners T (eds) Chemoecology of insect eggs and egg deposition. Blackwell, Berlin

Burgio G, Santi F, Maini S (2002) On intra-guild predation and cannibalism in *Harmonia axyridis* (Pallas) and *Adalia bipunctata* L. (Coleoptera: Coccinellidae). Biol Control 24:110–116

Cottrell TE (2004) Suitability of exotic and native lady beetle eggs (Coleoptera: Coccinellidae) for development of lady beetle larvae. Biol Control 31:362–371

Cottrell TE, Yeargan KV (1998) Intraguild predation between an introduced lady beetle, *Harmonia axyridis* (Coleoptera: Coccinellidae), and a native lady beetle, *Coleomegilla maculata* (Coleoptera: Coccinellidae). J Kan Entomol Soc 71:159–163

Dixon AFG (2000) Insect predator-prey dynamics: ladybird beetles and biological control. Cambridge University Press, Cambridge

Doumbia M, Hemptinne JL, Dixon AFG (1998) Assessment of patch quality by ladybirds: role of larval tracks. Oecologia 113:197–202

Dyson E (1996) An investigation of *Calvia quatuordecimguttata*: egg cannibalism, egg predation and evidence for a male-killer. Undergraduate Dissertation, University of Cambridge

Hemptinne JL, Dixon AFG, Gauthier C (2000a) Nutritive cost of intraguild predation on eggs of *Coccinella septempunctata* and *Adalia bipunctata* (Coleoptera: Coccinellidae). Eur J Entomol 97:559–562

Hemptinne JL, Lognay G, Dixon AFG (1998) Mate recognition in the two-spot ladybird beetle, *Adalia bipunctata*: role of chemical and behavioural cues. J Insect Physiol 44:1163–1171

Hemptinne JL, Lognay G, Gauthier C, Dixon AFG (2000b) Role of surface chemical signals in egg cannibalism and intraguild predation in ladybirds (Coleoptera: Coccinellidae). Chemoecology 10: 123–128

Hironori Y, Katsuiro S (1997) Cannibalism and interspecific predation in two predatory ladybirds in relation to prey abundance in the field. Entomophaga 42:153–163

Holloway GJ, de Jong PW, Brakefield PM, de Vos H (1991) Chemical defence in ladybird beetles (Coccinellidae). I. Distribution of coccinelline and individual variation in defence in 7-spot ladybirds (*Coccinella septempunctata*). Chemoecology 2:7–14

Holloway GJ, de Jong PW, Ottenheim M (1993) The genetics and cost of chemical defence in the two-spot ladybird *Adalia bipunctata* L. Evolution 47:1229–1239

King AG, Meinwald J (1996) Review of the defensive chemistry of coccinellids. Chem Rev 96:1105–1122

Lucas É (2005) Intraguild predation among aphidophagous predators. Eur J Entomol 102:351–364

Magro A, Téné JN, Bastin N, Dixon AFG, Hemptinne JL (2007) Assessment of patch quality by ladybirds: relative response to conspecific and heterospecific larval tracks a consequence of habitat similarity? Chemoecology 17:37–45

Majerus MEN (1991) Habitat and host plant preferences of British ladybirds. Entomol Mon Mag 127: 167–175

Majerus MEN (1994) Ladybirds. Harper Collins, London

Majerus MEN, Majerus TMO (1997) Cannibalism among ladybirds. B Amat Ent Soc 56:235–248

Marples NM (1993) Toxicity assays of ladybirds using natural predators. Chemoecology 4:33–38

Michaud JP (2002) Invasion of the Florida citrus ecosystem by *Harmonia axyridis* (Coleoptera: Coccinellidae) and asymmetric competition with a native species, *Cycloneda sanguinea*. Environ Entomol 31:827–835

Mills NJ (1982) Voracity, cannibalism and coccinellid predation. Ann Appl Biol 10:144–148

Obrycki JJ, Giles KL, Ormord AM (1998) Interactions between an introduced and indigenous coccinellid species at different prey densities. Oecologia 117:279–285

Omkar Pervez A, Gupta AK (2004) Role of surface chemicals in egg cannibalism and intraguild predation by neonates of two aphidophagous ladybirds *Propylea dissecta* and *Coccinella transversalis*. J Appl Entomol 128:691–695

Pasteels JM, Deroe C, Tursch B, Braekman JC, Daloze D, Hootele C (1973) Distribution et activités des alcaloïdes défensifs des Coccinellidae. J Insect Physiol 19:1771–1784

Pienkowski RL (1965) The incidence and effect of egg cannibalism in first-instar *Coleomegilla maculata lengi* (Coleoptera: Coccinellidae). Ann Entomol Soc Am 58:150–153

Polis GA, Myers CA, Holt RD (1989) The ecology and evolution of intraguild predation: potential competitors that eat each other. Annu Rev Ecol Syst 20:297–330

Roy HE, Brown P, Majerus MEN (2006) *Harmonia axyridis*: a successful biocontrol agent or an invasive threat? In: Eilenberg J, Hokkanen H (eds) An ecological and societal approach to biological control. Kluwer, The Netherlands

Sato S, Dixon AFG (2004) Effect of intraguild predation on the survival and development of three species of aphidophagous ladybirds: consequences for invasive species. Agric For Entomol 6:21–24

Snyder WE, Joseph SB, Preziosi RF, Moore AJ (2000) Nutritional benefits of cannibalism for the lady beetle *Harmonia axyridis* (Coleoptera: Coccinellidae) when prey quality is poor. Environ Entomol 29: 1173–1179

Takahashi K (1989) Intra- and inter-specific predations of lady beetles in spring alfalfa fields. Jpn J Entomol 57:199–203

Vanhove F (1998) Impact de la défense des oeufs sur la structure des communautes de Coccinellidae. Masters Dissertation, University of Gembloux

Ware RL, Evans N, Malpas L, Michie LJ, O'Farrell K, Majerus MEN (in press) Intraguild predation by the invasive ladybird *Harmonia axyridis*: 1: British & Japanese coccinellid eggs. Neobiota

Ware RL, Majerus MEN (2007) Intraguild predation of immature stages of British and Japanese coccinellids by the invasive ladybird *Harmonia axyridis*. BioControl (this issue). doi:10.1007/s10526-007-9135-8

Pollen as an alternative food for *Harmonia axyridis*

Nick Berkvens · Jochem Bonte · Dirk Berkvens · Koen Deforce · Luc Tirry · Patrick De Clercq

Abstract The current study examines the potential of the multicoloured Asian lady beetle *Harmonia axyridis* (Pallas) (Coleoptera: Coccinellidae) to use pollen as a food to sustain development and reproduction in the absence of insect prey. Three populations of *H. axyridis* were used in this study: a long-term laboratory population (since 1998) and a melanic and non-melanic population originating from field collected individuals in Belgium. The insects were allowed to develop and reproduce on frozen eggs of *Ephestia kuehniella* Zeller (Lepidoptera: Phycitidae), frozen moist bee pollen or an even mixture of the two. Females of the field population offered the mixed diet initiated oviposition sooner than those fed only *E. kuehniella* eggs, but other developmental and reproductive traits were similar on these diets. A diet of pollen alone allowed 35–48% of the larvae of the field population of *H. axyridis* to successfully reach adulthood. However, developmental time for these individuals was prolonged by 31–49% and adult body weight was reduced by 37–68%, compared to individuals offered the diets containing *E. kuehniella* eggs. When fed exclusively on pollen in their larval and adult life, about 40% of the adult females of

N. Berkvens (✉) · J. Bonte · L. Tirry · P. De Clercq
Laboratory of Agrozoology, Department of Crop Protection, Faculty of Bioscience Engineering, Ghent University, Coupure Links 653, 9000 Ghent, Belgium
e-mail: Nick.Berkvens@UGent.be

J. Bonte
e-mail: Jochem.Bonte@UGent.be

L. Tirry
e-mail: Luc.Tirry@UGent.be

P. De Clercq
e-mail: Patrick.DeClercq@UGent.be

D. Berkvens
Department of Animal Health, Prince Leopold Institute of Tropical Medicine, Antwerp, Belgium
e-mail: dberkvens@itg.be

K. Deforce
Flemish Heritage Institute, Brussels, Belgium
e-mail: koen.deforce@rwo.vlaanderen.be

Biological Control to Invasion: Ladybird Harmonia axyridis. H.E. Roy, E. Wajnberg (eds.), 201
doi: 10.1007/978-1-4020-6939-0_13, © IOBC 2007

either field population were able to produce a small number of viable eggs. The laboratory and field strains differed in their response to diet for a number of developmental and reproductive traits. The exploitation of pollen and other plant foods at times when insect prey is scarce, may offer a further competitive advantage to the non-indigenous coccinellid *H. axyridis* over native European predatory lady beetles that share the same niche and are less capable of using pollen as an alternative food.

Keywords Coccinellidae · Coleoptera · Development · *Ephestia kuehniella* · *Harmonia axyridis* · Invasive species · Morph type · Pollen · Reproduction

Introduction

Harmonia axyridis (Pallas) (Coleoptera: Coccinellidae) is an aphidophagous coccinellid, indigenous to central and eastern Asia (Coderre et al. 1995; Nalepa et al. 1996; Koch 2003). The species was used as an augmentative biological control agent of aphid pests in orchards, field crops and greenhouses in both North America and Western Europe. Most probably as a result of these augmentative releases *H. axyridis* established in both continents (Koch 2003; Coutanceau 2006; Brown et al. 2007). Accompanying the large-scale colonisation in North America, several harmful side-effects were recorded: a shift in the coccinellid communities due to the dominance of the invader along with a decline in population numbers of some indigenous lady beetle species, nuisance to humans during its dormancy phase and secondary damage to crops (Koch 2003; Kenis et al. 2007; Pell et al. 2007). The species is at present exponentially expanding its geographical range in Europe (Brown et al. 2007) and could in its aftermath have a similar impact as that in North America. The mechanisms driving the successful colonisation and the ecological consequences for Europe are, however, still not entirely clear.

The present direction in scientific research concerning invasive species is in understanding the mechanisms determining the success of invasions and identifying the traits of successful invaders and the conditions under which these traits promote invasion (Holway 1999; Byers 2000). An important factor in the establishment of an invading species is the food range (van Lenteren et al. 2003). The exploitation of available resources and the nature of mutual competitive interactions with native ecological analogs is an important determinant of the invasion success of a species (Byers 2000). *Harmonia axyridis* is a generalist predator attacking Tetranychidae, Psyllidae, Coccoidea and immature stages of Chrysomelidae, Curculionidae and Lepidoptera, but the main food source for adults consists of several aphid species (Tedders and Schaefer 1994; Hodek and Honek 1996; Koch 2003). Many predaceous arthropods, including coccinellids, are also facultative plant feeders and additionally feed on a wide variety of non-prey foods such as pollen, nectar, fungal spores, plant wound exudates and phloem sap (Gilbert 1981; Hodek and Honek 1996; Lamana and Miller 1996; Patt et al. 2003; Koch et al. 2004; Lundgren et al. 2004). The supplementary nutrients found in these foods can either sustain the predators in periods where nutritionally optimal prey is scarce or complement the nutrients obtained from suboptimal prey (Kiman and Yeargan 1985; Hodek and Honek 1996; Jørgenson and Toft 1997; Patt et al. 2003).

This study investigated whether pollen can be exploited by *H. axyridis* as an alternative food source. This was done by assessing the effect of pollen on the developmental and reproductive performance of melanic and non-melanic morphs of a European field population of *H. axyridis*. The results are discussed in view of the significance of non-prey food for the establishment potential of this species. Additionally, a long-term laboratory population originating from a commercial culture that was one of the sources of biological control releases in the area since 1997 (Brown et al. 2007) was included in the study to compare its performance with that of the field population.

Materials and methods

Populations of *H. axyridis*

All experiments were conducted in the laboratory using three populations: a laboratory population and two field populations.

Larvae acquired from Biobest NV (Westerlo, Belgium) were used to start the laboratory population in September 1998 at the Laboratory of Agrozoology of Ghent University. The laboratory colony was repeatedly infused with new adults obtained from the same source up to August 2000. The commercial supplier did not reveal the origin of its culture or the methods used for rearing. The current experiments were conducted using individuals of the 55th generation of rearing at Ghent University. At the time of this study, the laboratory population consisted of more than 99% melanic *spectabilis* and *conspicua* morphs.

Using visual, sweeping and beating techniques (Elliott et al. 1991) individuals from an established wild population were collected in October 2005 in a wild open space in Ghent, Belgium. Two populations were initiated using the collected individuals; these populations will be referred to as 'field populations'. The first field population consisted of non-melanic *succinea* individuals (the 'red field population') and the second of melanic *spectabilis* and *conspicua* individuals (the 'black field population'). Individuals of the 5th generation of these populations were used for the experiments.

All populations were reared on frozen *Ephestia kuehniella* Zeller (Lepidoptera: Phycitidae) eggs, as described by Ongagna et al. (1993) and De Clercq et al. (2003).

Experimental design

Experiments were conducted in Petri dishes (diameter 9 cm, height 2 cm) that were kept in incubators held at 23 ± 1°C, 65 ± 5% RH and a 16:08 h (L:D) photoperiod. Three diets were tested: frozen *E. kuehniella* eggs, frozen moist pollen collected by honeybees (i.e. fresh, non-dried pollen placed in the deepfreeze immediately upon collection) and a mixture of frozen bee pollen and *E. kuehniella* eggs in equal weight proportions (referred to as the mixed diet). Both frozen bee pollen and *E. kuehniella* eggs were obtained from Koppert BV (Berkel en Rodenrijs, The Netherlands). The composition of the pollen was determined after Erdtman's acetolysis and mounting on microscope slides in glycerol (Erdtman 1960). Identifications of the pollen were based on Moore et al. (1991) and Beug (2004) and a reference pollen collection. The honeybee collected pollen consisted mainly of pollen from Brassicaceae (74.9–79.2%) and *Verbascum* spp. (Scrophulariaceae) (9.7–16.2%). Additionally small percentages of pollen of *Cornus sanguinea* L., *Trifolium repens* L., *Ranunculus acris* L. and of *Prunus* and *Vitis* sp. were found. Pollen pellets were thawed and finely pulverized using a Nova type 20 grinder before being offered to the insects or being mixed with the *E. kuehniella* eggs.

Twenty pairs of adults (3–5 days old) were randomly taken from each population. The adult pairs were placed in individual Petri dishes and left to mate and reproduce for one week. During this period they were fed a mixture of *E. kuehniella* eggs and pollen to prevent food source conditioning. After this week an accordion-pleated carton board strip was placed in each Petri dish to serve as an oviposition substrate and all eggs deposited during the following day were collected. Upon hatching, nine larvae were randomly collected per adult pair and placed individually in a Petri dish. Three larvae from each female were allotted to each of the diets tested. In this manner each diet treatment started off with 60 first-instar larvae that were then left to develop to adulthood. Sex of the resulting adults was determined (McCornack et al. 2007) and twenty to thirty pairs were formed from individuals that had experienced the same diet. When composing adult pairs, the pairing of brothers and sisters was carefully avoided. Each pair was then placed in a Petri dish, subjected to the same food regimen as during immature development and left to reproduce. During the entire experiment the foods were offered ad libitum and were replaced every other day; water was provided by way of a moist paper plug fitted into a 1.5-cm plastic dish.

Development and reproduction

The influence of the tested diets on the developmental and reproductive performance of the three *H. axyridis* populations was assessed.

Developmental duration and survival of the immature stages were monitored on a daily basis. The presence of exuviae and colour patterns (Koch 2003) were used to identify larval instars. The total developmental period was defined as the period from oviposition to adult emergence.

Fresh body weight was determined at adult emergence. Adults were starved for one day and weighed on a semi-microbalance Sartorius Genius ME215P (Sartorius AG, Goettingen, Germany) (±0.01 mg). In addition, body length of adults was measured at the time of death. Using an ocular micrometer the distance between the anterior margin of the pronotum and the posterior margin of the elytra was measured.

Reproduction was characterised using four parameters: pre-oviposition period, number of oviposition days, egg batch size and hatching rate of eggs. The values of pre-oviposition period and number of oviposition days were determined by checking the Petri dishes daily for oviposited eggs. Egg batch size and hatching rate were determined on a weekly basis. In addition, longevity of female adults was determined on each diet.

Data analysis

Stata/SE 9.2 was used to perform the statistical data analysis (StataCorp 2005). A generalised linear model was used with the link function and error distribution depending on the nature of the data (Poisson distribution for count data and negative binomial distribution in case of over dispersion). This generalised linear model is an extension of the general linear model and allows the use of error distributions other than the normal distribution. The model uses the principle of regression analysis and thus each time tests the difference between the average of the reference group and the average of the group in question. Each test consists of a regression coefficient being calculated and tested for being significantly different from zero, for which *P*-values are presented (McCullagh and Nelder

1989). If none of the parametric models could be used, a bootstrapped median regression was applied. Analysis started with a saturated model and interactions and non-significant main factors were dropped at a significance level of 0.05. Each saturated model contained three factors: population (laboratory, red field or black field), diet (*E. kuehniella* eggs, pollen or the mixed diet) and sex (male or female). The most parsimonious model is reported, using likelihood ratios to assure model fit.

Results

Figures 1 and 2 show the values of the fitness parameters found during this study for the three populations reared on the three diets. Figure 1a–d present immature survival, developmental time, adult body weight and female longevity, respectively, while Fig. 2a–d present the reproductive parameters pre-oviposition period, number of oviposition days, egg batch size and egg hatch, respectively. There was a strong significant correlation between body weight of adults at emergence and body length of adults at the time of death ($r = 0.85$, $P < 0.001$). For this reason only data on body weight are presented.

When reared individually, a diet of moist pollen alone allowed 35–48% of the larvae to successfully reach adulthood, whereas this was 90–98% for larvae offered *E. kuehniella* eggs or a mixture of *E. kuehniella* eggs and pollen. About 40–43% of the females from the field populations maintained on pollen were able to produce a small number of viable eggs, whereas only 5% of the females from the laboratory population laid eggs, none of which hatched. On the other hand, beetles reared on pollen took about twice as long to develop ($P < 0.001$) and had considerably lower body weights ($P < 0.001$) than those reared on *E. kuehniella* eggs or the mixed diet. Overall, when compared with females fed *E. kuehniella* eggs or the mixed diet, females of the field populations maintained on pollen alone took longer to initiate oviposition ($P = 0.015$ for the black field population fed

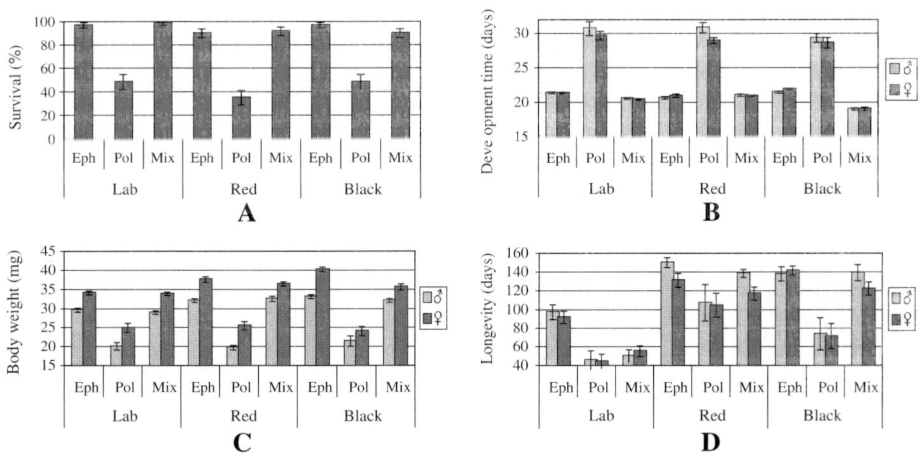

Fig. 1 Immature survival (means ± SE,%) (**A**), development time (means ± SE, days) (**B**), body weight (means ± SE, mg) (**C**), and longevity (means ± SE, days) (**D**) of a laboratory population (Lab) and a red (Red) and black (Black) field population of *H. axyridis* on frozen *E. kuehniella* eggs (Eph), frozen moist honeybee pollen (Pol), or a mixture of frozen moist bee pollen and frozen *E. kuehniella* eggs in equal weight proportions (Mix)

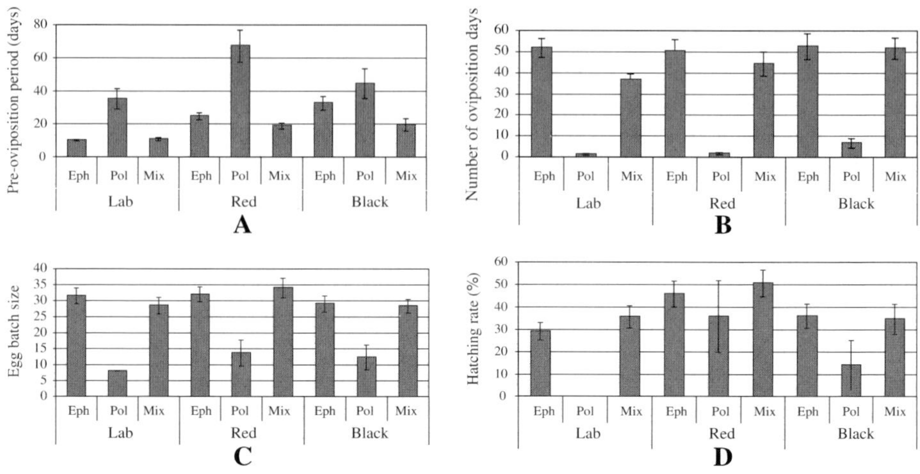

Fig. 2 Pre-oviposition period (means ± SE, days) (**A**), number of oviposition days (means ± SE) (**B**), egg batch size (means ± SE) (**C**), and egg hatch (means ± SE,%) (**D**) of a laboratory population (Lab) and a red (Red) and black (Black) field population of *H. axyridis* on frozen *E. kuehniella* eggs (Eph), frozen moist honeybee pollen (Pol), or a mixture of frozen moist bee pollen and frozen *E. kuehniella* eggs in equal weight proportions (Mix)

E. kuehniella and $P < 0.001$ in the other cases) and produced fewer eggs ($P < 0.001$ for number of oviposition days and egg batch size). Eggs laid by females of the field populations fed pollen had lower hatching rates than those laid by females maintained on *E. kuehniella* eggs and the mixed diet ($P = 0.045$ and 0.014, respectively).

Adding pollen to a diet of *E. kuehniella* eggs overall yielded similar developmental and reproductive rates compared to a diet of *E. kuehniella* eggs alone, but pre-oviposition period of females from the field populations was shorter when reared on the mixed diet ($P < 0.001$).

In the current study, melanic and non-melanic morphs responded differentially to diet for a number of traits. Immatures of the red field population developed faster on *E. kuehniella* eggs than on the mixed diet ($P = 0.009$). Individuals of the red field population also developed faster than those of the black field population on *E. kuehniella* eggs ($P < 0.001$). In contrast, immatures of the black field population developed faster on the mixed diet than on *E. kuehniella* alone ($P < 0.001$) and total developmental time of the black morphs was shorter than that of the red morphs when fed the mixed diet ($P < 0.001$). On a diet of pollen alone, the black morphs laid eggs on more days than the red ones ($P < 0.001$), whereas the latter were longer lived ($P = 0.012$).

Several biological differences were found between individuals from the long-term laboratory population and the two field populations. Overall, individuals of the laboratory population were smaller than those of the field populations when fed *E. kuehniella* eggs ($P < 0.001$, both sexes pooled) and the mixed diet ($P = 0.013$ compared to the black field population and $P < 0.001$ compared to the red field population, both sexes pooled). Further, females of the laboratory population were shorter lived ($P < 0.001$) and started ovipositing sooner ($P = 0.002$ compared to the red field population when fed pollen, $P < 0.001$ in all other cases), except when compared to those of the black field population reared on pollen alone.

Discussion

Harmonia axyridis demonstrated a lower developmental and reproductive fitness when fed on pollen than when offered diets containing *E. kuehniella* eggs. This may indicate a lack or shortage of certain nutritional components in the pollen used in the current study, which are essential for optimal development and reproduction of *H. axyridis*. The composition of pollen varies among plant species, but in general it contains high levels of proteins, amino acids and starch and to a lesser extent lipids, sugars and traces of vitamins, flavonoids, carotenoids, and minerals (Patt et al. 2003; Atrouse et al. 2004; Lundgren and Wiedenmann 2004). According to Hodek and Honek (1996) 'essential' foods support full development and reproduction of a predaceous coccinellid, whereas 'alternative' foods serve simply as a source of energy and thus increase survival of the predator. As pollen allows full development, albeit at lower rates, and the production of a small number of viable eggs in a proportion of *H. axyridis* females, it is difficult to categorize pollen as an essential or alternative food for this species based on the above definitions. However, we consider this food to be 'alternative' in a broader sense, as it is expected to be a temporary food source that will be used primarily in the absence of nutritionally superior insect prey.

Hodek and Honek (1996) suggested that many polyphagous coccinellids may have a 'mixed' feeding habit, in which they select a favourable balance of important nutrients from various foods, including plant materials. Besides the provision of supplementary nutrients, pollen may contain phagostimulating compounds that can influence food uptake. De Clercq et al. (2005) found that individuals of *Adalia bipunctata* (L.) fed on a mixed diet of frozen bee pollen and *E. kuehniella* eggs had a faster development, higher oviposition rate and egg hatch than on a diet of only *E. kuehniella* eggs. Likewise, Smith (1965a, b) found that *Coleomegilla maculata* (DeGeer) developed better on a mixture of aphids and corn pollen than on aphids or pollen alone. In the current study, however, supplementing a diet of *E. kuehniella* eggs with bee pollen only shortened the pre-oviposition period of *H. axyridis*, but had no effect on other developmental and reproductive parameters.

The availability and suitability of food is fundamental for the establishment of an exotic species in a new area (Byers 2000; Bartel and Nair 2003; Barlow and Kean 2004; Boivin et al. 2006). As a polyphagous predator, *H. axyridis* can feed on an extensive range of arthropod species (Tedders and Schaefer 1994; Hodek and Honek 1996; Koch 2003). Our findings indicate that the use of plant foods such as pollen may further sustain populations of the predator when insect prey are scarce or absent, albeit with reduced developmental and reproductive rates. Field research by Lundgren et al. (2004) demonstrated that even at times when aphids were at hand, *H. axyridis* larvae fed on pollen of corn (*Zea mays* L.). *Harmonia axyridis* has been reported to use other types of plant foods besides pollen. In autumn the species shows a tendency to aggregate on late season fruits, probably aiming to obtain carbohydrates to boost their reserves for winter (Koch et al. 2004).

The capacity of *H. axyridis* to exploit pollen when insect prey is scarce may offer this invasive species a competitive advantage over other native lady beetle species that are less capable in using this alternative food source. Blackman (1965) observed that *A. bipunctata* frequently fed on fresh hazel and willow pollen even when aphids were present. Hemptinne and Desprets (1986) and Hemptinne and Naisse (1987) reported that pollen was used as a spring food by the two-spotted ladybird. When De Clercq et al. (2005) reared *A. bipunctata* on moist bee pollen they reported that only 10% of the larvae successfully reached adulthood; the resulting adults failed to reproduce on pollen alone. Several other species of Coccinellidae have been observed to feed on pollen (Hagen 1962; Hemptinne and Desprets 1986; Hodek and Honek 1996); for these species, pollen was rarely sufficient to sustain larval development and it did not allow the females to initiate oogenesis. Watson and Thompson (1933) and Ibrahim (1955) reported that *Harmonia conformis* (Boisduval) and *Coccinella undecimpunctata aegyptiaca* Reiche, respectively, were observed to feed on pollen and other plant foods in the field during periods of low aphid densities. Adults of both species were however reproductively inactive during these periods. According to Hagen (1962) the physiological state of the females of *Hippodamia convergens* Guérin-Méneville, *Hippodamia quinquesignata punctulata* LeConte, *Hippodamia sinuata* Mulsant and *Hippodamia parenthesis* (Say) in western North America were influenced by the quantity and quality of food. If sufficient aphids were available reproduction occurred in these species; however, if the females had to feed on non-insect foods such as pollen, honeydew, nectar, etc. these foods were converted to fat and stored, but were inadequate for oogenesis. Smith (1961) performed laboratory feeding experiments where individuals of *Cycloneda sanguinea* (L.) and *Coccinella trifasciata* L. were fed pollen of several plant species. *Cycloneda sanguinea* completed three instars when fed on pollen from hemp (*Cannabis sativa* L.), two on pollen of corn (*Z. mays*), and one on pollen of grey birch (*Betula populifolia* Marsh), hemlock (*Tsuga canadensis* L.), hickory (*Carya ovata* Mill.), hornbeam (*Carpinus caroliniana* Walt) and red oak (*Quercus rubra* L.). *Coccinella trifasciata* completed only the first instar when fed pollen of corn and butternut (*Juglans cinerea* L.), but failed to develop on the other pollen tested in the study. These findings indicate that pollen from different plant species may differ in their nutritional value for coccinellids. *Coleomegilla maculata* is the only other species of Coccinellinae that is reportedly able to fully develop and reproduce on pollen (Lundgren and Wiedenmann 2004; Michaud and Grant 2005).

Soares et al. (2001, 2005) and Berkvens et al. (2007) showed that different morph types of *H. axyridis* responded dissimilarly in terms of their developmental and reproductive performance when offered various insect foods. The current study also reveals differential responses of *H. axyridis* morph types to a plant food. When fed on pollen alone, the black morphs produced more eggs than the red ones, whereas the red morphs lived longer than the black ones. This suggests that there is an energetic trade-off between reproduction and longevity and that the morph types involved may use opposite strategies for allocating their limited resources. However, more research is needed to substantiate this hypothesis. Further, individuals from a long term laboratory population and two field populations differed in their response to diet for several biological traits. As discussed in more detail by Berkvens et al. (2007) this could either be due to inter-population variation within *H. axyridis* or to genetic and non-genetic adaptation during long term laboratory rearing.

In conclusion, the current laboratory study has demonstrated that pollen can be used by *H. axyridis* as an alternative food, yielding full development and even some reproduction in part of the population. Zoophytophagous feeding habits may contribute to the establishment potential of a non-indigenous predator by sustaining its populations in times of prey scarcity. The extent to which pollinivory determines the outcome of competition among the invasive species *H. axyridis* and native coccinellid predators in Europe requires further investigation. Gut analysis of field collected lady beetles may indicate pollen consumption and digestion by *H. axyridis* in different periods of the year.

Acknowledgements This study was funded by grant number 53215 from the Institute for the Promotion of Innovation by Science and Technology in Flanders (IWT). We are grateful to Bea Roos and Maarten Bonte for invaluable help during the experiments.

References

Atrouse OM, Oran OM, Al-Abbadi SY (2004) Chemical analysis and identification of pollen grains from different Jordanian honey samples. Int J Food Sci Technol 39:413–417

Barlow ND, Kean JM (2004) Resource abundance and invasiveness: a simple model. Biol Invasions 6: 261–268

Bartel SM, Nair SK (2003) Establishment risks for invasive species. Risk Anal 24:833–845

Berkvens N, Bonte J, Berkvens D, Tirry L, De Clercq P (2007) Influence of diet and photoperiod on development and reproduction of European populations of *Harmonia axyridis* (Pallas) (Coleoptera: Coccinellidae). BioControl (this issue). doi:10.1007/s10526-007-9130-0

Blackman RL (1965) Studies on specificity in Coccinellidae. Ann Appl Biol 56:336–338

Beug H-J (2004) Leitfaden der Pollenbestimmung für Mitteleuropa und angrenzende Gebiete. Pfeil, Munich

Boivin G, Kölliker-Ott UM, Bale J, Bigler F (2006) Assessing the establishment potential of inundative biological control agents. In: Bigler F, Babendreier D, Kuhlmann U (eds) Environmental impact of invertebrates for biological control of arthropods. Methods and risk assessment. CABI Publishing, Wallingford

Brown PMJ, Adriaens T, Bathon H, Cuppen J, Goldarazena A, Hägg T, Kenis M, Kausnitzer BEM, Kovar I, Loomans AJM, Neved O, Pedersen J, Rabitsch W, Roy HE, Ternois V, Zakharov IA, Roy BD (2007) *Harmonia axyridis* in Europe: spread and distribution of a non-native coccinellid. BioControl (this issue). doi:10.1007/s10526-007-9132-y

Byers JE (2000) Competition between two estuarine snails: implications for invasions of exotic species. Ecology 81:1225–1239

Coderre D, Lucas É, Gagné I (1995) The occurrence of *Harmonia axyridis* (Pallas) (Coleoptera: Coccinellidae) in Canada. Can Entomol 127:609–611

Coutanceau JP (2006) *Harmonia axyridis* (Pallas, 1773): une coccinelle asiatique introduite, acclimaté et en extension en France. Bull Soc Entomol Fr 111:395–401

De Clercq P, Bonte M, Van Speybroeck K, Bolckmans K, Deforce K (2005) Development and reproduction of *Adalia bipunctata* (Coleoptera: Coccinellidae) on eggs of *Ephestia kuehniella* (Lepidoptera: Phycitidae) and pollen. Pest Manag Sci 61:1129–1132

De Clercq P, Peeters I, Vergrauwe G, Thas O (2003) Interaction between *Podisus maculiventris* and *Harmonia axyridis*, two predators used in augmentative biological control in greenhouse crops. BioControl 48:39–55

Eliott NC, Kieckhefer RW, Kauffman WC (1991) Estimating adult coccinellid populations in wheat fields by removal, sweepnet, and visual count sampling. Can Entomol 123:13–22

Erdtman G (1960) The acetolysis method. Svensk Bot Tidskr 54:561–564

Gilbert FS (1981) Foraging ecology of hoverflies: morphology of the mouthparts in relation to feeding on nectar and pollen in some common urban species. Ecol Entomol 6:245–262

Hagen KS (1962) Biology and ecology of predaceous Coccinellidae. Annu Rev Entomol 7:289–326

Hemptinne JL, Desprets A (1986) Pollen as a spring food for *Adalia bipunctata*. In: Hodek I (ed) Ecology of Aphidophaga. Academia, Prague and Dr W Junk, Dordrecht, pp 29–35

Hemptinne JL, Naisse J (1987) Ecophysiology of the reproductive activity of *Adalia bipunctata* L. (Co., Coccinellidae). Meded Fac Landbouww Rijksuniv Gent 52:225–233

Hodek I, Honek A (1996) Ecology of Coccinellidae. Kluwer Academic Publishers, Dordrecht

Holway DA (1999) Competitive mechanisms underlying the displacement of native ants by the invasive argentine ant. Ecology 80:238–251

Ibrahim MM (1955) Studies on *Coccinella undecimpunctata aegyptiaca* Rche. II. Biology and life history. Bull Soc Entomol Egypte 39:395–423

Jørgensen HB, Toft S (1997) Role of granivory and insectivory in the life cycle of the carabid beetle *Amara similata*. Ecol Entomol 22:7–15

Kenis M, Roy HE, Zindel R, Majerus ME (2007) Current and potential management strategies against *Harmonia axyridis*. BioControl (this issue). doi:10.1007/s10526-007-9136-7

Kiman ZB, Yeargan KV (1985) Development and reproduction of the predator *Orius insidiosus* (Hemiptera: Anthocoridae) reared on diets of selected plant material and arthropod prey. Ann Entomol Soc Am 78:464–467

Koch RL (2003) The multicolored Asian lady beetle, *Harmonia axyridis*: a review of its biology, uses in biological control, and non-target impacts. J Insect Sci 3:1–16

Koch RL, Burkness EC, Wold Burkness SJ, Hutchinson WD (2004) Phytophagous preferences of the multicolored Asian lady beetle (Coleoptera: Coccinellidae) from autumn-riping fruit. J Econ Entomol 97:539–544

Lamana ML, Miller JC (1996) Field observation on *Harmonia axyridis* Pallas (Coleoptera: Coccinellidae) in Oregon. Biol Control 6:232–237

Lundgren JG, Razzak AA, Wiedenmann RN (2004) Population responses and food consumption by predators *Coleomegilla maculata* and *Harmonia axyridis* (Coleoptera: Coccinellidae) during anthesis in an Illinois cornfield. Environ Entomol 33:958–963

Lundgren JG, Wiedenmann RN (2004) Nutritional suitability of field corn pollen as food for the predator, *Coleomegilla maculata* (Coleoptera: Coccinellidae). J Insect Physiol 50:567–575

McCornack BP, Koch RL, Ragsdale DW (2007) A simple method for in-field sex determination of the multicolored Asian lady beetle *Harmonia axyridis*. J Insect Sci 7:10

McCullagh P, Nelder J (1989) Generalized linear models. Chapman & Hall, London

Michaud JP, Grant AK (2005) Suitability of pollen sources for the development and reproduction of *Coleomegilla maculata* (Coleoptera: Coccinellidae) under simulated drought conditions. Biol Control 32:363–370

Moore PD, Webb JA, Collinson ME (1991) Pollen analysis, 2nd edn. Blackwell Science, Oxford

Nalepa CA, Kidd KA, Ahlstrom KR (1996) Biology of *Harmonia axyridis* (Coleoptera: Coccinellidae) in winter aggregations. Ann Entomol Soc Am 89:681–685

Ongagna P, Giuge L, Iperti G, Ferran A (1993) Cycle de développement d'*Harmonia axyridis* (Col., Coccinellidae) dans son aire d'introduction, le sud-est de la France. Entomophaga 38:125–128

Patt LM, Wainright SC, Hamilton GC, Whittinghill D, Bosley K, Dietrick J, Lashomb JH (2003) Assimilation of carbon and nitrogen from pollen and nectar by a predaceous larva and its effects on growth and development. Ecol Entomol 28:717–728

Pell J, Baverstock J, Roy HE, Ware RL, Majerus ME (2007) Intraguild predation involving *Harmonia axyridis*: a review of current knowledge and future perspectives. BioControl (this issue). doi:10.1007/s10526-007-9125-x

Smith BC (1961) Results of rearing some coccinellid (Coleoptera: Coccinellidae) larvae on various pollens. Proc Entomol Soc Ontario 91:270–271

Smith BC (1965a) Growth and development of coccinellid larvae on dry foods (Coleoptera, Coccinellidae). Can Entomol 97:760–768

Smith BC (1965b) Effects of food on the longevity, fecundity and development of adult coccinellids. Can Entomol 97:910–919

Soares AO, Coderre D, Schanderl H (2001) Fitness of two phenotypes of *Harmonia axyridis* (Coleoptera: Coccinellidae). Eur J Entomol 98:287–293

Soares AO, Coderre D, Schanderl H (2005) Influence of prey quality on the fitness of two phenotypes of *Harmonia axyridis* adults. Entomol Exp Appl 114:227–232

StataCorp (2005) Stata Statistical Software: Release 9. StataCorp LP, College Station

Tedders WL, Schaefer PW (1994) Release and establishment of *Harmonia axyridis* (Coleoptera: Coccinellidae) in the South-eastern United States. Entomol News 105:228–243

Watson JP, Thompson WL (1933) Food habits of *Leis conformis* Boisd. (Chinese ladybeetle). Florida Entomol 17:27–29

van Lenteren JC, Babendreier D, Bigler F, Burgio G, Hokkanen HMT, Kuske S, Loomans AJM, Menzler-Hokkanen I, van Rijn PCJ, Thomas MB, Tommasini MG, Zeng QQ (2003). Environmental risk assessment of exotic natural enemies used in inundative biological control. BioControl 48:3–38

Influence of diet and photoperiod on development and reproduction of European populations of *Harmonia axyridis* (Pallas) (Coleoptera: Coccinellidae)

Nick Berkvens · Jochem Bonte · Dirk Berkvens · Luc Tirry · Patrick De Clercq

Abstract The current study examines the effect of photoperiod (16:08 or 12:12 h L:D) and diet (eggs of *Ephestia kuehniella* Zeller (Lepidoptera: Phycitidae) or the pea aphid *Acyrthosiphon pisum* (Harris) (Hemiptera: Aphididae)) on the development and reproduction of the multicoloured Asian lady beetle *Harmonia axyridis* (Pallas) (Coleoptera: Coccinellidae). A long-term laboratory population of *H. axyridis* (since 1998) and a melanic and non-melanic population originating from field collected individuals of *H. axyridis* in Belgium were used in this study. Long day conditions (16 h photoperiod) shortened development of the field populations with 2–3 days when compared with short day conditions (12 h photoperiod). Oviposition in the field populations was delayed by 1–3 months when reared at a 12 h photoperiod. Dissections indicated that the females were in reproductive diapause. As compared with live pea aphids, a diet consisting of *E. kuehniella* eggs yielded heavier adult body weights (up to 12%) and increased the number of egg laying days (by 45–169%) for both field populations at a 16 h photoperiod and lengthened adult life span (by 45–92%) under both light regimens. The morph types differed in their response to the foods offered in terms of developmental rate, pre-oviposition period and number of oviposition days. The laboratory and field strains responded differentially to regimens of food and photoperiod. The study indicated a greater nutritional plasticity of the non-melanic morphs which may offer them a competitive

N. Berkvens (✉) · J. Bonte · L. Tirry · P. De Clercq
Laboratory of Agrozoology, Department of Crop Protection, Faculty of Bioscience Engineering, Ghent University, Coupure Links 653, 9000 Ghent, Belgium
e-mail: Nick.Berkvens@UGent.be

J. Bonte
e-mail: Jochem.Bonte@UGent.be

L. Tirry
e-mail: Luc.Tirry@UGent.be

P. De Clercq
e-mail: Patrick.DeClercq@UGent.be

D. Berkvens
Department of Animal Health, Prince Leopold Institute of Tropical Medicine, Antwerp, Belgium
e-mail: dberkvens@itg.be

doi: 10.1007/978-1-4020-6939-0_14, © IOBC 2007

advantage that may in part explain the predominance of non-melanic morphs in newly colonized areas.

Keywords *Harmonia axyridis* · Morph type · Establishment · Photoperiod · *Ephestia kuehniella* · *Acyrthosiphon pisum* · Development · Reproduction

Introduction

The multicoloured Asian lady beetle *Harmonia axyridis* (Pallas) (Coleoptera: Coccinellidae) is an aphidophagous coccinellid indigenous to Central and Eastern Asia (Coderre et al. 1995; Koch 2003). Because of its high predatory effectiveness, the species was introduced as a biological control agent in both North America and Europe. However, the species succeeded in establishing wild populations on both continents (Koch 2003; Cuppen et al. 2004; Coutanceau 2006, Brown et al. 2007). Several authors recorded harmful side effects accompanying the invasion by *H. axyridis* in the USA: a dominance of the invader in coccinellid communities along with a decline in population numbers of some lady beetle species, aggregating adults causing nuisance to humans when overwintering en masse in occupied buildings and the inflicting of damage to agricultural crops (Koch 2003; Koch et al. 2004; Kenis et al. 2007; Pell et al. 2007). Due to these harmful side-effects the ladybird has lost its environment friendly image as a biological control agent and is increasingly being stigmatized as a major invasive alien species. The ecological mechanisms driving the successive colonisation of the invader in Europe are, however, still unclear.

Harmonia axyridis shows a high level of polymorphism, with more than 100 elytral patterns having been recorded (Soares et al. 2003). The morph types are commonly classified into a 'red' non-melanic group, comprising the *succinea* subgroup (0 to 21 dots, h), and a 'black' melanic group comprising the *conspicua* (two red dots, h^C), *spectabilis* (four red dots, h^S) and *axyridis* (more than four red dots, h^X) subgroups (Osawa and Nishida 1992; Serpa et al. 2003). A multi-allelic gene determines the colour polymorphism. Hosino (1940) and Tan (1946) together reported 15 alleles determining the large variation in colours and patterns. Black is generally the dominant colour, with a dominance order of $h < h^X < h^S < h^C$ in the four subgroups (Hosino 1936; Komai 1956; Sasji 1971). Colouration and maculation can to some extent also be influenced by larval diet (Grill 1999), and by temperature and humidity during pupation (Komai et al. 1950), with a higher quality diet causing a fuller red colour and a low temperature and high humidity increasing the extension of the melanic pigment. True (2003) indicates that melanic alleles, in addition to determining colour, can have pleiotropic effects on the developmental processes and life history of an individual. Soares et al. (2001, 2003, 2005) and Serpa et al. (2003) compared ecological and physiological characteristics of *H. axyridis* morphs and determined mutual dissimilarities that could support this theory. These dissimilarities could be the basis for the selection of certain genotypes in certain environments (Serpa et al. 2003; Soares et al. 2003, 2005), and consequently be the reason for both the spatial and temporal variations in the occurrence of the morphs and the maintenance of polymorphism in this species (Komai 1956; Osawa and Nishida 1992; Hodek and Honěk 1996; Krafsur et al. 1997). In the Flanders region of Belgium, where established populations have been found since 2001, both melanic and non-melanic individuals occur, with a predominance of red *succinea* morph types (Adriaens et al. 2007). In Flanders, Hantson (2004) reported a proportion of 74.9% f. *succinea*, 23.1%

f. *spectabilis* and 2% f. *conspicua*. This indicates a potential role for polymorphism in the adaptation of the species to a new environment.

In the current study the effects of photoperiod and diet on the development and reproduction of European populations of *H. axyridis* were investigated. A first objective was to compare the responses of a melanic and non-melanic field population in order to judge possible differences in adaptiveness of the morph types. Second, performance of these field populations was compared to that of a long-term laboratory population that had originally been established using insects from a commercial culture that was one of the sources of biological control releases in the area since 1997 (Brown et al. 2007).

Materials and methods

Populations of *H. axyridis*

All experiments were conducted in the laboratory using three populations: a laboratory population and two field populations.

The laboratory population was started in September 1998 at the Laboratory of Agrozoology of Ghent University using larvae acquired from Biobest NV (Westerlo, Belgium). The origin of the commercial culture and the rearing methods used at its facilities were not disclosed by the company. The laboratory culture at Ghent University was infused with individuals from Biobest NV on several occasions thereafter; the last infusion was done in August 2000. Individuals of the 50th generation of laboratory rearing were used in the experiments. The population originally consisted of an even mixture of morphs, but at the time of this study, it consisted of more than 99% melanic *spectabilis* and *conspicua* morphs.

The field populations were started in April 2005 with individuals collected from an established wild population in an orchard in Wippelgem, near Ghent, Flanders. A first melanic population was established using *conspicua* and *spectabilis* individuals (referred to as the "black field population"), whereas the second non-melanic population only consisted of *succinea* individuals (referred to as the "red field population"). The very few (<5%) non-melanic offspring that occurred in the melanic field population during the process of laboratory rearing were discarded from the population upon adult emergence. Experiments were done using individuals of the 4th generation of both field populations.

All populations were reared on frozen eggs of the Mediterranean flour moth *Ephestia kuehniella* Zeller (Lepidoptera: Phycitidae), as described by Ongagna et al. (1993) and De Clercq et al. (2003).

Experimental design

To study the effects of diet and photoperiod on developmental time and reproductive capacity of *H. axyridis*, experiments were conducted in Petri dishes (diameter 9 cm, height 2 cm), kept in incubators held at 23 ± 1°C and 65 ± 5% RH. The insects were exposed to two diet regimens (deep-frozen *E. kuehniella* eggs or a mixture of live late instars and adults of the pea aphid, *Acyrthosiphon pisum* (Harris) (Hemiptera: Aphididae)) and two photoperiod regimens (16 and 12 h light). Hodek and Honěk (1996) mention *A. pisum* as a natural and essential prey of *H. axyridis*, whereas *E. kuehniella* eggs are considered a suitable substitute food source. Pea aphids, fed on faba bean *Vicia faba* L., and

E. kuehniella eggs were obtained from Biobest NV (Westerlo, Belgium) and Koppert BV (Berkel en Rodenrijs, The Netherlands), respectively. Foods were always offered ad libitum and were replaced every other day. Water was provided by way of a moist paper plug fitted into a 1.5-cm plastic dish. The 16 and 12 h photoperiods were chosen to simulate late spring/early summer and early spring/early autumn light conditions, respectively. A photoperiodicity of 16 h occurs in Belgium on 25 May and 18 July, while 12 h occurs on 18 March and 25 September (Anonymous 2007).

Twenty pairs of adults (3–5 days old) were randomly taken from stock cultures of each population. The pairs were placed in individual Petri dishes, held at 23 ± 1°C, 65 ± 5% RH and a 16 h photoperiod and left to mate and reproduce for 1 week on a mixed diet of *A. pisum* and *E. kuehniella*, to prevent food source conditioning. After this week, an accordion-pleated carton board strip (15 by 2 cm, folded 6–8 times along its length) was placed in each Petri dish to serve as an oviposition substrate and egg batches deposited during the following day were collected. The egg batches (F1) of each pair were divided into four equal sub-batches which were assigned to one of four experimental regimens: 16 h + *A. pisum*; 16 h + *E. kuehniella*; 12 h + *A. pisum*; 12 h + *E. kuehniella*. On the first day of hatching, three larvae were randomly collected per sub-batch. Each first-instar larva was then transferred to an individual Petri dish and allowed to develop to adulthood under the same regimen as experienced by the egg stage. By way of this design, 60 individuals of each population were subjected to each regimen and allowed to complete immature development. Sex of the resulting F1 adults was determined (McCornack et al. 2007) and per population 20–25 pairs were formed from the cohort of individuals that had been exposed to the same regimen. The pairing of brothers and sisters was carefully avoided when composing adult pairs. Each pair was then placed in an individual Petri dish and subjected to the same regimen as during immature development in order to monitor reproductive performance.

Fitness parameters

Several fitness parameters were assessed to evaluate effects of the tested regimens on the developmental and reproductive performance of the F1-individuals of the three *H. axyridis* populations.

Developmental duration of the immature stages was monitored on a daily basis. The total developmental period was defined as the period from oviposition to adult emergence. Newly emerged adults were starved for one day and weighed on a semi-microbalance Sartorius Genius ME215P (Sartorius AG, Goettingen, Germany) (±0.01 mg).

Body length of adults was measured at the time of death. The distance between the anterior margin of the pronotum and the posterior margin of the elytra was measured with an ocular micrometer.

Reproduction of females was characterized using four parameters: pre-oviposition period, number of oviposition days, size of egg batches and egg hatch. To determine the pre-oviposition period and the number of oviposition days, each Petri dish was checked daily for oviposited eggs. The values for egg batch size and egg hatch were determined on a weekly basis. Once a week a fresh accordion-pleated carton board was introduced into each Petri dish; the following day the carton board was removed and the eggs were collected and counted to calculate mean batch size. The collected eggs were transferred to a new Petri dish, placed under the same regimen as that experienced by the ovipositing female adult and were checked daily for hatching. In addition, longevity of female adults was determined. The experiment was terminated 210 days after F1 egg hatching. At that moment all individuals of the laboratory population had died, whereas about 15% of the individuals of both field populations fed *A. pisum* at 16 h light and between 0 and 5% of the individuals in the remaining treatment groups were still alive.

Dormancy

A significant delay in oviposition in the main experiment described above (see "Results") suggested the occurrence of dormancy in the field populations reared at a 12 h photoperiod. A second experiment was set up to confirm this finding. In this experiment a cohort of 60 larvae from either field population was allowed to develop to adulthood and reproduce, while fed *E. kuehniella* at a 12 h photoperiod. The state of dormancy of the resulting adult females was judged by performing dissections 4 and 7 weeks after emergence. Ten randomly chosen females of each population were dissected each time. The state of dormancy was evaluated in a qualitative manner, by visually inspecting gut content and the developmental status of the fat body and ovaria in comparison with reproducing females of the same populations fed *E. kuehniella* eggs at 23°C and a 16 h photoperiod. At the start of diapause, female adults of *H. axyridis* have a well developed fat body and a full gut. Fat body degenerates and the gut becomes emptied of food during dormancy. Ovarian development is also suppressed during dormancy (Sakurai et al. 1992; Iperti and Bertand 2001).

Data analysis

Data analysis was carried out in Stata/SE 9.2 (StataCorp 2005). A generalised linear model was used with the link function and error distribution depending on the nature of the data (Poisson distribution for count data and negative binomial distribution in case of over dispersion). This generalized linear model is an extension of the general linear model and allows the use of error distributions other than the normal distribution. The model uses the principle of the regression analysis and thus tests each time the difference between the average of the reference group and the average of the group in question. Thus, each test consists of a regression coefficient being calculated and tested for being significantly different from zero, for which *P*-values are presented (McCullagh and Nelder 1989). If none of the parametric models could be used, a bootstrapped median regression was applied. Analysis started with a saturated model and interactions and non-significant main factors were dropped at a significance level of 0.05. Each saturated model contained four factors: population (laboratory, red field or black field), photoperiod (16 or 12 h), diet (*E. kuehniella* or *A. pisum*) and sex (male or female). The most parsimonious model is reported, using likelihood ratios to assure model fit.

Results

Fitness parameters

The results of the fitness assessment for the three populations exposed to the four tested regimens are presented in Figs. 1 and 2. A significant correlation between body weight and

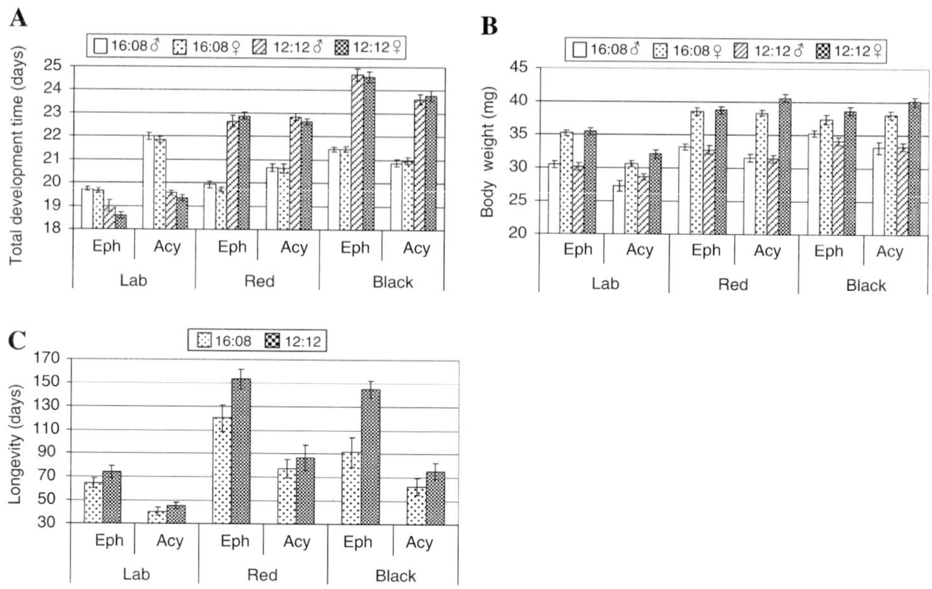

Fig. 1 Total development time (means ± SE, days) (**a**) and body weight (means ± SE, mg) (**b**) of males and females and longevity of females (means ± SE, days) (**c**) of a laboratory population (Lab) and a red (Red) and black (Black) field population of *Harmonia axyridis* at different photoperiods (16:08 and 12:12 h L:D) and diets (*Ephestia kuehniella* (Eph) and *Acyrthosiphon pisum* (Acy))

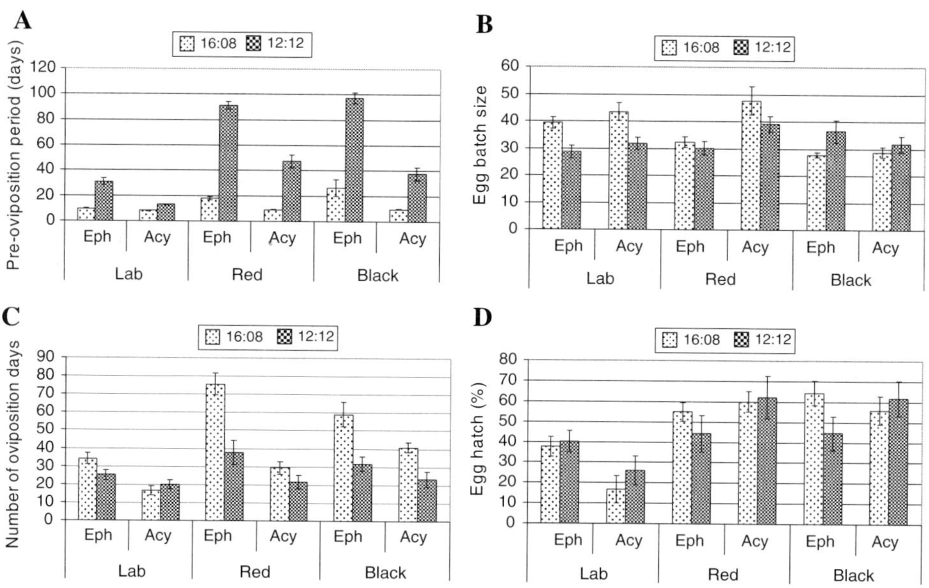

Fig. 2 Pre-oviposition period (means ± SE, days) (**a**) egg batch size (means ± SE) (**b**) number of oviposition days (means ± SE) (**c**) and egg hatch (means ± SE, %) (**d**) of a laboratory population (Lab) and a red (Red) and black (Black) field population of *Harmonia axyridis* at different photoperiods (16:08 and 12:12 h L:D) and diets (*Ephestia kuehniella* (Eph) and *Acyrthosiphon pisum* (Acy))

body length was found for adults of either sex ($r = 0.68$, $P < 0.001$ and $r = 0.475$, $P < 0.001$ for males and females, respectively). For this reason only data on body weight are presented.

A late spring/early summer photoperiod of 16 h shortened larval and pupal development ($P < 0.001$ for both sexes) compared to an early spring/early autumn photoperiod of 12 h (Fig. 1a). Equally, the pre-oviposition periods of all three populations were shortened at 16 h light ($P = 0.002$ for the laboratory population fed *A. pisum*, $P < 0.001$ for all other cases) (Fig. 2a). Females of all populations lived longer at 12 h than at 16 h light ($P = 0.002$) (Fig. 1c). The 16 h light regimen increased the number of oviposition days of both field populations fed *E. kuehniella* ($P < 0.001$ for the red population and $P = 0.009$ for the black one) and of the black field population fed *A. pisum* ($P < 0.001$) (Fig. 2c).

Although *E. kuehniella* eggs caused longer pre-oviposition periods than aphids ($P < 0.001$ in all cases except for the laboratory population reared at a 16 h photoperiod) (Fig. 2a), they yielded an increase in the number of egg laying days for both field populations at a 16 h photoperiod ($P < 0.001$ for the red field population, $P = 0.001$ for the black field population) (Fig. 2c). A diet of *E. kuehniella* eggs furthermore increased life span of the females under both light regimens ($P < 0.001$) (Fig. 1c) and yielded heavier adult body weights at long day conditions ($P < 0.001$ for both sexes) (Fig. 1b).

Two marked differences between the red and black morphs were found in this study. First, there was a differential effect of the diet on development and reproduction of the morphs. When fed *E. kuehniella* eggs, the red individuals developed faster than the black ones, a difference that was marginally significant at a 16 h photoperiod ($P = 0.055$ for both sexes), but significant at a 12 h photoperiod ($P = 0.016$ for both sexes) (Fig. 1a); there was, however, no difference in developmental time between the populations when fed *A. pisum* (Fig. 1a). The black field population had a shorter pre-oviposition period than the red field population when reared at a 12 h photoperiod on *A. pisum* ($P = 0.044$), but the inverse was the case at 16 h light on *E. kuehniella* ($P = 0.003$) (Fig. 2a). When fed *A. pisum* under long day conditions the black field population had a higher number of oviposition days than the red population ($P = 0.035$) (Fig. 2c). Furthermore, the difference in the number of oviposition days between diets was greater for the red field population than for the black field population (Fig. 2c), whereas the opposite was found for the pre-oviposition period (Fig. 2a). Second, in all treatment groups the red morphs were longer lived than the black morphs ($P = 0.043$) (Fig. 1c).

Several differences were found between the performances of the field populations and the laboratory population. In contrast to the field populations, longer day length resulted in a significantly longer developmental time for the laboratory population ($P = 0.008$ for both sexes) and *E. kuehniella* eggs shortened the developmental time of the laboratory population compared with *A. pisum* ($P = 0.017$ for both sexes) (Fig. 1a). Both field populations had a longer developmental time ($P < 0.001$ for both sexes) (Fig. 1a) and pre-oviposition period ($P < 0.001$) (Fig. 2a) than the laboratory population when reared at a 12 h light regimen, irrespective of diet, and a longer pre-oviposition period at 16 h light on *E. kuehniella* ($P < 0.001$). When fed *E. kuehniella* eggs at a 16 h photoperiod both field populations oviposited on more days than did the laboratory population ($P < 0.001$ for the red field population and $P = 0.006$ for the black field population) (Fig. 2c). On *A. pisum*, egg hatch in the laboratory population was lower than that in both field populations ($P < 0.001$) (Fig. 2d). Females of both field populations lived longer than those of the laboratory population in all cases ($P < 0.001$) (Fig. 1c). Males and females of the field populations weighed more than those of the laboratory population under all diet and light regimens ($P \leq 0.001$ for both sexes) (Fig. 1b).

Dormancy

As females of the field populations showed prolonged pre-oviposition periods at a 12 h photoperiod in the main experiment, a second experiment was done to confirm if rearing at 12 h light on *E. kuehniella* eggs induced a state of reproductive diapause in the red and black field populations. Females of either morph again went through a prolonged pre-oviposition period like those in the first experiment. Dissections done 4 and 7 weeks after adult emergence revealed that females reared at 12 h light had a lower amount of fat body compared to those kept at a 16 h photoperiod. All females maintained under short day conditions had empty ovaries and empty or quasi empty guts. Suppressed activity, aggregation behaviour and a minimal food uptake of male and female adults reared at a 12 h photoperiod were also observed during the prolonged pre-oviposition period.

Discussion

Photoperiod had a significant influence on developmental and reproductive performance of *H. axyridis*. Ongagna and Iperti (1994) also observed a significant decrease in the developmental duration of *H. axyridis* at a 16 h photoperiod compared to shorter photoperiods varying between 9 and 12 h of light. Omkar and Pathak (2006) reported similar findings for the lady beetle *Coelophora saucia* (Mulsant) (Coleoptera: Coccinellidae) and ascribed the greater fitness of the insect at a 16 h photoperiod to the feeding behaviour of lady beetles. Coccinellidae are diurnal organisms and thus most likely only feed during light conditions (Hodek and Honěk 1996). The individuals can therefore consume more prey under a 16 h light regimen and ad libitum food conditions, which in turn benefits development and reproduction.

In the field populations, oviposition was delayed by 1–3 months when held at short day conditions. A shortening of day length in the autumn, prey scarcity and low temperatures are cues for adult *H. axyridis* and other lady beetles to prepare for their winter dormancy (Ongagna and Iperti 1994; Hodek and Honěk 1996). The prolongation of the pre-oviposition period found in this study indicates a state of reproductive diapause induced by the short day length during larval development. This is supported by the physiological state (gut content, fat body and ovarian development) of the dissected adult individuals and the behaviour observed during the prolonged pre-oviposition period (low activity, aggregation and minimal food uptake). The reproductive diapause in the field populations lasted longer when *E. kuehniella* was provided as food. These observations may indicate that *E. kuehniella* eggs are not nutritionally optimal for maturing eggs directly out of dormancy. The delay of oviposition at short day conditions was much less pronounced for the laboratory population. This can be due to the loss of photoperiodic response after prolonged rearing at a 16 h light regimen. As also reported by Michaud and Qureshi (2006) for *Hippodamia convergens* Guérin-Méneville (Coleoptera: Coccinellidae), the adults of *H. axyridis* undergoing a reproductive diapause lived longer in our study, suggesting a trade-off between the reproductive capacities and survival.

Ephestia kuehniella eggs appeared to be a better food for *H. axyridis* than live pea aphids based on the results of this study. Eggs of *E. kuehniella* have been shown to be an effective unnatural food for the production of different insects, including coccinellids (Specty et al. 2003; De Clercq et al. 2005). Specty et al. (2003) found that *E. kuehniella* eggs were nutritionally superior to *A. pisum* in terms of amino acid and fatty acid content and composition.

Better performance of the *H. axyridis* populations tested in this study on *E. kuehniella* eggs under long day conditions may be the result of artificial selection. The laboratory population was reared on *E. kuehniella* and at long day photoperiods for ca. 50 generations at our facilities and originated from a commercial source that most probably used the same conditions during mass culturing (Kabiri 2006). Also the field populations were reared on *E. kuehniella* eggs and at a 16 h photoperiod in the laboratory for three generations prior to the experiments and may originally have sprung from the same commercial source.

Like in the current study and in Berkvens et al. (2007), Soares et al. (2001, 2005) also found dissimilar reproductive and food assimilating responses among morph types of *H. axyridis* to various food sources. The findings imply that red morphs may be nutritionally more adaptive and may be more able to exploit new and unknown food sources, such as the factitious food *E. kuehniella* in our study. Their greater nutritional plasticity may offer non-melanic morphs a competitive advantage for the exploitation of food sources during establishment and may in part explain the predominance of non-melanic morphs in newly colonized areas (Koch 2003; Hantson 2004). Surprisingly, however, a laboratory population of *H. axyridis* was predominated by black morphs after some years of laboratory rearing on a factitious host, suggesting that other factors are involved in morph determination. Selective factors that could have played a role in this increase in melanics during laboratory culturing include the genetic dominance of melanism in *H. axyridis* (Komai 1956), a potentially higher mating success of *H. axyridis* melanics in the summer conditions used in the laboratory (Osawa and Nishida 1992), and the reported increase in melanism at high population densities in several insects (True 2003).

Serpa et al. (2003) also observed a difference in adult longevity between certain morph types of *H. axyridis*. They found that the redder morph, *aulica*, had a longer life span compared to a completely black morph, *nigra*. One of the explanations might be thermal melanism, which suggests that when insolated, dark ectothermic insects heat up faster and reach a higher equilibrium temperature. This results in a higher level of activity and a reproductive advantage, but equally a lower longevity (Serpa et al. 2003). However, since our experiments were carried out in climate chambers using fluorescent lamps with low radiation, thermal melanism is not a valid explanation for the shorter life span observed in the black morphs.

A laboratory population of *H. axyridis* was included in the experiments in order to compare its performance with that of the field populations. Individuals of the laboratory population showed different developmental and reproductive responses to the tested regimens of diet and photoperiod from those of the field populations. The observed differences between the field and laboratory populations may be due to inter-population variation within *H. axyridis*. However, the broad consistency in the morph frequency data from Europe provides some evidence of the genetic similarity of the populations found in different European countries and suggests that the species has spread from limited points of origin in Europe (Brown et al. 2007). If the field and laboratory populations used in this study would thus have the same origin, the differences observed here may also indicate that prolonged rearing may have altered the biological characteristics of the insect as a result of genetic and non-genetic adaptation (Mackauer 1972, 1976; Chambers 1977; Grenier and De Clercq 2003). As noted above, the long term use of an unnatural food like *E. kuehniella* eggs and long day photoperiods in the laboratory may have led to unintentional selection in *H. axyridis*.

In conclusion, this laboratory study has demonstrated differential effects of photoperiod and diet on the developmental and reproductive fitness of the morph types of a European population of *H. axyridis*. A different capability to exploit available food sources may lead to a competitive advantage that may in part explain the predominance of non-melanic morphs in newly colonized areas. Further research is needed to fully understand the bearing of the ecophysiological differences among the morph types of *H. axyridis* on the establishment potential of the species.

Acknowledgements We are grateful to Dr. Jonathan G. Lundgren for his valuable comments on a first draft of this paper. This study was funded by grant number 53215 from the Institute for the Promotion of Innovation by Science and Technology in Flanders (IWT). Bea Roos is thanked for excellent technical assistance.

References

Adriaens T, San Martin y Gomez G, Maes D (2007) Invasion history, habitat preferences and phenology of the invasive ladybird *Harmonia axyridis* in Belgium. BioControl (this issue). doi:10.1007/s10526-007-9137-6

Anonymous (2007) Volkssterrenwacht Urnia. Onze "Zon en duisternis" efemeriden-fiches. http://www.urania.be/sterrenkunde/zon-en-duisternis.php (consulted on 10/05/2007)

Berkvens N, Bonte J, Berkvens D, Deforce K, Tirry L, De Clercq P (2007) Pollen as an alternative food for *Harmonia axyridis*. BioControl (this issue). doi:10.1007/s10526-007-9128-7

Brown PMJ, Adriaens T, Bathon H, Cuppen J, Goldarazena A, Hägg T, Kenis M, Klausnitzer BEM, Kovář I, Loomans AJM, Majerus MEN, Nedved O, Pedersen J, Rabitsch W, Roy HE, Ternois V, Zakharov IA, Roy DB (2007) *Harmonia axyridis* in Europe: spread and distribution of a non-native coccinellid. BioControl (this issue). doi:10.1007/s10526-007-9132-y

Chambers DL (1977) Quality control in mass rearing. Annu Rev Entomol 22:289–308

Coderre D, Lucas É, Gagné I (1995) The occurrence of *Harmonia axyridis* (Pallas) (Coleoptera: Coccinellidae) in Canada. Can Entomol 127:609–611

Coutanceau JP (2006) *Harmonia axyridis* (Pallas, 1773) une coccinelle asiatique introduite, acclimatée : une Coccinelle asiatique introduite et en extension en France. B Soc Entomol Fr 111:395–401

De Clercq P, Bonte M, Van Speybroeck K, Bolckmans K, Deforce K (2005) Development and reproduction of *Adalia bipunctata* (Coleoptera: Coccinellidae) on eggs of *Ephestia kuehniella* (Lepidoptera: Phycitidae) and pollen. Pest Manage Sci 61:1129–1132

Cuppen J, Heijerman T, Van Wielink P, Loomans A (2004) Het lieveheersbeestje *Harmonia axyridis* in Nederland: een aanwinst voor onze fauna of een ongewenste indringer (Coleoptera: Coccinellidae)? Nederl Faun Meded 20:1–12

De Clercq P, Peeters I, Vergauwe G, Thas O (2003) Interaction between *Podisus maculiventris* and *Harmonia axyridis*, two predators used in augmentative biological control in greenhouse crops. BioControl 48:39–55

Grenier S, De Clercq P (2003) Comparison of artificially vs. naturally reared natural enemies and their potential for use in biological control. In: van Lenteren JC (ed) Quality control of biological control agents. Theory and testing procedures. CABI publishing, Wallingford, pp 115–132

Grill CP (1999) Development of colour in an aposematic lady beetle: The role of environmental conditions. Evol Ecol Res 1:651–662

Hantson E (2004) Kleurverschillen bij het veelkleurig Aziatisch lieveheersbeestje in Vlaanderen. Coccinula 10:16–19

Hodek I, Honěk A (1996) Ecology of Coccinellidae. Kluwer Academic Publishers, Dordrecht

Hosino Y (1936) Genetical studies of the lady-bird beetle, *Harmonia axyridis* Pallas. Rep II. Jpn J Genet 12:307–320

Hosino Y (1940) Genetical studies on the pattern types of the ladybird beetle, *Harmonia axyridis* Pallas. J Genet 40:215–228

Iperti G, Bertand E (2001) Hibernation of *Harmonia axyridis* (Coleoptera: Coccinellidae) in South-Eastern France. Acta Soc Zool Bohem 65:207–210

Kabiri F (2006) *Harmonia*: a problem or a distraction? REBECA Macrobials Workshop: WICC Wageningen, The Netherlands, 5–6 April 2006. http://www.rebeca-net.de/downloads/Risk%20assessment%20macrobials%20Kabiri.pdf (consulted on 17/09/2007)

Kenis M, Roy HE, Zindel R, Majerus MEN (2007) Current and potential management strategies against *Harmonia axyridis*. BioControl (this issue). doi:10.1007/s10526-007-9136-7

Koch RL (2003) The multicolored Asian lady beetle, *Harmonia axyridis*: a review of its biology, uses in biological control, and non-target impacts. J Insect Sci 3:1–16
Koch RL, Burkness EC, Would Burkness SJ, Hutchinson WD (2004) Phytophagous preferences of the multicolored Asian lady beetle (Coleoptera: Coccinellidae) from autumn-riping fruit. J Econ Entomol 97:539–544
Komai T (1956) Genetics of lady-beetles. Adv Genet 8:155–185
Komai T, Chio M, Hosino Y (1950) Contributions to the evolutionary genetics of the lady-beetle, *Harmonia axyridis*. I. Geographic and temporal variations in the relative frequencies of the elytral pattern types and in the frequency of elytral ridge. Genetics 35:589–601
Krafsur ES, Kring TJ, Miller JC, Nariboli P, Obrycki JJ, Ruberson JR, Schaefer PW (1997) Gene flow in the exotic colonizing lady beetle *Harmonia axyridis* in North America. Biol Control 8:207–214
Mackauer M (1972) Genetic aspects of insect production. Entomophaga 17:27–48
Mackauer M (1976) Genetic problems in the production of biological control agents. Annu Rev Entomol 21:369–385
McCornack BP, Koch RL, Ragsdale DW (2007) A simple method for in-field sex determination of the multicolored Asian lady beetle *Harmonia axyridis*. J Insect Sci 7:10
McCullagh P, Nelder J (1989) Generalized linear models. Chapman & Hall, London
Michaud JP, Qureshi JA (2006) Induction of reproductive diapause in *Hippodamia convergens* (Coleoptera: Coccinellidae) hinges on prey quality and availability. Eur J Entomol 102:483–487
Omkar, Pathak S (2006) Effects of different photoperiods and wavelengths of light on the life-history traits of an aphidophagous lady beetle, *Coelophora saucia* (Mulsant). J Appl Entomol 130:45–50
Ongagna P, Iperti G (1994) Influence de la température et de la photopériode chez *Harmonia axyridis* Pall. (Col., Coccinellidae): obtention d'adultes rapidement féconds ou en dormance. J Appl Entomol 117:314–317
Ongagna P, Giuge L, Iperti G, Ferran A (1993) Cycle de développement d'*Harmonia axyridis* (Col., Coccinellidae) dans son aire d'introduction, le sud-est de la France. Entomophaga 38:125–128
Osawa N, Nishida T (1992) Seasonal variation in elytral colour polymorphism in *Harmonia axyridis* (the lady-bird beetle): the role of non-random mating. Heredity 69:297–307
Pell JK, Baverstock J, Roy HE, Ware RL, Majerus MEN (2007) Intraguild predation involving *Harmonia axyridis*: a review of current knowledge and future perspectives. BioControl (this issue). doi:10.1007/s10526-007-9125-x
Sakurai H, Kawai T, Takeda S (1992) Physiological changes related to diapause of the lady beetle *Harmonia axyridis* (Coleoptera: Coccinellidae). Appl Entomol Zool 27:479–487
Sasji H (1971) Coccinellidae. Fauna Japonica. Academic Press of Japan, Tokyo
Serpa L, Schanderl H, Brito C, Soares AO (2003) Fitness of five phenotypes of *Harmonia axyridis* Pallas (Coleoptera: Coccinellidae). In: Soares AO, Ventura MA, Garcia V, Hemptinne J-L (eds) Proceedings of the 8th International Symposium on Ecology of Aphidophaga: Biology, ecology and behaviour of aphidophagous insects. Arquipélago, Life and Marine Sciences, Supplement 5, pp 43–49
Soares AO, Coderre D, Schanderl H (2001) Fitness of two phenotypes of *Harmonia axyridis* (Coleoptera: Coccinellidae). Eur J Entomol 98:287–293
Soares AO, Coderre D, Schanderl H (2003) Effect of temperature and intraspecific allometry on predation by two phenotypes of *Harmonia axyridis* Pallas (Coleoptera: Coccinellidae). Environ Entomol 32. 939–944
Soares AO, Coderre D, Schanderl H (2005) Influence of prey quality on the fitness of two phenotypes of *Harmonia axyridis* adults. Entomol Exp Appl 114:227–232
Specty O, Febvay G, Grenier S, Delobel B, Piotte C, Pageaux J, Ferran A, Guillaud J (2003) Nutritional plasticity of the predatory lady beetle *Harmonia axyridis* (Coleoptera: Coccinellidae): comparison between natural and substitution prey. Arch Insect Biochem Physiol 52:81–91
StataCorp (2005) Stata Statistical Software: Release 9. StataCorp LP, College Station
Tan CC (1946) Mosaic dominance in the inheritance of color patterns in the lady-bird beetle *Harmonia axyridis*. Genetics 31:195–210
True JR (2003) Insect melanism: the molecules matter. Trends Ecol Evol 18:640–647

Ladybird population dynamics in potato: comparison of native species with an invasive species, *Harmonia axyridis*

Jean Pierre Jansen · Louis Hautier

Abstract Following the detection of the harlequin ladybird, *Harmonia axyridis*, in 2003 in potato crops in Belgium, a study was carried out between 2004 and 2006 on the phenology of this species compared to native species in potato. The results confirmed the success of *H. axyridis*, with high population levels in 2004 and 2005. In 2006, aphid populations were very low and no *H. axyridis* larvae were sampled in potato, but the indigenous species *Coccinella septempunctata* and *Propylea quatuordecimpunctata* were detected. A species by species comparison of the date of first larvae detection, the larvae population peak, and the difference between this peak and the aphid population peak was performed. Results showed a clear correlation between *C. septempunctata* and *P. quatuordecimpunctata* and potato aphids, with a delay of 3.5 and 6.5 days between the aphid and ladybird population peaks for the two native species. *H. axyridis* arrived 7–8 days after the two indigenous species and the larval peak population occurred 15.8 days after the aphid population peak. This meant that *H. axyridis* had to complete its larval development with very low aphid populations or even with no aphids at all. The reason for its late arrival and the possible food resources used by *H. axyridis* larvae are discussed.

Keywords *Adalia bipunctata* · *Coccinella septempunctata* · *Harmonia axyridis* · Invasive species · Intraguild predation · *Propylea quatuordecimpunctata*

Introduction

Despite the reported impact of the harlequin ladybird, *Harmonia axyridis* (Pallas) (Coleoptera: Coccinellidae), in the USA on indigenous ladybird populations (LaMana and Miller 1998; Colunga-Garcia and Gage 1998; Michaud 2002; Brown 2003; Alyokhin and Sewell 2004), this species was commercially introduced in large numbers into Belgium between 1997 and 2003 for aphid biological control in glasshouses. This

J. P. Jansen (✉) · L. Hautier
Department of Biological Control and Plant Genetic Resources, Walloon Agricultural Centre, Chemin de Liroux, 2, 5030 Gembloux, Belgium
e-mail: labecotox@cra.wallonie.be

doi: 10.1007/978-1-4020-6939-0_15,

ladybird species rapidly became invasive and was first reported outdoors in 2001 in northern Belgium, where glasshouse production is widespread. After this first report, *H. axyridis* populations gradually increased and they were found in spring and summer 2003 on *Acer* and *Tilia* tree species and in large numbers during winter in 2003 (Adriaens et al. 2003). In parks and gardens in Brussels, *H. axyridis* soon became the dominant species on *Tilia*, *Acer* and *Pinus* trees (San Martin 2003; Ottart 2005). In 2003 it was first recorded on arable crops in Belgium, on potato (Jansen and Warnier 2004). Eggs, larvae, nymphs and adults were found at various sites, indicating that the species could complete its development on potato, as it does in the USA (Nault and Kennedy 2003; Alyokhin and Sewell 2004).

Several authors have emphasised the risk of introducing exotic species for indigenous biodiversity (van Lenteren et al. 2003). In this context, *H. axyridis* appears to be a great threat for indigenous aphid predator species. Studies have shown that it is very aggressive and able to dominate *Coccinella septempunctata* L. and *Adalia bipunctata* (L.) (Coleoptera: Coccinellidae) in terms of competition for food resources and direct predation (Yasuda et al. 2001; Kajita et al. 2000; Hautier 2003; Hokkanen et al. 2003). This could explain why this species rapidly became dominant in several habitats when it was introduced in the USA, leading to major changes in aphidophagous ladybird diversity in natural and agricultural ecosystems.

In potato, in particular, change in the ladybird populations could disrupt aphid biological control and lead to an increase in insecticide use. Aphids can sometimes be a problem in potato for markets and industry, but aphid surveys in Belgium showed that populations rarely reached high levels, due mainly to the high activity of beneficial arthropods (Jansen 2002). In the 1994–2001 period, insecticide applications were required only in one field out of eight, on average. A dramatic decrease in aphid populations is regularly observed in mid-July, due to plant maturation and the activity of aphid-specific predators and parasitoids (Karley et al. 2003). Potato aphids are an important food source for at least three species: *C. septempunctata*, *Propylea quatuordecimpunctata* (L.) and *A. bipunctata*. These species completed their development in the potato crop between the end of June and mid-July (Jansen and Warnier 2004).

The aim of this study was to examine the establishment and phenology of *H. axyridis* in potato compared with native species, and to detect possible changes in aphidophagous ladybird populations in the crop, in order to be able to anticipate major changes in aphid control by beneficial arthropods and retain the biological control of potato aphids.

Material and methods

Field sites

Ladybirds and aphids were sampled in the 2004–2006 period in commercial potato fields at Corroy-le-Château (2004, cv Asterix), Florennes (2005 and 2006, cv Bintje), Gembloux (2004 and 2005, cv Bintje, 2006, cv Nicola) and Nivelles (2004, 2005 and 2006, cv Bintje). Potato fields were all located in typically intensive agriculture production area in Belgium. Neighbouring crops were mainly cereals (winter wheat and barley) and sugar beet. Field margins, hedgerows, tree rows, grasslands or fallows were nearly absent in these landscapes. Potato fields were at least 4 ha large and regularly received fungicide applications at intervals of 7–15 days, depending on disease pressure, humidity and temperature. In 2005, the Nivelles and Florennes fields received one insecticide application (Plenum

0.3 kg/ha, WG 50% pymetrozine) at the beginning of August to control the buckthorn aphid, *Aphis nasturtii* Kaltenbach (Homoptera; Aphididae). No other insecticide treatments were applied.

Aphid and ladybird sampling

Sampling was performed weekly, from mid-June, when the first aphids were observed, until mid-August, when no aphids or ladybirds were observed. The aphids were counted in situ on four samples of 50 leaves on each occasion. The samples were taken randomly along four transect lines in the field, using 25 leaves in lower part of the plants and 25 in the upper part.

The ladybirds were sampled using a beating method. Plants were shaken 30 s above a plastic tray (40 × 50 × 18 cm) and the insects collected were transferred to plastic vials and taken to the laboratory for counting and identification. All the samples were kept at 5–10°C during transportation to limit insect activity and avoid ladybird larvae cannibalism. Each week, ladybirds were sampled by shaking 4 × 10 plants in each field, taken randomly along four transects, as for the aphid counting. In 2006, due to the very low aphids and ladybird population levels, the sample size was increased from 10 plants to 15 per transect. The results obtained in 2006 were divided by 1.5 so that a comparison between years could be made.

Statistical analysis

The dates of the first larvae and adult detections and of the peak larvae and adult populations, and the differences between the peak larvae and adult populations and the peak aphid numbers, were compared by species and by field, using a Generalised Linear Model (GLM) with species ($p = 4$) and field × year ($q = 8$) as variables. The dates were compared two by two with the help of a Student t test based on Least Square Means at $P = 0.05$ level, using SAS software. Before analysis, the dates were converted into numerical values, taking 30 June as 0 for larvae data and 31 May as 0 for adult data.

Results

The results of the aphid population counts and the ladybird larvae and adult samplings are presented in Fig. 1 (2004), Fig. 2 (2005) and Fig. 3 (2006). The mean of the three sites is given, except for 2006 when it was calculated for two of the three sites, as no ladybirds were caught at the third site (Gembloux). Only species found at both larval and adult stages, and considered relevant to potato, were taken into account in the figures: the three native species *C. septempunctata, P. quatuordecimpunctata* and *A. bipunctata*, and the invasive species *H. axyridis.* The other species collected from time to time only at adult instar stage were the indigenous species *Hippodamia variegata* (Goeze) (1 specimen in 2004, 8 in 2005, none in 2006) and *Adalia decimpunctata* (L.) (1 specimen in 2004).

The aphid populations in 2004 and 2006 followed a typical pattern observed in potato in Belgium, with a progressive increase in populations towards the end of June, a peak between 6 and 12 July and then a rapid decline due to the activity of natural enemies of aphids and to plant physiology changes which slow down aphid population growth. The peak aphid populations in 2004 and 2006 were below the economic threshold of 10 aphids

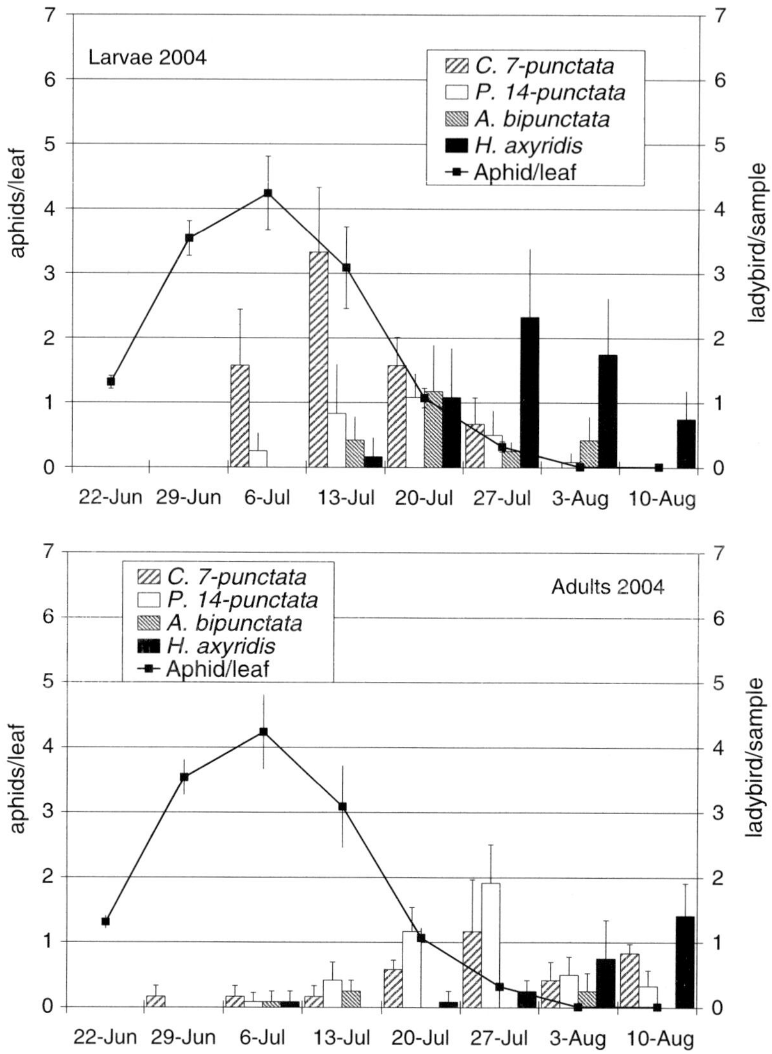

Fig. 1 Seasonal abundance of aphids and ladybird larvae and adult ladybird populations in potato in Belgium in 2004 (Corroy-le-Chateau, Gembloux and Nivelles). Mean aphids per leaf ± sd and mean ladybird per sample ± sd

per leaf, with moderate aphid populations in 2004 (a mean of 4.2 aphids per leaf, range 3.3–5.6) and low levels in 2006 (a mean of 0.7 aphids per leaf, range 0.1–1.3). The aphids observed belonged to three species of equal importance numerically: *Myzus persicae* Sulzer, *Macrosiphum euphorbiae* (Thomas) and *A. nasturtii*. Both *Aulacorthum solani* (Kaltenbach) and *Aphis fabae* (Scopoli) were also regularly found, but not in large colonies. In 2005, the aphid populations were much higher, with a mean of 18.0 aphids per leaf (range 1.3–43.2) and it was necessary to apply insecticide in two of the three fields. These high levels resulted from an *A. nasturtii* outbreak, with the populations of other aphid species being similar to the 2004 and 2006 counts. The aphid population peak in 2005 also occurred later than normal because of a second and late growing period at the beginning of

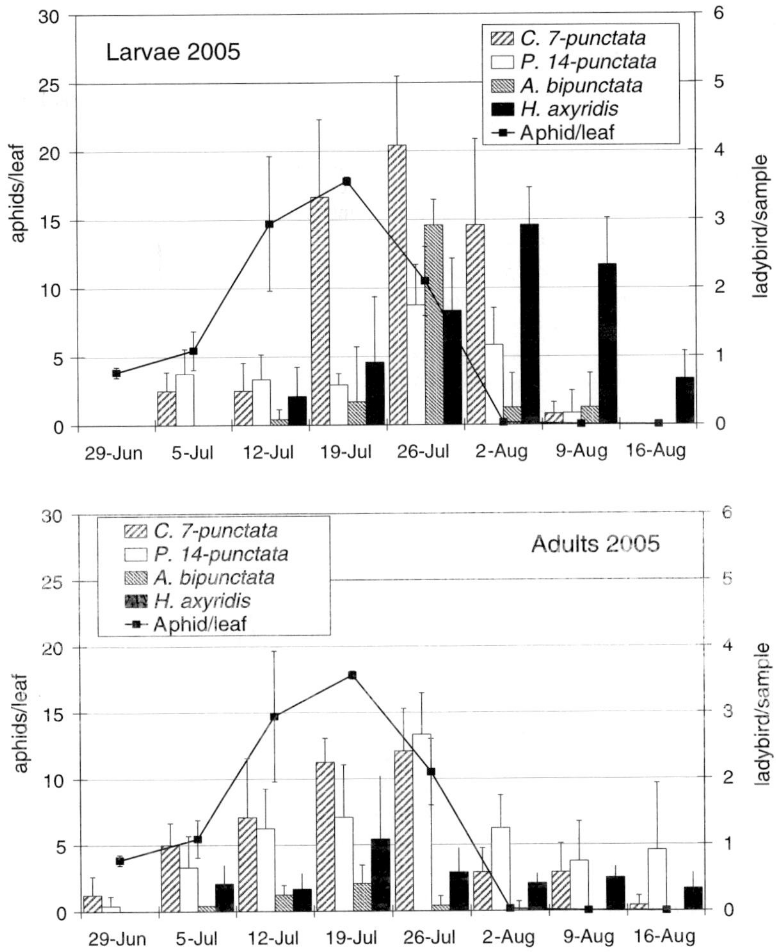

Fig. 2 Seasonal abundance of aphids and ladybird larvae and adult ladybird populations in potato in Belgium in 2005 (Florennes, Gembloux, Nivelles). Mean aphids per leaf ± sd and mean ladybird per sample ± sd

July due to abundant rainfall after 3–4 weeks of drought that stopped foliage development completely. The decline in aphid numbers observed after mid-July was due to the use of pymetrozine in two of the three fields. This insecticide is know to be selective for ladybirds and has no acute toxicity on ladybird populations (Hautier et al. 2006). However, the drastic reduction in aphid populations probably affected the ladybirds by reducing food availability and increasing cannibalism and intraguild predation.

The ladybird sampling results showed a clear population trend in 2004 and 2005 for larvae and adults. Two species, *C. septempunctata* and *P. quatuordecimpunctata*, arrived significantly earlier than *H. axyridis* and *A. bipunctata* and their population dynamics were more closely related to aphid population abundance (Table 1). The ladybird populations could be split into two groups: (1) *C. septempunctata* and *P. quatuordecimpunctata* and (2) *A. bipunctata* and *H. axyridis* which arrived 7–8 days later. The same delay was observed for the larva population peak, with a difference of 13 days between

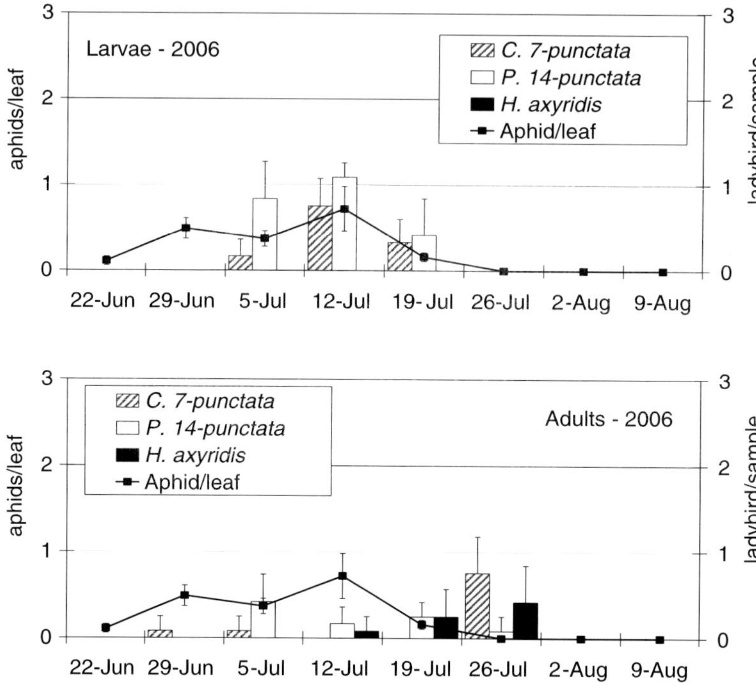

Fig. 3 Seasonal abundance of aphids and ladybird larvae and adult ladybird populations in potato in Belgium in 2006 (Florennes, Nivelles). Mean aphids per leaf ± sd and mean ladybird per sample ± sd

C. septempunctata and *H. axyridis*. This delay was also observed with regard to the date of the aphid population peak, with a delay of only 3.5 days for *C. septempunctata* but 15.8 days for *H. axyridis*.

The sampling method used did not allow data on ladybird pupae to be collected but visual observations and collecting pupae manually in order to study the parasitism of *C. septempunctata* and *H. axyridis* in potato were performed at the same time as the beating sampling in 2005. Pupae were identified by the presence or absence of spines at the base and identification was confirmed when the adults emerged. There was evidence that

Table 1 Date of first larva detection, peak larval population and delay between aphid and ladybird larval population peaks in 8 potato in Belgium (3 in 2004, 3 in 2005 and 2 in 2006). Numbers in the same columns followed by different letters are different at $P = 0.05$ level

Ladybird species	Occurence (2004, 2005, 2006)	First larval detection	Peak larval population	Peak larval – Peak aphid population
C. 7-punctata	3, 3, 2	7.1 July (a)	16.4 July (a)	3.5 days (a)
P. quaturodecimpunctata	3, 3, 2	8.0 July (a)	19.4 July (ab)	6.5 days (ab)
A. bipunctata	3, 2, 0	15.7 July (b)	22.5 July (bc)	9.6 days (bc)
H. axyridis	3, 3, 0	15.1 July (b)	28.7 July (c)	15.8 days (c)
F		7.62	7.42	7.42
df		3, 7	3, 7	3, 7
P		0.0022	0.0025	0.0025

Table 2 Date of first adult detection, peak adult population and delay between peak aphid and ladybird adult population in potato in 8 potato in Belgium (3 in 2004, 3 in 2005 and 2 in 2006). Numbers in the same columns followed by different letters are different at $P = 0.05$ level

Ladybird species	Occurence (2004, 2005, 2006)	First adult detection	Peak adult population	Peak adult—Peak aphid population
C. 7-punctata	3, 3, 2	8.4 July (a)	18.8 July (a)	24.3 days (a)
P. quatuordecimpunctata	3, 3, 2	8.1 July (a)	19.4 July (a)	24.9 days (a)
A. bipunctata	3, 2, 0	15.3 July (a)	24.6 July (a)	30.0 days (a)
H. axyridis	3, 3, 2	15.9 July (a)	29.0 July (a)	32.0 days (a)
F		2.30	2.14	1.62
df		3, 7	3, 7	3, 7
P		0.1115	0.1306	0.2198

H. axyridis larvae reached the pupa stage in a same extend as *C. septempunctata*, with 134 *C. septempunctata* and 83 *H. axyridis* pupae collected at Nivelles on 3 August 2005, and 36 *C. septempunctata* and 14 *H. axyridis* larvae collected by beating in the same field one week earlier.

The same trends as for larvae results were observed for the adult ladybird samplings, with an apparent late detection date of *H. axyridis* and a longer delay between peak aphid and adult ladybird populations compared to *C. septempunctata* and *P. quatuordecimpunctata* (Table 2). However, the statistical analysis conclude that they were no differences concerning adult first detection date, adult peak population date and delay between peak aphid and adult ladybird population (see Table 1 for details). This can be explain by the fact that several adult catches did not correlate with the larva data in terms of dates and probably related to adult ladybirds searching for food and not becoming successfully established in the visited field. The direct consequence of this is an increase of the data variability that made the statistical analysis of the adult observations more difficult than for larvae.

The relative abundance of the different species is illustrated in Fig. 4 (2004) and Fig. 5 (2005). The minimum and maximum records of larvae were 9 (10 August) and 59 (20 July) in 2004 and 8 (16 August) and 125 (26 July) in 2005. The minimum and maximum records of adults were 5 (6 July) and 40 (27 July) in 2005 and 16 (16 August) and 69 (26 July) in 2005. At the beginning of July, only *C. septempunctata* larvae and *P. quatuordecimpunctata* larvae were found. Subsequently, *A. bipunctata* and *H. axyridis* arrived. The relative abundance of *H. axyridis* increased to a point where it completely dominated the other species by the end of the aphid potato season in August, when no aphids were observed. The results with the adults were less clear, probably because of the possible 'background noise' in adult catches and the low adult densities in August.

Discussion

This 3-year study showed that the harlequin ladybird, *H. axyridis*, has become successfully established in potato fields in Belgium and that the first detection in 2003 (Jansen and Warnier 2004) was not accidental. The first *H. axyridis* larvae were detected, on average, 7–8 days later than *C. septempunctata* and *P. quatuordecimpunctata*, the two most important species found in those fields. Compared to these two native species, the

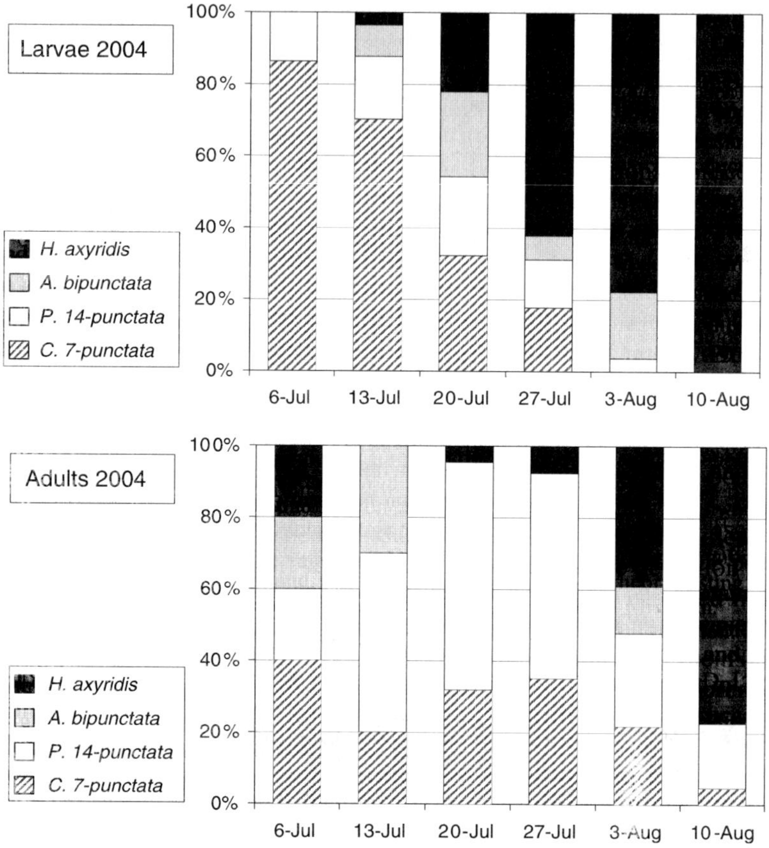

Fig. 4 Relative abundance of ladybird larvae and adults sampled in potato fields in 2004 (sum of Corroy-le-Chateau, Gembloux and Nivelles)

H. axyridis larval population peak occurred 12 and 9 days later, respectively. As aphid resources are very limited in time (Jansen 2005), the harlequin ladybird is not at all synchronised with aphids and that the main preimaginal population development occurred with very few aphids. This suggests that *H. axyridis* is able to complete its larval development feeding on prey other than aphids. This has been observed in the USA where *H. axyridis* larvae were found in great numbers in potato fields where they were supposed to feed on Colorado beetle larvae (Nault and Kennedy 2003), and in Japan on *Hibiscus* plants (Hironori and Katsuhiro 1997). In terms of the phenology of *H. axyridis*, *C. septempunctata* and *P. quatuordecimpunctata* appear to be the prey and *H. axyridis* the predator, with a delay in arrival times and peak population levels between prey and predator. This hypothesis is also supported by the fact that no *H. axyridis* larvae were collected in 2006, when the population levels of European species were much lower than in 2004 and 2005, due to low aphid populations.

Many laboratory studies clearly show that *H. axyridis* larvae are able to reach the imago stage only when the ladybird larvae of other species are available as food, such as *C. septempunctata* and *A. bipunctata* (Yasuda and Ohnuma 1999; Yasuda et al. 2001; Sato and Dixon 2004). When placed in competition, *H. axyridis* easily dominates *A. bipunctata*

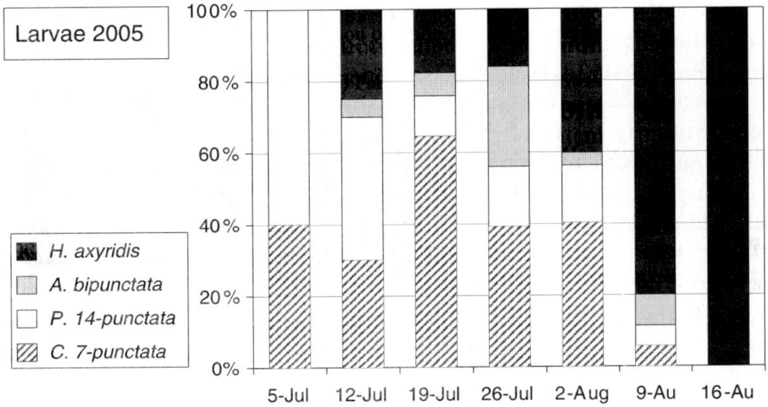

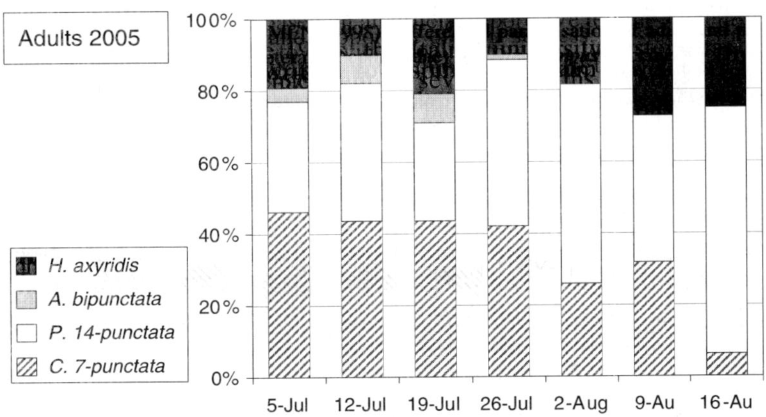

Fig. 5 Relative abundance of ladybird larvae and adults sampled in potato fields in 2005 (sum of Florennes, Gembloux and Nivelles)

in terms of direct predation and competing for food resources (Hautier 2003). The hypothesis that *H. axyridis* acts as an intraguild predator rather than an aphid predator has often been put forward but never completely verified in situ. Although intraguild predation was regularly observed in the field during this study and in previous studies (Hironori and Katsuhiro 1997; Nault and Kennedy 2003), no real assessment of the consequences of this predation was carried out because of the lack of quantification techniques in practical conditions.

Another explanation for the late arrival of *H. axyridis* is that it is known to be principally an arboreal species, at least in its area of origin (Iablokoff-Khnzorian 1982), and is commonly found on trees in Belgium, as indicated by the records on urban trees (San Marin 2003; Ottart 2005). One hypothesis is that *H. axyridis* colonizes herbaceous plants, as potato and wheat fields for example, only when it cannot find food resources on trees. This hypothesis is strengthened by the fact that *A. bipunctata*, generally considered as an arboreal species with the ability to extend to other ecosystems when there is a lack of food (Iperti 1965; Iablokoff-Khnzorian 1982), arrives at the same time as *H. axyridis*, as indicated by the results of this study. The great difference between *A. bipunctata* and

H. axyridis is that the developmental success of *A. bipunctata* when the aphid population is low or null is limited.

Whatever the reason for the late arrival of *H. axyridis* in potato fields compared with indigenous species, the direct consequence is the same: *H. axyridis* has to complete its development with very low aphid populations present, or even with no aphids at all, and it is able to do it successfully. The principal question, therefore, is what do *H. axyridis* larvae feed on? As very few insects other than ladybirds were observed in the potato fields at the end of July and the beginning of August and no alternative food resources such as pollen were available, conspecific and intraguild predation on the last indigenous ladybird larvae and pupae seems to be the most realistic hypothesis. The long-term consequences of this could be dramatic for the European species. However, no clear evolution of the dominance of *H. axyridis* was observed during this 3-year study.

References

Adriaens T, Maes D, Branquart E (2003) The multicoloured Asian ladybird *Harmonia axyridis* Pallas (Coleoptera: Coccinellidae), a threat for native aphid predators in Belgium? Belgian J Zool 133:195–196

Alyokhin A, Sewell G (2004) Changes in a ladybeetle community following the establishment of three alien species. Biol Invasions 6:463–471

Brown MW (2003) Intraguild responses of aphid predators on apple to the invasion of an exotic species, *Harmonia axyridis*. BioControl 48:141–153

Colunga-Garcia M, Gage SH (1998) Arrival, establishment and habitat use of the multicolored Asian ladybeetle (Coleoptera: Coccinellidae) in a Michigan landscape. Environ Entomol 27:1574–1580

Hautier L (2003) Impact sur l'entomofaune indigène d'une coccinelle exotique utilisée en lutte biologique. DES en Gestion de l'Environnement, Université Libre de Bruxelles, Belgium

Hautier L, Jansen JP, Mabon N, Schiffers B (2006) Building a selectivity list of plant protection products on beneficial arthropods in open field: a clear example with potato crop. IOBC/WPRS bulletin 29(10): 21–32

Hironori Y, Katsuhiro S (1997) Cannibalism and interspecific predation in two predatory ladybirds in relation to prey abundance in the field. Entomophaga 42:153–164

Hokkanen HM, Barbendreier D, Bigler F, Burgio B, Kuske S, Van Lenteren JC, Loomans A, Menzler-Hokkanen I, Van Rijn P, Thomas M, Tommasini M, Zeng QQ (2003) Evaluating environmental risks of biological control introductions into Europe (ERBIC), final report

Iablokoff-Khnzorian SM (1982) Les coccinelles, Coleoptères – Coccinellidae. ed. Société Nouvelle des éditions Boubée, Paris, 517 pp

Iperti G (1965) Contribution à l'étude de la spécificité chez les principales coccinelles aphidiphages des Alpes-Maritimes et des Basses-Alpes. Entomophaga 10:159–178

Jansen JP (2002). Pucerons et auxiliaires de lutte en pomme de terre de consommation : synthèse des observations réalisées entre 1994 et 2001 en Belgique. 2ème conférence internationale sur les moyens alternatifs de lutte contre les organismes nuisibles aux végétaux, Lille – 4, 5, 6 et 7 mars 2002

Jansen JP (2005) Pucerons en pomme de terre de consommation: bilan de 12 années d'observation. Journée d'étude Pomme de terre, édition 2005, pp 41–50. éd. Centre wallon de recherches Agronomiques, Gembloux, Belgique

Jansen JP, Warnier AM (2004) Aphid specific predators in potato in Belgium. Commun Agric Appl Biol Sci 69(3):151–156

Kajita Y, Takano F, Yasuda H, Agarwala BK (2000) Effects of indigenous ladybird species on the survival of an exotic species in relation to prey abundance. Appl Entomol Zool 35:473–479

Karley AJ, Pitchford JW, Douglas AE, Parker WE, Howard JJ (2003) The causes and process of the mid-summer population crash of the potato aphids *Macrosiphum euphorbiae* and *Myzus persicae* (Hemiptera: Aphididae). B Entomol Res 93:425–437

LaMana ML, Miller JC (1998) Temperature-dependant development in an Oregon population of *Harmonia axyridis* (Coleoptera: Coccinellidae). Environ Entomol 27:1001–1005

Michaud JP (2002) Invasion of the Florida citrus ecosystem by *Harmonia axyridis* (Coleoptera: Coccinellidae) and asymmetric competition with a native species, *Cycloneda sanguinea*. Environ Entomol 31:827–835

Nault BA, Kennedy GG (2003) Establishment of multicolored Asian lady beetle in Eastern North Carolina: Seasonal abundance and crop exploitation within an agricultural landscape. BioControl 48:363–378

Ottart N (2005) Impact de la coccinelle invasive *Harmonia axyridis* sur les populations de coccinelles indigènes à Bruxelles. Travail de fin d'études, Université Libre de Bruxelles, 84 pp

San Martin G (2003) Etude de l'impact de l'urbanisation sur les populations de coccinelles à Bruxelles. Mémoire de licence en biologie, Université Libre de Bruxelles, 64 pp

Sato S, Dixon AF (2004) Effects of intraguild predation on the survival and development of three species of aphidophagous ladybirds: consequences for invasive species. Agri For Entomol 6:21–24

van Lenteren JC, Babendreier D, Bigler F, Burgio G, Hokkanen HM, Kuske S, Loomans AJ, Menzler-Hokkanen I, Van Rijn PC, Thomas MB, Tommasini MG, Zeng QQ (2003) Environmental risk assessment of exotic natural enemies used in inundative biological control. BioControl 48:3–38

Yasuda H, Ohnuma N (1999) Effects of cannibalism and predation on the larval performance of two ladybird beetles. Entomol Exp Appl 93:63–67

Yasuda H, Kikuchi T, Kindlmann P, Sato S (2001) Relationships between attack and escaped rates, cannibalism and intraguild predation in larvae of two predatory ladybirds. J Ins Behavior 14:373–384

Current and potential management strategies against *Harmonia axyridis*

Marc Kenis · Helen E. Roy · Renate Zindel · Michael E. N. Majerus

Abstract This paper reviews the current and potential methods to control the harlequin ladybird, *Harmonia axyridis* (Pallas) (Coleoptera: Coccinellidae), an Asian predatory beetle invasive in Europe and the Americas where it has become a human nuisance, a grape and wine pest and a threat to native biodiversity. Current methods to manage this invasive species include: techniques to mechanically prevent adult beetles from entering buildings in autumn or to remove aggregates of beetles inside buildings, e.g. using various trapping methods; the use of insecticides on buildings or in vineyards to prevent aggregation in houses or on grapes; cultivation practices in vineyards to lower the impact of the ladybird on grape production and wine quality; remedial treatments for wine tainted by the ladybird. Other methods are presently being developed or considered. Semiochemicals could be used as deterrents or as attractants to develop more efficient trapping systems in buildings and open fields. Natural enemies include pathogens, parasitoids, predators and a parasitic mite but few of them show potential as biological control agents. While management methods presently used or under development may eventually solve the problems caused by beetles aggregating in buildings or vineyards, the issue of *H. axyridis* populations outcompeting native species is much more challenging. Only the sudden adaptation of a native natural enemy or the importation of a natural enemy from the area of origin of the ladybird may

M. Kenis (✉) · R. Zindel
CABI Europe-Switzerland, 1, Rue des Grillons, 2800 Delemont, Switzerland
e-mail: m.kenis@cabi.org

H. E. Roy
NERC Centre for Ecology and Hydrology Monks Wood, Huntingdon PE28 2LS, UK
e-mail: hele@ceh.ac.uk

R. Zindel
Department of Biology, Unit of Ecology and Evolution, University of Fribourg, Chemin du Musée 10, 1700 Fribourg, Switzerland
e-mail: renatezindel@gmail.com

M. E. N. Majerus
Department of Genetics, University of Cambridge, Downing Street, Cambridge CB2 3EH, England, UK
e-mail: m.majerus@gen.cam.ac.uk

Biological Control to Invasion: Ladybird Harmonia axyridis. H.E. Roy, E. Wajnberg (eds.), 235
doi: 10.1007/978-1-4020-6939-0_16, © IOBC 2007

ultimately lower population densities. The problems linked to the importation of an Asian natural enemy of *H. axyridis* are discussed.

Keywords *Harmonia axyridis* · Integrated pest management · Biological control · Mechanical control · Chemical control · Cultivation practices · Parasitoids · Entomopathogenic fungi · Parasitic mites

Introduction

Harmonia axyridis (Pallas) (Coleoptera: Coccinellidae), the harlequin ladybird, or multicoloured Asian ladybeetle, is a polyphagous predatory coccinellid, native to central and eastern Asia. It has been widely used as a biological control agent of aphids in the Americas and Europe, in greenhouses as well as in outdoor crops (Koch 2003; Pervez and Omkar 2006). In North America, since its establishment in the 1980s, it has spread and increased dramatically so that it is now the dominant ladybird species in much of the USA and Canada (Koch and Galvan 2007). Three categories of hazards caused by this invasive ladybird are reported (Koch et al. 2006). First, it is considered as a human nuisance because it aggregates in buildings when seeking overwintering sites, causing cosmetic damage and, occasionally, biting humans and causing allergic reactions (McCutcheon and Scott 2001; Huelsman and Kovach 2004; Kovach 2004; Ray and Pence 2004). Second, *H. axyridis* has been reported to damage fruit crops in late summer and to taint wine when harvested and crushed with grapes (Koch et al. 2004; Kovach 2004; Pickering et al. 2005). Finally, due to its predatory and competitive abilities, *H. axyridis* may have strong negative effects on biodiversity, affecting many non-target species, including native ladybirds and other aphidophagous insects but also non-pest aphids or other herbivorous insects (see reviews by Majerus et al. 2006 and Koch and Galvan 2007).

In Europe, *H. axyridis* has been released in the wild since 1990 (Brown et al. 2007). Since the first observations of feral populations were reported in 1999 in Germany and in Belgium in 2001, numbers have increased exponentially and it is now regarded as established in 12 European countries (Brown et al. 2007). In Belgium, recent surveys showed that *H. axyridis* has become by far the most abundant ladybird on broadleaved trees (Adriaens et al. 2007).

Since the establishment of *H. axyridis* in North America, this invasive ladybird has been the target of numerous research programmes, which have been summarised in several review articles (e.g. Adriaens et al. 2003; Koch 2003; Majerus and Roy 2005; Koch et al. 2006; Pervez and Omkar 2006). None of these reviews, however, cover the currently used and potential management methods to control *H. axyridis*.

Several methods are presently implemented or being studied to control the ladybird in North America. These methods differ according to the three main types of hazards caused by the beetle (aggregation in buildings, damage to fruits and wine and competition with the native fauna). Most management methods, proposed so far, concern preventive or curative control against ladybird aggregation in houses or in vineyards. No method is presently available to lower population densities in natural environments and to limit the impact of the ladybird on native species.

Mechanical control and trapping methods

Numerous extension documents in North America describe methods to mechanically prevent adult beetles from entering buildings in autumn or to remove aggregates of beetles inside buildings (e.g. McCutcheon and Scott 2001; Jones and Boggs 2002; Baniecki et al. 2004; Potter and Townsend 2005). Recommendations to prevent the entrance of beetles in buildings include: checking for all possible entry sites; sealing windows, cracks and small holes throughout the house with weather stripping; placing insect screening over windows, attic, exhaust vents, etc. To remove large aggregates of beetles in buildings, the use of a broom and dust pan or, better, a vacuum cleaner is recommended (Jones and Boggs 2002; Baniecki et al. 2004; Potter and Townsend 2005). Jones and Boggs (2002) describe how a vacuum cleaner can be modified to capture ladybirds and avoid them entering the vacuum by using a sweeper attachment with a knee-high nylon hose inserted in the end.

Various traps are recommended and commercially available to catch adult beetles in buildings (Banieki et al. 2004). Black light traps and other light traps can be used for capturing beetles in dark confined places such as attics. Window traps may be effective during daytime. Although it has been suggested that these traps could reduce beetle populations by 50–70% (Jones and Boggs 2002; Banieki et al. 2004), thorough testing is required to ascertain whether they provide a solution for large infestations of *H. axyridis* (Jones and Boggs 2002; Potter and Townsend 2005). Furthermore, it is highly unlikely that light traps could efficiently control ladybirds in outdoor conditions, e.g. in vineyards or orchards. The efficiency of trapping methods in buildings and open field could be greatly enhanced by coupling with semiochemicals. There are many examples in the literature of insects using olfactory communication for dispersal, migration, reproduction, aggregation and alarm signalling (reviewed in a pest control context by Van Emden and Service 2004). Over the last 30 years the behavioural manipulation of insects through semiochemical techniques (for example: push–pull strategies) has been considered for the control of insect populations, with varying success (Cook et al. 2007). However, our theoretical understanding of insect semiochemistry has increased enormously in recent years (reviewed by Hassanali et al. 2005; Cook et al. 2007; Roy et al. 2008) and applications exploiting this knowledge, within pest control strategies, are being developed further. Pheromones have been used within integrated pest management (IPM) strategies for: monitoring pest populations, trapping out pest populations ("lure and kill"), confusing pests (saturating a crop environment with synthetic pheromones to prevent males locating females), deterring oviposition and manipulating or encouraging natural enemies.

Information on semiochemically-mediated behavioural mechanisms of coccinellids is limited (Pettersson et al. 2005) but there are sufficient recent studies to suggest that, like other insects, chemical communication is an important component of ladybird behaviour (Burns et al. 1998; Hemptinne et al. 2004; Pettersson et al. 2005). The use of autumn aggregation pheromones for the location of suitable overwintering sites perhaps has the most significant potential for use in a control strategy. It has been widely speculated that many species of ladybirds use species-specific aggregation chemicals in preparation for dormancy (Majerus 1994). Hills (1969) suggested that *Adalia bipunctata* (L.) (Coleoptera: Coccinellidae) used pheromonal volatiles from the excretal remains of individuals overwintering the previous year. Majerus (1994) demonstrated that *A. bipunctata* were less likely to locate sites that had been washed between overwintering events, providing strong circumstantial evidence of a long lasting, aggregation pheromone. In a more recent study, Pettersson et al. (2005) demonstrated that *Coccinella septempunctata* L. (Coleoptera: Coccinellidae) use an autumn aggregation pheromone to locate overwintering sites. There is no current information on the chemical nature of these aggregation pheromones or whether they are species specific or, indeed, seasonally specific (active only in autumn). Nalepa et al. (2000, 2005) investigated factors influencing aggregation behaviour and found little evidence for volatile aggregation pheromones. Their results suggest that beetles are rather attracted by visual intensity contrast and by contact chemoreception with conspecifics or the faeces and residues that persist in aggregation sites from previous years. However, if an aggregation semiochemical specific to *H. axyridis* could be produced, it would be possible to utilise this in a "lure and kill" strategy for controlling this species in sensitively determined areas. A recent paper suggests that two volatile sesquiterpenes, (E)-β-farnesene (aphid alarm pheromone) and (−)-β-caryophyllene (isolated from the headspace volatiles above *H. axyridis* aggregations in winter), are ecologically relevant to *H. axyridis* (Verheggen et al. 2007). (E)-β-farnesene elicited electroantennogram (EAG) and behavioural responses within an olfactometer in both male and female *H. axyridis*. In contrast, (−)-β-caryophyllene resulted in EAG and behavioural activity (attraction) in male *H. axyridis* but only EAG activity in female beetles (no associated behavioural response within an olfactometer). However, interestingly, Verheggen et al. (2007) demonstrated in a bioassay that both male and female *H. axyridis* aggregated in the (−)-β-caryophyllene-treated side of a ventilated plastic box as opposed to the untreated side. Perhaps the presence of the male *H. axyridis* evokes the response of the females. The authors conclude that (E)-β-farnesene and (−)-β-caryophyllene could have potential use within a push–pull control strategy for *H. axyridis*, but it is possible that *H. axyridis* exhibits different antennal sensitivity to (−)-β-caryophyllene in relation to season, responding more in winter. The complexity of insect behavioural responses to semiochemicals should be assessed thoroughly when considering manipulating insects as a component of an IPM strategy.

Trapping methods based on aggregative pheromones are commonly used for the monitoring and control of scolytid bark beetles (Byers 2004). A sap beetle, *Carpophilus lugubris* Murray (Coleoptera: Nitidulidae) was successfully controlled at overwintering locations, with minimal impacts on other insect species, using an autoinoculative device baited with aggregation pheromone and the fungal pathogen *Beauveria bassiana* (Balsamo) Vuillemin (Ascomycota: Hypocreales) (Dowd and Vega 2003). A combination of the fungal pathogen *Metarhizium anisopliae* (Metsch.) Sorokin (Ascomycota: Hypocreales) and an attractant (phenethyl propionate:eugenol:geraniol) has shown potential for controlling Japanese beetle *Popillia japonica* Newman (Coleoptera: Scarabaeidae) (Klein and Lacey 1999). These systems could be appropriately designed for controlling *H. axyridis*, however, recent studies have shown that *H. axyridis* is relatively resilient to fungal (*B. bassiana*) mortality (Roy et al. 2007). It is possible that a synthetic insecticide could be used in an attractant trap which would provide a targeted method of control with limited impacts on non-target insects and the environment.

Chemical control

Many insecticides have been tested against *H. axyridis*, but rather to test the non-target impact of insecticides used against aphids and other pests than to assess their potential use against the ladybird (see Koch 2003, for a review, and Galvan et al. 2005a, b, 2006d for the most recent investigations). These studies showed that most insecticides commonly used in agricultural environments are toxic to *H. axyridis*. Larvae were more sensitive than adults and, in general, the toxicity of the insecticides was lower to the ladybird than to aphids. Galvan et al. (2006a), Inclan et al. (2006) and Williams et al. (2006) recently carried out field and laboratory evaluation of the efficacy and suitability of insecticides for management of *H. axyridis* in vineyards. One of the limiting factors of using insecticides is that many of them, e.g. most pyrethroids, have a pre-harvest interval of several weeks whereas, to be efficient, treatments should be applied within a week before harvest (Galvan et al.

2006a). Among the most efficient compounds, only carbaryl and imidacloprid were labelled for use on wine grape within 7 days of harvest in USA at the time of the study (Galvan et al. 2006a). Insecticide treatments against *H. axyridis* in vineyards should not be carried out preventively but should rather follow decision protocols based on rigorous sampling plans and well defined action thresholds (Galvan et al. 2006a, 2007). In addition, the potential effect of such treatments on non-target beneficial organisms should be investigated.

In North America, it is sometimes advised to apply insecticides, as a preventive method, to the exterior of buildings that are regularly infested year after year (McCutcheon and Scott 2001; Jones and Boggs 2002; Baniecki et al. 2004; Potter and Townsend 2005). Application of a residual formulation of synthetic pyrethroids (e.g. bifenthrin, cyfluthrin, cypermethrin, deltamethrin, etc.) to walls and sidings, around eaves, attic vents, windows, doors and other likely points of entry, in late-September–October, i.e. just before beetles enter the structure is recommended. Insecticide use inside buildings is usually not advised (McCutcheon and Scott 2001; Potter and Townsend 2005) and the application of chemicals in other overwintering sites has never been considered.

Deterrents

Deterrents could be employed to reduce *H. axyridis* populations in locations where their presence is undesired, for example, houses used as overwintering locations and vineyards. *Harmonia axyridis* reduces rates of feeding and oviposition in the presence of conspecific faeces but not with the faeces of *Propylea japonica* (Thunberg) (Coleoptera: Coccinellidae); whereas *P. japonica* reduces feeding and oviposition in response to faeces of conspecifics and *H. axyridis* (Agarwala et al. 2003). This is hypothesised to reduce intra-guild predation (Agarwala et al. 2003). Further work could investigate the use of these pheromonal faecal volatiles as deterrents for *H. axyridis*. However, it is possible that these could act as deterrents to other ladybird species and so be detrimental by impacting on beneficial insects in agricultural systems. Some plant compounds, such as camphor or menthol have proved to be repellent to *H. axyridis* in field and laboratory tests and could be used to protect buildings (Riddick et al. 2000). However, under field conditions, camphor was ineffective after 48 h post-application (Riddick and Aldrich 2004). Laboratory bioassays also showed that *H. axyridis* is repelled by DEET (*N,N*-diethyl-3-methylbenzamide), a widely used insect repellent. A formulation of DEET plus paraffin had repellent activity for 23 days in the laboratory, but has not been field-tested (Riddick et al. 2004).

Cultivation practices in fruit production and remedial treatment for tainted wine

H. axyridis is often reported as a pest of fruit production in North America (Koch 2003; Kovach 2004). Nevertheless, the few studies on feeding activity suggest that it rarely causes primary injury on fruits, but rather feeds on damaged fruits (Koch et al. 2004; Kovach 2004; Galvan et al. 2006b). The main impact of the ladybird as a fruit pest is in vineyards when it aggregates in clusters of grapes, in particular those containing damaged berries, and is processed with the grapes, resulting in tainted, unmarketable wine or juice (Pickering et al. 2005). Several recommendations on cultivation practices in vineyards have been suggested to lower the impact of the ladybird in regions where *H. axyridis* causes recurrent problems to fruits, although these recommendations are not always based on rigorous studies. Key components of an IPM strategy against *H. axyridis* in vineyards include proper surveys for beetle densities before harvest and the determination of a threshold density, to assist in management decisions. Galvan et al. (2007) have described various sampling plans and assessed their usefulness. They found binomial sampling plans to be more accurate methods to measure beetle densities than enumerative plans. Kovach (2004) and Pickering et al. (2007) evaluated the threshold density for wine contamination to be about 0.9 and 1.3–1.5 beetle per kg of grapes, respectively, but the latter authors recommend a more conservative limit of 0.2–0.4 beetles per kg of grapes above which interventions in the field or in the winery should be considered. Including berry injuries in the sampling procedures may also be useful since ladybirds are primarily found on damaged fruits (Galvan et al. 2007). Such damage is caused by a variety of mechanisms including by splitting, feeding by birds or other insects, disease (rot) etc. (Galvan et al. 2007). Growers could reduce berry injury by using irrigation to avoid long periods of drought and by avoiding injuring to berries when pruning or spraying. Selecting varieties with higher resistance or tolerance to splitting may also be envisaged, as a potential long-term measure, when vineyards are replanted through the normal process of renewing stock.

Harvesting methods may have an impact on the density of beetles in harvested grapes. Beetles may be more likely to leave the grapes during day harvesting rather than during night harvesting (Anonymous 2005). Hand harvesting may be more favourable than mechanical harvesting because aggregations of beetles in grape clusters can be monitored during harvesting and infested grapes can be discarded. Beetles can be removed by shaker clusters, by hand or by using shaker tables, and by floating clusters in water or vacuum clusters (Galvan et al. 2006c). Shaker tables could also be used to monitor populations just before harvest. However, these methods result in significant increase in time, labour and costs of harvest (Galvan et al. 2006c). Managing the surrounding crops and vegetation could also be critical because beetles found in vineyards and orchards in autumn probably come from nearby aphid infestations. In North America, damage in vineyards is often associated with the presence of soybean infested by the soybean aphid, *Aphis glycines* Matsumura (Hemiptera: Aphididae), one of the preferred preys of *H. axyridis*, although the exact relationship between the proximity of soybean, soybean aphids and *H. axyridis* populations has not as yet been rigorously monitored.

Studies are being made on refining agents and other technologies to restore the quality of wine tainted by *H. axyridis* (Pickering et al. 2006). The main compound responsible for the taint is 2-isopropyl-3-methoxypyrazine (IPMP), a component of the reflex bleeding fluid of *H. axyridis* (Pickering et al. 2005). IPMP concentrations in wine could be lowered by activated charcoal in white wine and deodorized oak in red wine, although these treatments do not translate into lower intensity of the sensory attributes associated with the taint. In contrast, oak chips were successful in reducing the intensity of the taint characteristics, probably through a masking effect (Pickering et al. 2006). However, this method does not completely remove the taint from contaminated wine and, furthermore, it cannot be used for all wine types, since many of them are not normally oaked. Thus, other solutions that may be suitable for all wine types need to be sought. Pickering et al. (2006) suggest searching for a technology to target and selectively bind IPMP, with subsequent removal of the complex from the juice or wine.

Biological control

Natural enemies can be used in various ways to control an invasive insect such as *H. axyridis* (Van Driesche and Bellows 1996). First, a species could be introduced from the region of origin of the pest to the region of introduction for permanent establishment, a method often called "classical biological control". Second, in an "augmentative biological control" programme, natural enemies can be increased through regular inundative or inoculative releases. Finally, in a "conservation biological control" strategy, natural enemies already present on site can be favoured, e.g. through various cultural practices, to increase their impact on the target pest. All these methods could be developed and employed for controlling *H. axyridis*, albeit not with the same agents.

Natural enemies reported, or tested against *H. axyridis* comprise pathogens, predators, parasitoids, nematodes and a parasitic mite. There is little information available on the importance of natural enemies in the population dynamics of *H. axyridis* in field conditions, particularly in its region of origin. Research on mortality factors in field conditions has been carried out in Japan (Osawa 1992, 1993) and South Korea (Park et al. 1996), but natural enemies in the largest part of its distribution range, i.e. China and Russia, are largely unknown. In Europe and North America, observations suggest that natural enemies are of little importance in the population dynamics of the ladybird.

Pathogens

There have been a number of studies focussing on the impact of entomopathogens on coccinellids. Undoubtedly, the literature on fungal entomopathogens dominates this field and is mainly from the perspective of natural mortality in overwintering populations or, more usually, the effect of biorational pesticides on non-target coccinellids. Entomopathogenic fungi are common natural enemies of insects and other arthropods worldwide (Roy et al. 2006) and have considerable potential as biological control agents of pest insects.

Two Hypocrealean fungi have been found infecting ladybirds: *M. anisopliae* (Ginsberg et al. 2002) and *B. bassiana* (James et al. 1995; Cottrell and Shapiro-Ilan 2003; Roy et al. 2007). Most research on the direct interactions between fungi and ladybirds has concentrated on *B. bassiana* because of the potential non-target impact of this fungus when used as a mycoinsecticide against various pest insects. Susceptibility of various ladybirds (Coleoptera: Coccinellidae) to *B. bassiana* has been demonstrated: *Hippodamia convergens* Guérin-Méneville (James et al. 1995), *A. bipunctata* (Roy et al. 2007), *C. septempunctata* (Roy et al. 2007), *Coleomegilla maculata lengi* Timberlake (Todorova et al. 1996, 2000), *Serangium parcesetosum* Sicard (Poprawski et al. 1998), *Olla v-nigrum* (Mulsant) (Cottrell and Shapiro-Ilan 2003) and *Cryptolaemus montrouzieri* Mulsant (Smith and Krischik 2000). In contrast, a number of studies have demonstrated the resilience of *H. axyridis* to *B. bassiana* (Cottrell and Shapiro-Ilan 2003; Roy et al. 2007). In North America one of the overwintering habitats of *H. axyridis* is under bark in pecan orchards (Cottrell and Shapiro-Ilan 2003). Adults of the native ladybird *O. v-nigrum* are also found in this overwintering location and are commonly recorded infected with *B. bassiana* (Cottrell and Shapiro-Ilan 2003); whereas *H. axyridis* are not. Cottrell and Shapiro-Ilan (2003) demonstrated in laboratory studies that *H. axyridis* is extremely resistant to *B. bassiana* infection. Further studies using various isolates of *B. bassiana* support this notion. Roy et al. (2007) showed that *A. bipunctata* and *C. septempunctata* are highly susceptible to doses of 10^7 spores ml^{-1} but even doses of *B. bassiana* exceeding 10^9 spores ml^{-1} have limited impact on *H. axyridis*. However, *B. bassiana* appears to have a sublethal effect on *H. axyridis* and even low doses (10^5 spores ml^{-1}) reduce egg production significantly (Roy et al. 2007).

There is one further fungal entomopathogen of coccinellids worthy of note: *Hesperomyces virescens* Thaxter (Ascomycota: Laboulbeniales). This intriguing fungus is an obligate ectoparasite that infects many coccinellids worldwide (Nalepa and Weir 2007). *Hesperomyces virescens* has been reported infecting *Chilocorus stigma* (Say) (Coleoptera: Coccinellidae) (Thaxter 1891), *A. bipunctata* (Welch et al. 2001), *H. axyridis* (Garcés and Williams 2004), *Chilocorus bipustulatus* (L.) (Coleoptera: Coccinellidae) (Kamburov et al. 1967; Applebaum et al. 1971), and *Coccinula crotchi* (Lewis) (Coleoptera: Coccinellidae) and *Coccinula sinensis* Weise (Coleoptera: Coccinellidae) (M. Majerus and R. Ware personal observation). This fungus is often reported on the ventroposterior of males and the dorsoposterior of females; a sexual dimorphism that reflects the major transmission mechanism which is thought to be via direct contact during mating (Weir and Beakes 1996; Welch et al. 2001; Riddick and Schaefer 2005). Nalepa and Weir (2007) have studied the interactions between *H. virescens* and *H. axyridis* in detail. In a field survey in North Carolina they demonstrated infection levels of 22–38% at the beginning of winter increasing to 62% by the end of winter. These overwintering ladybirds did not exhibit the characteristic elytral distribution of *H. virescens* as hypothesised by the sexual transmission theory. In contrast, fungal thalli were distributed in accordance with direct contact through aggregation and orientation behaviour, notably on the anterior part of the body. A similar pattern was reported for *A. bipunctata* which, when collected from overwintering aggregations, had fungal thalli distributed at the margins and front angles of the elytra (Weir and Beakes 1996). Nalepa and Weir (2007) conclude that direct transmission through contact with conspecifics is the most important mechanism for transmission of *H. virescens* and sexual contact is of primary importance in the mating season but aggregation in winter also plays a significant role.

Laboulbeniales characteristically do not penetrate the insect cuticle, however, *H. virescens* is one of the few species, in this family, to exhibit rhizoidal penetration into the host body by production of a circular appressorium, which attaches and penetrates the host cuticle (Weir and Beakes 1996). Even though this fungus invades the host there are no known deleterious impacts on the beetle (Nalepa and Weir 2007). However, *H. virescens* is suspected to have a negative effect on *C. bipustulatus*, through increased mortality (Applebaum et al. 1971; Kamburov et al. 1967). Heavy infections could also impede flight, foraging and mating but this requires further investigation.

It is also of note that some Asian populations of *H. axyridis* are strongly female biased, due to the presence of a male-killing bacterium of the genus *Spiroplasma* (Majerus et al. 1998). The bacterium is maternally inherited in the cytoplasm of eggs, but not sperm. As a consequence, it kills male hosts early in embryogenesis. In doing so, the male-killer makes the soma of these dead male eggs available as a food source to infected female sibling neonate larvae, providing them with a significant fitness advantage (Majerus and Hurst 1997; Majerus 2006). Prevalence of this *Spiroplasma* in populations of *H. axyridis* in its native range is variable, but can reach 80%. The male-killer has not been reported from *H. axyridis* in its introduced range. Intuitively, it seems likely that introduction of male-killer infected females into populations of *H. axyridis* in Europe and North America would depress *H. axyridis*, because half the offspring (the males) of infected females would die in the eggs. The male-killer would spread naturally through the population to an equilibrium prevalence dictated by the vertical transmission efficiency of the male-killer (>0.99 for this *Spiroplasma*), any direct fitness effects that the male-killer has on females and the level of benefit gained by infected females as a result of the death and consumption of their male siblings (Hurst et al. 1997). The death of half the infected female's offspring would seem to be beneficial in reducing host population size, and consequently both intraguild competition and predation imposed by *H. axyridis*. However, this may not be the case in practice. As male *H. axyridis* can mate with many females, a bias in the sex ratio towards females is unlikely to lead to reduced fertility in females. Moreover, as *H. axyridis* larvae are highly cannibalistic, loss of a significant proportion of male progeny during embryogenesis may simply lead to a reduction in cannibalism of immatures, and not to a reduction in the adult population. Finally, as female ladybirds eat considerably more than males (Hodek 1996), the level of intraguild competition may actually be increased as a result of biasing the population sex ratio in favour of females. Due to these uncertainties, use of a male-killer in an attempt to reduce *H. axyridis* populations should not be undertaken until a full analysis of the consequences of such an action has been conducted.

Current research indicates that, under current knowledge, pathogens are unlikely to be suitable biological control agents for *H. axyridis*. The fungi examined so far (mainly *B. bassiana*) lack efficacy and specificity towards *H. axyridis* and could impact on non-target coccinellids and other insects. Nevertheless, the development of a novel strategy utilising pathogens as a component of a complex control strategy should not be ruled out. The sublethal effects of disease require further research and could provide a solution to controlling *H. axyridis*. Furthermore, *H. axyridis* should be screened throughout its native and non-native distribution to thoroughly assess the diversity of pathogens (fungi, protists, bacteria and viruses) and other "hitchhikers" (nematodes and other parasites) associated with it. One of these could be the key to control.

Predators

As other ladybirds, *H. axyridis* is considered to be largely protected from generalist predators thanks to its alkaloid-rich defence secretions and its aposematic colouration. Nevertheless, Nechayev and Kuznedsov (1973) mention eight species of birds feeding on *H. axyridis* in Russia. Although Kuznetsov (1997) reports that the two bird species *Picus canus* Gmelin (Piciformes: Picidae) and *Sitta europaea* L. (Passeriformes: Sittidae) will feed on *H. axyridis* in mass overwintering aggregations, he concludes that predation of coccinellids, including *H. axyridis*, by birds is generally rare. Eggs and larvae may also be preyed upon by other coccinellids (Pell et al. 2007; Ware and Majerus 2007) and by predatory bugs (De Clercq et al. 2003), although, in general, *H. axyridis* has better physical and chemical defences against predation than most ladybirds (see Pell et al. 2007, for review). It is also less attacked by ants and spiders than other ladydirds (Dutcher et al. 1999; Yasuda and Kimura 2001). It is very unlikely that any invertebrate or vertebrate predator will ever play a significant role in the natural control of *H. axyridis* in Europe or the Americas.

Parasitoids

Only two parasitoid species were found on *H. axyridis* in its introduction range and these have never been reported as being a significant source of mortality. The tachinid fly *Strongygaster triangulifera* (Loew) (Diptera: Tachinidae) was found in adult beetles in North Carolina (Nalepa et al. 1996; Nalepa and Kidd 2002) and Oregon (Katsoyannos and Aliniazee 1998). All were reared from overwintering beetles. Parasitism was very variable between sites and from year to year but, on average, only 2–4% of the beetles were parasitized. In Oregon, Katsoyannos and Aliniazee (1998) observed parasitism rates of

6.7–15.4% by dissection and 4.8% by rearing. *Strongygaster triangulifera* is a small parasitic fly that is widely distributed in North America and very polyphagous. It is most frequently reported as a parasitoid of adult Coleoptera and there are scattered reports of parasitism within the Dermaptera, Hemiptera, Orthoptera and Lepidoptera (Reeves and O'Hara 2004).

The braconid wasp *Dinocampus coccinellae* (Schrank) (Hymenoptera: Braconidae) is a common endoparasitoid of ladybirds in Europe, Asia and the Americas. Females usually parasitise adults, although larvae and pupae can also be attacked, in which case the development continues in the adult beetle (Obrycki et al. 1985). *Dinocampus coccinellae* is recorded from many ladybirds of the subfamily Coccinellinae but shows strong preferences for some large species. For example, in the UK it mainly attacks *C. septempunctata*, *Coccinella undecimpunctata* L. (Coleoptera: Coccinellidae) and *Harmonia quadripunctata* Pontoppidan (Coleoptera: Coccinellidae) (Majerus 1997). In North and South America it is most often recorded from *Coleomegilla maculata* (DeGeer) (e.g. Obrycki et al. 1985; Berti Filho and Costa 1995; Hoogendoorn and Heimpel 2002). It is occasionally observed on *H. axyridis* in North America (Hoogendoorn and Heimpel 2002; Firlej et al. 2005), Europe (Koyama and Majerus 2007) and East Asia (Liu 1950; Maeta 1969; Park et al. 1996). In Europe and North America, *H. axyridis* is considered as a marginal host for *D. coccinellae.* Hoogendoorn and Heimpel (2002) compared parasitism in *H. axyridis* and the native ladybird *C. maculata* in the USA. They found similar parasitism rates by dissection of field-collected beetles, varying from 9% to 24%. But the emergence rate was much higher in *C. maculata* than in *H. axyridis*, hardly any parasitoid larvae being able to complete their development in *H. axyridis*. In laboratory assays, female wasps attack the two hosts readily, but larvae were found in 90% of the attacked *C. maculata* and in only 7% of the *H. axyridis*. Similar observations were made in Canada (Firlej et al. 2005) and, in the UK (Koyama and Majerus 2007). Koyama and Majerus (2007) have shown that while the attack rates of *D. coccinellae* on *H. axyridis* and *C. septempunctata* are similar, successful development in *H. axyridis* is much lower than in *C. septempunctata*. Firlej (2006) observed that, in *H. axyridis*, parasitism by *D. coccinellae* was successful only when the ladybird larvae were parasitised, whereas adults were attacked but no parasitoid developed in these adults. Firlej (2006) suggested that the poor performance of *D. coccinellae* in *H. axyridis* may be at least partly due to the lower number of teratocytes produced by the parasitoid in *H. axyridis* compared to suitable hosts. In Asia, *D. coccinellae* is less commonly reared from *H. axyridis* than from, e.g. *Coccinella septempunctata brucki* Mulsant (Coleoptera: Coccinellidae) (Maeta 1969). However, successful parasitism is higher in Asia compared to Europe (Liu 1950; Maeta 1969; Park et al. 1996), suggesting either that Asian populations of *D. coccinellae* are better adapted to *H. axyridis* than European populations, or that invasive populations of *H. axyridis* are more resistant to *D. coccinellae*.

Other parasitoids of *H. axyridis* are known only from their area of origin, but details on their impact, biology and ecology are scarce. The tachinid fly *Medina* (=*Degeeria*) *luctuosa* (Meigen) (Diptera: Tachinidae) was found in adult beetles collected in October–November in Korea (Park et al. 1996). Parasitism rates fluctuated between 0.7% and 21.1%. In laboratory rearing, parasitism by *M. luctuosa* shortened the life span and prevented egg development and oviposition. The true identity of this parasitoid is questionable because, in Europe, *M. luctuosa* is specific to adult chrysomelids of the genus *Haltica*, whereas many Coccinellidae, but not yet *H. axyridis*, are parasitized by a closely-related species, *Medina separata* (Meigen) (Diptera: Tachinidae), which was sometimes wrongly identified as *M. luctuosa* (Tschorsnig and Herting 1994). A gregarious pupal parasitoid, the phorid fly *Phalacrotophora philaxyridis* Disney (Diptera: Phoridae) was described from Japan (Disney 1997). A congeneric species, *Phalacrotophora fasciata* (Fallén) (Diptera: Phoridae) is reported from *H. axyridis* pupae in the Russian Far East by Kuznetsov (1997) and Park et al. (1996) cite a *Phalacrotophora* sp. parasitizing 0.4–6.7% of the pupae in Korea, 1–8 adults emerging from a single host pupa. The host range of *P. philaxyridis* is unknown but, in Europe, *P. fasciata* and another species, *Phalacrotophora berolinensis* Schmitz (Diptera: Phoridae), parasitize several coccinellid species (Hodek 1973; Majerus and Kearns 1989).

From this survey, it appears that the potential of using parasitoids for the biological control of *H. axyridis* in its area of introduction is limited. The parasitoids found on *H. axyridis* in Europe and North America are of marginal importance and, in the case of *D. coccinellae*, not well adapted to *H. axyridis*. Thus, there is little prospect for their use in conservation or augmentative biological control programmes. The specificity of the parasitoids found only in Asia remains to be investigated but they belong to genera that are known to comprise rather polyphagous species, at least within the family Coccinellidae. It must be noted, however, that our knowledge on the parasitoid complex of *H. axyridis* in its native range is based on a couple of studies in Japan and Korea, whereas the largest part of the native range has never been investigated for parasitism. The parasitoid complex of indigenous ladybirds often comprises more than three species. For example, at least seven parasitoids are known from *C. septempunctata* in Europe (Klausnitzer and Klausnitzer 1997). Larval parasitism has never been recorded on *H. axyridis*, although several parasitoids of the genera *Aprostocetus*, *Oomyzus* and *Tetrastichus* (Hymenoptera: Eulophidae), *Homalotylus* (Hymenoptera: Encyrtidae) and *Pachyneuron* (Hymenoptera: Pteromalidae) are common larval parasitoids of Coccinellidae (Klausnitzer and Klausnitzer 1997; Noyes 2007). Larval parasitism by Encyrtidae or Eulophidae may kill up to 90–95% of ladybird populations and severely affect their impact as aphid or coccid predators (Iperti 1964; Hodek 1973). Thus, it would be worthwhile investigating parasitism in *H. axyridis* in regions which have never or rarely been surveyed, such as China and Siberia. It remains to be seen whether a parasitoid may be found that is totally specific to *H. axyridis*, as this would be a pre-requirement for introduction into Europe or America. Most ladybird parasitoids are specific to Coccinellidae or to coccinellid sub-families, but often rather polyphagous at family or sub-family level (Klausnitzer and Klausnitzer 1997). An exception may be *Homalotylus platynaspidis* Hofer (Hymenoptera: Encyrtidae), which, in Europe, may be specific to *Platynaspis luteorubra* (Goeze) (Coleoptera: Coccinellidae) (Hodek 1973), although it has been recently reported from *Scymnus subvillosus* (Goeze) (Coleoptera: Coccinellidae) in Russia (Noyes 2007).

Nematodes

To our knowledge, nematodes have never been recorded from wild populations of *H. axyridis*. Parasitic nematodes of the families Allantonematidae and Mermitidae are able to infest ladybird populations (Iperti 1964). For example, Iperti (1964) found a parasitic nematode (later described as *Parasitylenchus coccinellinae* Iperti & Van Waerebeke (Tylenchida: Allantonematidae) in Iperti and van Waerebeke (1968)) parasitizing the intestine of adults of several ladybirds, particularly *Propylea quatuordecimpunctata* (L.) (Coleoptera: Coccinellidae) and *Oenopia conglobata* (L.) (Coleoptera: Coccinellidae). Parasitism rates in *P. quatuordecimpunctata* reached up to 46% in early autumn. The infection was not lethal, but strongly reduced egg maturation (Iperti 1964). Commercially available entomopathogenic nematodes (*Heterorhabditis bacteriophora* Poinar (Nematoda: Heterorhabditidae) and *Steinernema carpocapsae* (Weiser) (Nematoda: Steinernematidae)) were tested on *H. axyridis* and other ladybirds by Shapiro-Ilan and Cottrell (2005) in the USA. They demonstrated that while *H. axyridis* and *C. septempunctata* (invasive alien species in USA) are susceptible to nematodes they were both more resistant than two native species (*C. maculata* and *O. v-nigrum*).

Parasitic mite

The mite *Coccipolipus hippodamiae* (McDaniel & Moril) (Acari: Podapolipidae) is a sexually transmitted ectoparasite of adult coccinellids in Europe (McDaniel and Morrill 1969; Webberley et al. 2004). It lives on the ventral surface of the elytra. Larval *C. hippodamiae* pass between male and female hosts during copulation, in either direction (Hurst et al. 1995). Once on a new host, larval mites embed their mouthparts into the ventral surface of the host elytra and feed on haemolymph, metamorphosing into adults. Within a few weeks, much of the ventral surface of the elytra is covered with mite adults, eggs and larvae (Webberley et al. 2002). In *A. bipunctata*, the mite has strong negative effects on male and female hosts, increasing overwintering mortality, particularly in males (Webberley et al. 2002) and reducing both fecundity and egg viability in females, the latter eventually to zero (Hurst et al. 1995). Similar, but marginally less extreme, negative effects of the mite have been found on female *Adalia decempunctata* (L.) (Coleoptera: Coccinellidae) and *O. conglobata* (Webberley et al. 2004). *Coccipolipus hippodamiae* is found on a variety of coccinellids, but shows highest prevalence on *A. bipunctata*, where over 90% of some populations are infected by late summer (Webberley et al. 2004). Comparative analysis of prevalence levels of the mite, on four European coccinellids, has indicated that prevalence is positively correlated to promiscuity and voltinism (Webberley et al. 2006).

Due to the high negative effects that *C. hippodamiae* imposes on its hosts, coupled to the behaviour and ecology of *H. axyridis* in Europe, this mite is a possible candidate for use as a biological control agent of *H. axyridis*. Importantly, *H. axyridis* has behavioural characteristics (high promiscuity, no requirement for dormancy before reproducing, multivoltinism, over-lapping generations) to facilitate both horizontal transmission and transmission down the generations of the mite, leading to high prevalence once the mite has established. What is currently not known is whether *C. hippodamiae* can use *H. axyridis* as a host, and, if it can, whether the strong negative effects that the mite imposes on *A. bipunctata* will also be imposed on *H. axyridis*. Encouragingly, preliminary tests in which *C. hippodamiae* have been artificially transferred from wild French *Harmonia quadripunctata* to *H. axyridis* have shown that the mite can feed and survive on *H. axyridis* (M. Majerus unpublished data). Moreover, transfer to *H. axyridis* might come about naturally, for inter-specific hybrid matings between *H. axyridis* and *A. bipunctata* has been recorded in the field in Europe (Pascoe et al. 2007), and Webberley et al. (2004) report that mites can successfully transfer between species during interspecific hybrid matings. If it is proven that *C. hippodamiae* is detrimental to *H. axyridis* and if the transfer from European ladybirds does not occur naturally, artificially infested *H. axyridis* adults could be inoculated to field populations for further propagation. *Coccipolipus hippodamiae* also occurs in North America (McDaniel and Morrill 1969), and may thus have a role to play there. However, given the wide host range of the mite, an appropriate risk assessment would be required before integrating this biological control agent into management programmes. An introduction to continents from which *C. hippodamiae* has not been reported (South America, Africa) should not be considered, although other congeneric mites with similar ecologies that occur on these continents may be considered as alternatives.

The efficacy of sexually transmitted diseases (STD) such as *C. hippodamiae* and *H. virescens* as controlling agents for *H. axyridis* might be improved in an IPM strategy that also involves the male-killer of *H. axyridis*. In species in which both sexes are highly promiscuous, such as *H. axyridis*, the rate of spread of an STD will be greater in populations in which sex ratios are female biased because males, due to their scarcity, will have more mating partners and so more opportunity to both contract and to pass on the disease (Majerus 2003). Thus, in a population of *H. axyridis* that is female biased due to the presence of a male-killer, an STD, such as *C. hippodamiae*, would reach higher prevalence than in a population with a 1:1 sex ratio.

Conclusions

Developing control methods against the harlequin ladybird remains a challenge, for several reasons. It is the first time a predatory ladybird, or any other predatory beetle, requires control. Thus, there is no previous experience on which to base new management strategies, and research to date has been very limited. Furthermore, as most other ladybirds are considered beneficial insects, control methods against *H. axyridis* will have to be highly specific to avoid non-target effects. Suitable control methods will continue to be developed to prevent or control aggregations in houses and fruit damage. Besides the use of chemical insecticides, which are efficient but should preferably be avoided to prevent their negative impact on human health and the environment, the most promising new strategies are the development of new cultural practices in vineyards and, especially of new trapping methods based on semiochemicals. Our current understanding of pheromonal and kairomonal communication by coccinellids, and specifically *H. axyridis*, is still limited. However, there are many effective model systems using semiochemical methods for controlling pest insects, which could prove to be informative in developing strategies for controlling *H. axyridis*.

While control methods presently used or under development may eventually solve the problems in buildings, orchards or vineyards, it is very unlikely that any of these methods will ever limit the impact of outbreak *H. axyridis* populations on native biodiversity. Unless a native European or American natural enemy (e.g. *C. hippodamiae*) adapts itself to the new invader and provides control, the only method that may eventually lower invasive population densities in a sustainable way is the importation of natural enemies from the region of origin. However, *H. axyridis* is a difficult target for classical biological control, partly because the invasion of *H. axyridis* is, in itself, most probably the result of bad biological control practices. Thus convincing the public, authorities and a part of the scientific community that a biological control agent needs to be introduced to control another biological control agent that turned bad may be challenging. In addition, a candidate biological control agent should be specific to *H. axyridis*. The chance of finding a specific pathogen or parasitoid in Asia is low but not negligible. As in all classical biological control programmes, the risk of introducing an exotic agent has to be weighted against the risk of doing nothing or the risk related to traditional control measures. In the case of *H. axyridis*, the introduction of a biological control agent should be considered only if there are clear signs that native ladybirds, or other aphidophagous insects, are seriously threatened over a significant part of their distribution range. At present, the priority for research is to gather more reliable and quantitative data on the exact impact of *H. axyiridis* on the native fauna in the invaded regions. In the meantime, comparative assessments of mortality factors affecting populations of *H. axyridis* in the native range and the areas of introduction should be carried out, e.g. through life table analysis, to understand which factors limit population densities in the native range and which do not occur in the regions of introduction. These studies would also allow the identification of potential classical biological control agents that could be considered for introduction, should the situation in the invaded regions become dramatic for native biodiversity.

Finally, it must be noted that, in North America, there are various agricultural contexts where *H. axyridis* is highly valued as a biological control agent, particularly in corn, soybeans, citrus and apples (Koch 2003; Pervez and Omkar 2006). A control method that would provide area wide reductions in *H. axyridis* populations might also result in reduced levels of aphid biocontrol and, thus, create a conflict of interest between the agricultural sector and conservationists.

Acknowledgements Marc Kenis and Renate Zindel are supported by a grant from the Swiss Federal Office for the Environment (F232-0377) and the EU FP6 project ALARM (GOCE-CT-2003-506675).

References

Adriaens T, Branquart E, Maes D (2003) The multicoloured Asian ladybird *Harmonia axyridis* Pallas (Coleoptera : Coccinellidae), a threat for native aphid predators in Belgium? Belg J Zool 133:201–287

Adriaens T, San Martin y Gomez G, Maes D (2007) Invasion history, habitat preferences and phenology of the invasive ladybird *Harmonia axyridis* in Belgium. BioControl (this issue). doi:10.1007/s10526-007-9137-6

Agarwala BK, Yasuda H, Kajita Y (2003) Effect of conspecific and heterospecific feces on foraging and oviposition of two predatory ladybirds: role of fecal cues in predation avoidance. J Chem Ecol 29: 357–376

Anonymous (2005) Viticulture pest alert. Multicolored Asian ladybeetle (*Harmonia axyridis*) on grapes. Missoury State University. http://www.mtngrv.missouristate.edu/Publications/PestAlert1.htm. Cited 20 Aug 2007

Applebaum S, Kfir R, Gerson U, Tadmor U (1971) Studies on the summer decline of *Chilocorus bipustulatus* in citrus groves in Israel. Entomophaga 16:433–444

Baniecki J, Dabaan ME, Freeborn J, Cheves B, Richmond D (2004) Multicolored Asian ladybird (*Harmonia axyridis*). West Virginia University extension Service. http://www.wvu.edu/~agexten/ipm/insects/ladybeetle.htm. Cited 20 Aug 2007

Berti Filho E, Costa VA (1995) *Perilitus coccinellae* (Schrank, 1802) (Hymenoptera, Braconidae, Euphorinae), a parasitoid of *Coleomegilla maculata* (DeGeer) (Coleoptera, Coccinellidae). Rev Agric (Piracicaba) 70:80

Brown PMJ, Adriaens T, Bathon H, Cuppen J, Goldarazena A, Hägg T, Kenis M, Klausnitzer BEM, Kovář I, Loomans AJM, Majerus MEN, Nedved O, Pedersen J, Rabitsch W, Roy HE, Ternois V, Zakharov IA, Roy DB (2007) *Harmonia axyridis* in Europe: spread and distribution of a non-native coccinellid. BioControl (this issue). doi:10.1007/s10526-007-9132-y

Burns DA, McDonnell JJ, Hemptinne J-L, Lognay G, Dixon AFG (1998) Mate recognition in the two-spot ladybird beetle, *Adalia bipunctata*: role of chemical and behavioural cues. J Insect Physiol 44:1163–1171

Byers JA (2004) Chemical ecology of bark beetles in a complex olfactory landscape. In: Lieutier F, Day KR, Battisti A, Grégoire J-C, Evans HF (eds) Bark and wood boring insects in living trees in Europe, a synthesis. Kluwer Academic Publishers, Dordrecht

Cook SM, Khan ZR, Pickett JA (2007) The use of push-pull strategies in integrated pest management. Annu Rev Entomol 52:375–400

Cottrell TE, Shapiro-Ilan DI (2003) Susceptibility of a native and an exotic lady beetle (Coleoptera: Coccinellidae) to *Beauveria bassiana*. J Invertebr Pathol 84:137–144

De Clercq P, Peeters I, Vergauwe G, Thas O (2003) Interactions between *Podisus maculiventris* and *Harmonia axyridis*, two predators used in augmentative biological control in greenhouse crops. BioControl 48:39–55

Disney RHL (1997) A new species of Phoridae (Diptera) that parasitises a wide-spread Asian ladybird beetle (Coleoptera: Coccinellidae). Entomologist 116:163–168

Dowd PF, Vega FE (2003) Autodissemination of *Beauveria bassiana* by sap beetles (Coleoptera: Nitidulidae) to overwintering sites. Biocontrol Sci Technol 13:65–76

Dutcher JD, Estes PM, Dutcher MJ (1999) Interactions in entomology: aphids, aphidophaga and ants in pecan orchards. J Entomol Sci 34:40–56

Firlej A (2006) Interaction du parasitoïde *Dinocampus coccinellae* avec la coccinelle exotique *Harmonia axyridis* Pallas et la coccinelle indigène *Coleomegilla maculata lengi* Timberlake. PhD Thesis, Université du Québec à Montréal, Canada

Firlej A, Boivin G, Lucas E, Coderre D (2005) First report of parasitism of *Harmonia axyridis* parasitism by *Dinocampus coccinellae* schrank in Canada. Biol Invasions 7:553–556

Galvan TL, Koch RL, Hutchison WD (2005a) Effects of spinosad and indoxacarb on survival, development and reproduction of the multicolored Asian lady beetle (Coleoptera: Coccinellidae). Biol Control 34:108–114

Galvan TL, Koch RL, Hutchison WD (2005b) Toxicity of commonly used insecticides in sweet corn and soybean to the multicolored Asian lady beetle (Coleoptera: Coccinellidae). J Econ Entomol 98:780–789

Galvan TL, Burkness EC, Hutchison WD (2006a) Efficacy of selected insecticides for management of the multicolored Asian lady beetle on wine grapes near harvest. Plant Health Prog. doi:10.1094/PHP-2006-1003-01-RS

Galvan TL, Burkness EC, Hutchison WD (2006b) Influence of berry injury on infestations of the multicolored Asian lady beetle in wine grapes. Plant Health Prog. doi:10.1094/PHP-2006-0607-01-BR

Galvan TL, Burkness EC, Hutchison WD (2006c) Wine grapes in the Midwest: reducing the risk of the multicolored Asian lady beetle. Public. 08232. University of Minnesota Extension Service, St. Paul

Galvan TL, Koch RL, Hutchison WD (2006d) Toxicity of indoxocarb and spinosad to the multicolored Asian lady beetle (Coleoptera: Coccinellidae) via three routes of exposure. Pest Manag Sci 62:797–804

Galvan TL, Burkness EC, Hutchison WD (2007) Enumerative and binomial sequential sampling plans for the multicolored Asian lady beetle (Coleoptera: Coccinellidae) in wine grapes. J Econ Entomol 100:1000–1010

Garcés S, Williams R (2004) First record of *Hesperomyces virescens* Thaxter (Laboulbeniales: Ascomycetes) on *Harmonia axyridis* (Pallas) (Coleoptera: Coccinellidae). J Kansas Entomol Soc 77:156–158

Ginsberg HS, Lebrun RA, Heyer K, Zhioua E (2002) Potential nontarget effects of *Metarhizium anisopliae* (Deuteromycetes) used for biological control of ticks (Acari: Ixodidae). Environ Entomol 31:1191–1196

Hassanali A, Njagi PGN, Bashir MO (2005) Chemical ecology of locusts and related acridids. Annu Rev Entomol 50:223–245

Hemptinne J-L, Lognay G, Doumbia M, Dixon AFG (2004) Chemical nature and persistence of the oviposition deterring pheromone in the tracks of the larvae of the two spot ladybird, *Adalia bipunctata* (Coleoptera: Coccinellidae). Chemoecology 11:43–47

Hills LD (1969) Biological pest control: report 3. Henry Doubleday Research Association, Essex

Hodek I (1973) Biology of coccinellidae. Dr. W. Junk N. V., Publishers, The Hage

Hodek I (1996) Food relationships. In: Hodek I, Honek A (eds) Ecology of coccinellidae. Kluwer Academic Publishers, Dordrecht

Hoogendoorn M, Heimpel GE (2002) Indirect interactions between an introduced and a native ladybird beetle species mediated by a shared parasitoid. Biol Control 25:224–230

Huelsman MF, Kovach J (2004) Behavior and treatment of the multicolored Asian lady beetle (*Harmonia axyridis*) in urban environments. Am Entomol 50:163–164

Hurst GDD, Sharpe RG, Broomfield AH, Walker LE, Majerus TMO, Zakharov IA, Majerus MEN (1995) Sexually transmitted disease in a promiscuous insect, *Adalia bipunctata*. Ecol Entomol 20:230–236

Hurst GDD, Hurst LD, Majerus MEN (1997) Cytoplasmic sex-ratio distorters. In: O'Neill SL, Hoffmann AA, Werren JH (eds) Influential passengers. Oxford University Press, UK, pp 125–154

Inclan DJ, Williams RN, Ellis MA, Fickle DS, McClure KB (2006) Managing the multicoloured Asian lady beetle in grapes, 2005. Arthrop Manag Tests 31:L7

Iperti G (1964) Les parasites des coccinelles aphidiphages dans les Alpes-Maritimes et les Basses-Alpes. Entomophaga 9:153–180

Iperti G, van Waerebeke D (1968) Description, biologie et importance d'une nouvelle espèce d'Allantonematidae (Nématoda), parasite des Coccinelles aphidiphages: *Parasitylenchus coccinellinae*, n. sp. Entomophaga 13:107–119

James RR, Shaffer BT, Croft B, Lighthart B (1995) Field evaluation of *Beauveria bassiana*: its persistence and effects on the pea aphid and a non-target coccinellid in alfalfa. Biocontrol Sci Technol 5:425–438

Jones CS, Boggs J (2002) Multicolored Asian lady beetle. Ohio State University Extension Fact Sheet: Entomology HSE-1030-01. http://www.ohioline.osu.edu/hse-fact/1030.html. Cited 20 August 2007

Kamburov SS., Nadel DJ, Kenneth R (1967) Observations on *Hesperomyces virescens* Thaxter (Laboulbeniales), a fungus associated with premature mortality of *Chilocorus bipustulatus* L. in Israel. Israel J Agric Res Rehovot 17:131–134

Katsoyannos P, Aliniazee MT (1998) First record of *Strongygaster triangulifera* (Loew) (Diptera: Tachinidae) as a parasitoid of *Harmonia axyridis* (Pallas) (Coleoptera: Coccinellidae) in western North America. Can Entomol 130:905–906

Klausnitzer B, Klausnitzer H (1997) Marienkäfer. Westarp Wissenschaften, Magdeburg

Klein MG, Lacey LA (1999) An attractant trap for autodissemination of entomopathogenic fungi into populations of the Japanese beetle *Popillia japonica* (Coleoptera: Scarabaeidae). Biocontrol Sci Technol 9:151–158

Koch RL (2003) The multicoloured Asian lady beetle, *Harmonia axyridis*: a review of its biology, uses in biological control and non target impacts. J Insect Sci 3:32

Koch RL, Galvan TL (2007) Bad side of a good beetle: the North American experience with *Harmonia axyridis*. BioControl (this issue). doi:10.1007/s10526-007-9121-1

Koch RL, Burkness EC, Wold Burkness SJ, Hutchison WD (2004) Phytophagous preferences of the multicolored Asian lady beetle (Coleoptera: Coccinellidae) to autumn ripening fruit. J Econ Entomol 97:539–544

Koch RL, Venette RC, Hutchison WD (2006) Invasions by *Harmonia axyridis* (Pallas) (Coleoptera: Coccinellidae) in the Western Hemisphere: implications for South America. Neotrop Entomol 35:421–434

Kovach J (2004) Impact of the multicolored Asian lady beetle as a pest of fruit and people. Am Entomol 50:165–167

Koyama S, Majerus MEN (2007) Interactions between the parasitoid wasp *Dinocampus coccinellae* and two species of coccinellid from Japan and Britain. BioControl (this issue). doi:10.1007/s10526-007-9138-5

Kuznetsov VN (1997) Lady beetles of the Russian far east, (Memoir series no. 1). Center for Systematic Entomology, Gainesville

Liu CL (1950) Contribution to the knowledge of Chinese Coccinellidae. X. Occurrence of *Perilitus coccinellidae* (Schrank), a parasite of adult Coccinelidae in North China (Hymenoptera, Braconidae). Entomol News 61:207–208

Maeta Y (1969) Biological studies on the natural enemies of some coccinellid beetles. I. On *Perilitus coccinellae* (Schrank). Kontyu 37:147–1166

Majerus MEN (1994) Ladybirds new naturalist series. Harper Collins, UK

Majerus MEN (1997) Parasitization of British ladybirds by *Dinocampus coccinellae* (Schrank) (Hymenoptera: Braconidae). Br J Entomol Nat Hist 10:15–24

Majerus MEN (2003) A new dimension to sex wars: microbes that benefit female hosts. Microbiol Today 30:68–70

Majerus MEN (2006) The impact of male-killing bacteria on the evolution of aphidophagous coccinellids. Eur J Entomol 103:1–7

Majerus MEN, Hurst GDD (1997) Ladybirds as a model system for the study of male-killing symbionts. Entomophaga 42:13–20

Majerus MEN, Kearns P (1989) Ladybirds. Naturalists' handbooks n° 10. Richemond publ., Slough

Majerus TMO, Majerus MEN, Knowles B, Wheeler J, Bertrand D, Kuznetzov VN, Ueno H, Hurst GDD (1998) Extreme variation in the prevalence of inherited male-killing microorganisms between three populations of the ladybird *Harmonia axyridis* (Coleoptera: Coccinellidae). Heredity 81:683–669

Majerus MEN, Roy HE (2005) Scientific opportunities presented by the arrival of the harlequin ladybird, *Harmonia axyridis*, in Britain. Antenna 29:196–208

Majerus MEN, Strawson V, Roy HE (2006) The potential impacts of the arrival of the harlequin ladybird, *Harmonia axyridis* (Pallas) (Coleoptera: Coccinellidae), in Britain. Ecol Entomol 31:207–215

McCutcheon TW, Scott HR (2001) Observations of cosmetic damage on a house caused by the multicolored Asian lady beetle, *Harmonia axyridis* (Coleoptera: Coccinellidae). West Virginia University Extension Service. http://www.wvu.edu/~agexten/ipm/insects/beetle.pdf. Cited 20 August 2007

McDaniel B, Morrill W (1969) A new species of *Tetrapolipus* from *Hippodamia convergens* from South Dakota (Acarina: Podopolipidae). Ann Entomol Soc Am 62:1456–1458

Nalepa CA, Kidd KA (2002) Parasitism of the multicolored Asian lady beetle (Coleoptera: Coccinellidae) by *Strongygaster triangulifera* (Diptera: Tachinidae) in North Carolina. J Entomol Sci 37:124–127

Nalepa CA, Weir A (2007) Infection of *Harmonia axyridis* (Coleoptera: Coccinellidae) by *Hesperomyces virescens* (Ascomycetes: Laboulbeniales): role of mating status and aggregation behavior. J Invertebr Pathol 94:196–203

Nalepa CA, Kidd KA, Ahlstrom KR (1996) Biology of *Harmonia axyridis* (Coleoptera: Coccinellidae) in winter aggregations. Ann Entomol Soc Am 89:681–685

Nalepa CA, Kidd KA, Hopkins DI (2000) The multicolored Asian lady beetle (Coleoptera: Coccinellidae): orientation to aggregation sites. J Entomol Sci 35:150–157

Nalepa CA, Kennedy GG, Brownie C (2005) Role of visual contrast in the alighting behavior of *Harmonia axyridis* (Coleoptera: Coccinellidae) at overwintering sites. Environ Entomol 34:425–431

Nechayev VA, Kuznetsov VN (1973) Avian predation on coccinellids in Primorie region. Entomofauna Sovetskogo Dalnego Vostoka, Vladivostok

Noyes J (2007) Universal chalcidoidea database. Natural history museum. http://www.nhm.ac.uk/research-curation/projects/chalcidoids/. Cited 20 August 2007

Obrycki JJ, Tauber MJ, Tauber CA (1985) *Perilitus coccinellae* (Hymenoptera: Braconidae): parasitization and development in relation to host-stage attacked. Ann Entomol Soc Am 78:852–854

Osawa N (1992) A life table of the ladybird beetle *Harmonia axyridis* Pallas (Coleoptera, Coccinellidae) in relation to the aphid abundance. Jpn J Entomol 60:575–579

Osawa N (1993) Population field studies of the aphidophagous ladybird beetle *Harmonia axyridis* (Coleoptera: Coccinellidae): life tables and key factor analysis. Res Popul Ecol 35:335–348

Park H, Park YC, Hong OK, Cho SY (1996) Parasitoids of the aphidophagous ladybeetles, *Harmonia axyridis* (Pallas) (Coleoptera: Coccinellidae) in Chuncheon areas, Korea. Korean J Entomol 26:143–147

Pascoe S, Brown PM, Majerus MEN (2007) Hybrid mating between *Harmonia axyridis* and *Adalia bipunctata*. Bull Amat Entomol Soc 66:156–159

Pell JK, Baverstock J, Roy HE, Ware RL, Majerus MEN (2007) Intraguild predation involving *Harmonia axyridis*: a review of current knowledge and future perspectives. BioControl (this issue). doi: 10.1007/s10526-007-9125-x

Pervez A, Omkar A (2006) Ecology and biological control application of multicoloured Asian ladybird, *Harmonia axyridis*: a review. Biocontrol Sci Technol 16:111–128

Pettersson J, Ninkovic V, Glinwood R, Birkett MA, Pickett JA (2005) Foraging in a complex environment—semiochemicals support searching behaviour of the seven spot ladybird. Eur J Entomol 102:365–370

Pickering GJ, Lin JY, Reynolds A, Soleas G, Riesen R, Brindle I (2005) The influence of *Harmonia axyridis* on wine composition and aging. J Food Sci 70:S128–S135

Pickering GJ, Lin JY, Reynolds A, Soleas G, Riesen R (2006) The evaluation of remedial treatments for wine affected by *Harmonia axyridis*. Int J Food Sci Technol 41:77–86

Pickering GJ, Ker K, Soleas GJ (2007) Determination of the critical stages of processing and tolerance limits for *Harmonia axyridis* for 'ladybug taint' in wine. Vitis 46:85–90

Poprawski TJ, Legaspi JC, Parker PE (1998) Influence of entomopathogenic fungi on *Serangium parcesetosum* (Coleoptera: Coccinellidae), an important predator of whiteflies (Homoptera: Aleyrodidae). Environ Entomol 27:785–795

Potter MF, Townsend L (2005) Asian lady beetle invasion of structures. University of Kentucky, College of Agriculture, Department of Agriculture.ENTFACT-416. http://www.ca.uky.edu/entomology/entfacts/ef416.asp. Cited 20 Aug 2007

Ray JN, Pence HL (2004) Ladybug hypersensitivity: report of a case and review of literature. Allergy Asthma Proc 25:133–136

Reeves WK, O'Hara JE (2004) First report of *Strongygaster triangulifera* (Diptera: Tachinidae) as a par asitoid of a cantharid beetle, *Chauliognathus pennsylvanicus* (Coleoptera: Cantharidae). Can Entomol 136:661–662

Riddick EW, Aldrich JR (2004) Search for repellents, attractants and pheromones of the multicolored Asian lady beetle. Instant symposium—multicolored Asian lady beetles in agriculture and urban environments. Am Entomol 50:167–168

Riddick EW, Schaefer PW (2005) Occurrence, density and distribution of parasitic fungus *Hesperomyces virescens* (Laboulbeniales: Laboulbeniaceae) on multicoloured Asian beetle (Coleoptera: Coccinellidae). Ann Entomol Soc Am 98:615–624

Riddick EW, Aldrich JR, De Milo A, Davis JC (2000) Potential for modifying the behavior of the multicolored Asian lady beetle (Coleoptera: Coccinellidae) with plant-derived natural products. Ann Entomol Soc Am 93:1314–1321

Riddick EW, Aldrich JR, Davis JC (2004) Deet repels *Harmonia axyridis* (pallas) (Coleoptera: Coccinellidae) adults in laboratory bioassays. J Entomol Sci 39:373–386

Roy HE, Steinkraus D, Eilenberg E Hajek A, Pell JK (2006) Bizarre interactions and endgames: entomopathogenic fungi and their arthropod hosts. Annu Rev Entomol 51:331–357

Roy HE, Baverstock J, Pell JK (2008) Manipulating behaviour: a strategy for pest control? In: Ekesi S, Maniania NK (eds) Use of entomopathogenic fungi in biological pest management. Research Signpost 37/661 (2), Fort P.O., Trivandrum-695 023, Kerala, India

Roy HE, Brown PMJ, Rothery P, Ware RL, Majerus MEN (2007) Interactions between the fungal pathogen *Beauveria bassiana* and three species of coccinellid: *Harmonia axyridis, Coccinella septempunctata* and *Adalia bipunctata*. BioControl (this issue). doi:10.1007/s10526-007-9122-0
Shapiro-Ilan DI, Cottrell TE (2005) Susceptibility of lady beetles (Coleoptera: Coccinellidae) to entomopathogenic nematodes. J Invertebr Pathol 89:150–156
Smith SF, Krischik VA (2000) Effects of biorational pesticides on four coccinellid species (Coleoptera: Coccinellidae) having potential as biological control agents in interiorscapes. J Econ Entomol 93: 732–736
Thaxter R (1891) Supplementary note on North American Laboulbeniaceae. Proc Am Acad Arts Sci 25:261–270
Todorova SI, Cote J-C, Coderre D (1996) Evaluation of the effects of two *Beauveria bassiana* (Balsamo) Vuillemin strains on the development of *Coleomegilla maculate lengi* Timberlake (Col, Coccinellidae). J Appl Entomol 120:159–163
Todorova SI, Coderre D, Cote J-C (2000) Pathogenicity of *Beauveria bassiana* isolates toward *Leptinotarsa decemlineata* (Coleoptera: Chrysomelidae), *Myzus persicae* (Homoptera: Aphididae) and their predator *Coleomegilla maculate lengi* (Coleoptera: Coccinellidae). Phytoprotection 81:15–22
Tschorsnig H-P, Herting B (1994) Die Raupenfliegen (Diptera: Tachinidae) Mitteleuropas: Bestimmungstabellen und Angaben zur Verbreitung und Ökologie der Einzelnen Arten. Stuttgarter Beitr Naturk Ser A (Biol) 506:1–170
Van Driesche RG, Bellows TS Jr (1996) Biological control. Chapman & Hall, New York
Van Emden HF, Service MW (2004) Pest and vector control. Cambridge University Press, UK
Verheggen FJ, Fagel Q, Heuskin S, Lognay G, Francis F, Haubruge E (2007) Electrophysiological and behavioral responses of the multicolored Asian lady beetle, *Harmonia axyridis* Pallas, to sesquiterpene semiochemicals. J Chem Ecol (in press). doi:10.1007/s10886-007-9370-6
Ware RL, Majerus MEN (2007) Intraguild predation of immature stages of British and Japanese coccinellids by the invasive ladybird *Harmonia axyridis*. BioControl (this issue). doi:10.1007/s10526-007-9135-8
Webberley KM, Hurst GDD, Buzcko J, Majerus MEN (2002) Lack of parasite mediated sexual selection in an insect-STD system. Anim Behav 63:131–141
Webberley KM, Hurst GDD, Husband RW, Schulenburg JHGVD, Sloggett JJ, Isham V, Buzcko J, Majerus MEN (2004) Host reproduction and a sexually transmitted disease: causes and consequences of *Coccipolipus hippodamiae* distribution on coccinellid beetles. J Anim Ecol 73:1195–1200
Webberley KM, Tinsley M, Sloggett JJ, Majerus MEN, Hurst GDD (2006) Spatial variation in the incidence of sexually transmitted parasites of the ladybird beetle *Adalia bipunctata*. Eur J Entomol 103:793–797
Weir A, Beakes GW (1996) Correlative light- and scanning electron microscope studies on the developmental morphology of *Hersperomyces virescens*. Mycologia 88:677–693
Welch VL, Sloggett JJ, Webberley KM, Hurst GDD (2001) Short-range clinal variation in the prevalence of a sexually transmitted fungus associated with urbanisation. Ecol Entomol 26:547–550
Williams RN, Fickle DS, McClure KB, Inclan DJ (2006) Chemical evaluations for control of the multicoloured Asian lady beetle on grapes, 2005. Arthropod Manag Tests 31:C11
Yasuda H, Kimura T (2001) Interspecific interactions in a tri-trophic arthropod system: effects of a spider on the survival of larvae of three predatory ladybirds in relation to aphids. Entomol Exp Appl 98:17–25

Interactions between the parasitoid wasp *Dinocampus coccinellae* and two species of coccinellid from Japan and Britain

Satoshi Koyama · Michael E. N. Majerus

Abstract The establishment, spread and increase of the invasive coccinellid *Harmonia axyridis* Pallas in North America and Britain have coincided with declines in native ladybirds. In Britain, this pattern was predicted soon after *H. axyridis* was first recorded. However, predictions of the impact that *H. axyridis* may have on the parasites, parasitoids and pathogens of native coccinellids have been less certain, largely because of uncertainty over whether *H. axyridis* would become an alternate host for such agents. *Dinocampus coccinellae* (Schrank) is a braconid parasitoid of ladybirds of the sub-family Coccinellinae. In Japan, this wasp uses both *Coccinella septempunctata brucki* Mulsant and *H. axyridis* as hosts, but successfully parasitizes a higher proportion of the former species. Data are here presented that indicate the same is true in Britain, but to a greater extent. This study investigates the interactions of *D. coccinellae* with *C. septempunctata* L. and *H. axyridis* from Japan and Britain. We show that coccinellid activity affects encounter rates with the wasp, with *H. axyridis* being more active than *C. septempunctata* in the tests. Escalation rates from investigation to ovipositional attack were higher for *C. septempunctata* than *H. axyridis* for both Japanese and British wasps. Wasp emergence was higher, relative to ovipositional attack, from *C. septempunctata* than from *H. axyridis*. Wasps did not discriminate between Japanese and British ladybirds of the same species. British wasp eclosion rates were higher from Japanese than from British *H. axyridis*. We consider both the effect *D. coccinellae* may have on *H. axyridis* population demography, and the effect *H. axyridis* may have on *D. coccinellae* in Europe.

S. Koyama · M. E. N. Majerus (✉)
Department of Genetics, Downing Street, Cambridge CB2 3EH, UK
e-mail: m.majerus@gen.cam.ac.uk

Present Address:
S. Koyama
Department of Veterinary Medicine, Faculty of Agriculture, Tokyo University of Agriculture and Technology, 3-5-8 Saiwai, Fuchu, Tokyo, Japan
e-mail: skoyama@cc.tuat.ac.jp

Biological Control to Invasion: Ladybird Harmonia axyridis. H.E. Roy, E. Wajnberg (eds.),
doi: 10.1007/978-1-4020-6939-0_17, © IOBC 2007

Keywords Braconidae · *Coccinella septempunctata* · Coccinellidae · *Dinocampus coccinellae* · Harlequin ladybird · *Harmonia axyridis* · Host favourability · Host selection · Parasitoid wasp

Introduction

The invasive alien coccinellid, *Harmonia axyridis* Pallas, has been widely released as a biological control agent of aphids (Gordon 1985; Koch 2003). It became established in North America in the mid 1980s (Chapin and Brou 1991) and in mainland Europe in 1999 (Brown et al. 2007a). More recently, it has spread to Britain, where it became established in 2004 (Majerus et al. 2006a). In all three regions it has subsequently spread and increased rapidly (Koch 2003; Brown et al. 2007a, b), threatening biodiversity through predation of non-target aphids and coccids (Majerus and Roy 2005), intra-guild competition and intra-guild predation (Majerus et al. 2006b; Roy et al. 2006; Ware and Majerus 2007). One intriguing and unanswered question is why *H. axyridis* appears to co-exist with other coccinellids in its native range, but has a tendency to displace other coccinellids in parts of its introduced range (Majerus et al. 2006b; Roy et al. 2006; Snyder and Evans 2006). Although a variety of possible reasons for this situation have been suggested (large size, better attack capabilities, shape of mandibles, better chemical and physical defences of immature stages, phenotypic plasticity, multivoltinism, high fecundity, low prevalence of parasitoids, parasites and pathogens in introduced range) (Koch 2003; Pell et al. 2007 for reviews), experimental evidence in support of any of these speculations is lacking.

Dinocampus coccinellae (Schrank) (Hymenoptera: Braconidae) is a solitary parasitoid of coccinellids of the sub-family Coccinellinae (Ceryngier and Hodek 1996). Within this sub-family, it has been reported from over 50 species. The species reproduces by thelytokous parthenogenesis. Female wasps lay a single egg into hosts during an oviposition attack (Balduf 1926). However, females do not discriminate against hosts into which they or other females have already oviposited (Ceryngier and Hodek 1996; Geoghegan et al. 1998; Majerus et al. 2000). This is despite the fact that only a single wasp can develop in a host (Ceryngier and Hodek 1996; Geoghegan et al. 2000).

Various studies have shown that *D. coccinellae* females have a range of host preferences. Richerson and DeLoach (1972) noted that the wasp usually prefers larger coccinellid species, with recorded prevalences (proportion infected) being highest in *Coleomegilla maculata* (DeGeer) and *Hippodamia convergens* (Guérin) in North America (Balduf 1926; Obrycki 1989), *Coccinella septempunctata* L. in continental Europe (Iperti 1964), *C. septempunctata brucki* Mulsant in Japan (Koide 1961; Maeta 1969; Kawauchi 1984) and *Coccinella transversalis* (Mulsant) in Australia (Anderson et al. 1986). Within a host species, field prevalence levels are higher in females than males (Maeta 1969; Parker et al. 1977; Cartwright et al. 1982). In laboratory tests, Davis et al. (2006) showed that female *D. coccinellae* have a preference to oviposit into female rather than male adults of *C. septempunctata*, and that this preference was in addition to the preference the wasp shows for larger hosts in choice tests. As female coccinellids eat more than males (Hodek 1996; Dixon 2000), Davis et al. (2006) argue that this preference is adaptive, providing developing wasp larvae with increased resources. *Dinocampus coccinellae* also has a preference to oviposit into adult compared to immature coccinellids (second, third and fourth instar larvae or pupae), in both choice and no choice tests (Geoghegan et al. 1998). This preference is also adaptive, as the level of successful wasp emergence was higher from adult coccinellids than from fourth instar larvae (Geoghegan et al. 1998). Finally,

D. coccinellae oviposits preferentially into young (pre-winter) rather than old (overwintered) hosts of the same sex and species, conferring increased fitness on offspring because recently eclosed hosts are more likely to survive to allow full parasitoid development than are old adults (Majerus et al. 2000).

In Britain, highest prevalence of *D. coccinellae* has been recorded in *C. septempunctata*, *Coccinella undecimpunctata* L. and *Harmonia quadripunctata* (Pontoppidan) (Geoghegan et al. 1997; Majerus 1997). We here give comparative data on the prevalence of *D. coccinellae* from wild, British samples of *H. axyridis* and *C. septempunctata*. Further, we present findings of choice tests to determine whether *D. coccinellae* shows any preference to attack *C. septempunctata* or *H. axyridis* in a choice situation, and whether the origin of wasps used (Japanese or British) had an effect on the outcome.

Materials and methods

Collection and treatment of field prevalence samples

Samples of *C. septempunctata* and *H. axyridis* were collected in south-west London (Battersea and Clapham districts), in March 2007. The samples were fed on an artificial diet (Majerus et al. 1989) for 1 day, and thereafter on *Acyrthosiphon pisum* (Harris) (pea aphids), at 21°C. The samples of both species were sexed, and then divided into two sub-samples, A and B, at a ratio of 2:1 for each sex. Sub-sample A beetles were retained on a diet of excess *A. pisum*, daily for 45 days. All ladybirds were dissected after 45 days, or when they died if earlier, and were examined for evidence of parasitization. Larval emergence, pupation and adult eclosion of *D. coccinellae* were recorded. Sub-sample B beetles were anaesthetized with CO_2, killed and dissected within a week of collection. Immature stages of *D. coccinellae* that were found were recorded. The two treatments, A and B, were undertaken to test whether prevalence levels in adults retained under laboratory conditions were not underestimated as a result of incomplete ascertainment when parasites failed to complete development.

Experimental stocks

Ladybirds used in choice tests were of four types, British *H. axyridis*, Japanese *H. axyridis*, *C. septempunctata septempunctata* L. and *C. septempunctata brucki*. British *H. axyridis* were either field-collected or F1 adults from samples collected in London in November 2005, May 2006 or September 2006. Japanese *H. axyridis* were F1 adults from samples collected in Fuchu, Honshu, in September 2005 or May 2006. *Coccinella septempunctata* were taken from overwintering sites around Cambridge and Thetford Forest, in February, March and October 2006. *Coccinella septempunctata brucki* were either field-collected or F1 adults from samples collected in Fuchu, Honshu, in May 2006. Field-collected ladybirds were kept in culture on *A. pisum*, at 21°C, for a minimum of 28 days before use to ensure that they were not previously infected with *D. coccinellae*. F1 *H. axyridis* were between 21 and 56 days old. F1 *C. septempunctata* and *C. septempunctata brucki* were between 21 and 56 days old and had not experienced a diapause. Ladybirds were sexed under CO_2 anaesthetic using previously described criteria (for *C. septempunctata* see Randall et al. 1992; for *H axyridis* see Majerus et al. 1998). All ladybird stocks were maintained at 21°C, 16L:8D, on *A. pisum*, with a once weekly provision of artificial food as described by Geoghegan et al. (2000).

The *D. coccinellae* adults used in tests were either of British origin from *C. septempunctata* from Cambridgeshire or Suffolk, or of Japanese origin from *C. septempunctata brucki* from the Fuchu stock. They were fed on an agar based artificial diet (Geoghegan et al. 2000).

Test procedures

Experiment 1

Does *D. coccinellae* prefer hosts from its own region?

Two female ladybirds of the same species (either *H. axyridis* or *C. septempunctata*), but from different countries, were placed in a 9 cm Petri dish. A single *D. coccinellae* of known origin (Japanese or British) was introduced into the dish at least 10 min after the ladybirds. The ladybirds used in each test were chosen randomly from a sample: they were not match-paired for size or other phenotypic traits apart from sex. All *H. axyridis* used were f. *succinea*. The two sub-species of *C. septempunctata* were easily distinguishable. When Japanese and British *H. axyridis* were used in the same test, patterning details were used to distinguish individual origin. Each ladybird was used only in a single test. Each wasp was used for a single test only. All four possible combinations of wasp and ladybird species were tested. Each type of test was replicated 20 times.

All interactions between wasp and ladybirds were recorded over the 20-min period immediately following the introduction of the wasp. Behavioural observations were categorized as: pursuance and investigation of the host (PI); ovipositional stance (OS); or ovipositional attack (OA) (after Richerson and DeLoach 1972). These categories were invariably consequent on one another, with OA following from OS and OS following from PI (Richerson and DeLoach 1972; Davis et al. 2006; this study). The origin of the ladybird involved in each interaction with a wasp was recorded. Whether a coccinellid involved in an interaction with *D. coccinellae* was moving at the moment that the interaction was initiated was recorded. As *D. coccinellae* does not avoid superparasitism (Ceryngier and Hodek 1996; Geoghegan et al. 1998; Majerus et al. 2000), comparative tests of total number of OAs on ladybirds could be used.

Experiment 2

Do Japanese *D. coccinellae* discriminate between *H. axyridis* and *C. septempunctata*?

The protocol used was broadly as that described for experiment 1. However, only Japanese wasps were used, and the two ladybirds used in a choice test were one *H. axyridis* female and one *C. septempunctata* female, with both ladybirds originating from the same country (Japan or Britain). Each type of test was replicated ten times.

Experiment 3

Do British *D. coccinellae* discriminate between *H. axyridis* and *C. septempunctata*?

The protocol was as used for experiment 2, but with British wasps used in place of Japanese wasps. Each type of test was replicated ten times.

When possible, ladybirds that were used in tests and had been the subject of OA were thereafter maintained singly in culture on a diet of *A. pisum* for 5 weeks, and wasp cocoons resulting were recorded.

Statistical analysis

For the field samples, an orthogonal analysis of *G*-test for goodness-of-fit was used to compared numbers recorded, with the level of prevalence of the parasite being considered against the variables: host species, host sex and ladybird sub-sample (A or B). Interaction between these factors was obtained from the residual *G*-value. In addition, the *G* statistic was calculated to compare the success of parasitism between host species, using the sub-set comprising only those ladybirds that were infected.

In the preference experiments, *G*-test goodness-of-fit tests were used to compare results. Where a series of comparisons were made, a fully orthogonal analysis of *G* was used to avoid repeat use of numerical data. A similar approach was used to analyse data on the development of *D. coccinellae* in different hosts following observed OA.

Results

Field prevalences

Details of the samples collected to assess field prevalence levels are given in Table 1. All larvae that emerged from both coccinellid species in the A sub-samples successfully pupated and eclosed. Total *G*-test for goodness-of-fit, of parasitized or not, with variables host species, host sex, and sub-sample, was highly significant ($G = 31.16$, $df = 9$, $p < 0.001$). The overall prevalences of *D. coccinellae* in the two species were 0.122 ± 0.018 for *C. septempunctata* and 0.055 ± 0.009 for *H. axyridis*, and were significantly different from one another ($G = 11.79$, $df = 1$, $p < 0.001$). There was no significant difference in the prevalence levels in male compared to female hosts, either in the two species combined ($G = 1.4$, $df = 1$, $p > 0.05$), or either species individually (for *C. septempunctata*, $G = 2.62$, $df = 1$, $p > 0.05$; for *H. axyridis*, $G = 0.02$, $df = 1$, $p > 0.05$). There was no significant difference in prevalence between sub-samples A and B for the two species combined ($G = 0.44$, $df = 1$, $p > 0.05$) or for either species individually (for *C. septempunctata*, $G = 0.72$, $df = 1$, $p > 0.05$; for *H. axyridis*, $G = 2.92$, $df = 1$, $p > 0.05$). The residual *G*-value representing the interaction between variables with respect to the numbers parasitized or not was significant ($G = 11.25$, $df = 2$, $p < 0.05$).

Within the sub-set of parasitized ladybirds, the proportion of infected individuals in the A sub-samples that produced adult wasps was significantly greater for *C. septempunctata* than for *H. axyridis* ($G = 41.66$, $df = 1$, $p < 0.001$).

Experiment 1

A summary of the interactions is given in Table 2. The numbers of OA did not differ significantly between test types ($G = 4.12$, $df = 3$, $p < 0.05$). There was no significant difference in the level of escalation from PI through OS to OA for wasps of either origin

Table 1 Details of field samples of *C. septempunctata* and *H. axyridis* from South-west London

Coccinella septempunctata Sub-sample A					
Males			Females		
Uninfected	Produced adult wasp	Dissection revealed infection	Uninfected	Produced adult wasp	Dissection revealed infection
102	10	2	88	16	1
Coccinella septempunctata Sub-sample B					
Males			Females		
Uninfected	Produced adult wasp	Dissection revealed infection	Uninfected	Produced adult wasp	Dissection revealed infection
53	N/A	4	46	N/A	7
Harmonia axyridis Sub-sample A					
Males			Females		
Uninfected	Produced adult wasp	Dissection revealed infection	Uninfected	Produced adult wasp	Dissection revealed infection
189	1	11	264	0	9
Harmonia axyridis Sub-sample B					
Males			Females		
Uninfected	Produced adult wasp	Dissection revealed infection	Uninfected	Produced adult wasp	Dissection revealed infection
96	N/A	5	123	N/A	13

The numbers of uninfected and infected males and females of each host species are given. Infected hosts are sub-divided into those in which wasps emerged and those revealed by dissection. For differences in treatments of sub-samples A and B, see text

Table 2 Results of experiment 1, in which the interactions of both Japanese and British *D. coccinellae* to sympatric and allopatric *C. septempunctata* and *H. axyridis* were tested

Origin of wasps	Host species	Origin of hosts	Interactions observed		
			PI	OS	OA
Japan	*C. septempunctata*	Japan	137	74	49
	C. septempunctata	UK	119	57	47
Japan	*H. axyridis*	Japan	128	72	34
	H. axyridis	UK	145	82	27
British	*C. septempunctata*	Japan	85	52	28
	C. septempunctata	UK	88	60	35
British	*H. axyridis*	Japan	205	96	32
	H. axyridis	UK	198	75	30

Interactions: PI pursuance and investigation, OS ovipositional stance, OA ovipositional attack, after Richerson and DeLoach 1972 towards allopatric compared with sympatric ladybirds for either *C. septempunctata* or *H. axyridis* (for Japanese wasp with *C. septempunctata* $G = 0.66$, $df = 2$, $p > 0.05$; with *H. axyridis* $G = 1.68$, $df = 2$, $p > 0.05$; for UK wasps with *C. septempunctata* $G = 0.46$, $df = 2$, $p > 0.05$; with *H. axyridis* $G = 3.4$, $df = 2$, $p > 0.05$).

Experiment 2

The results (Table 3) show that Japanese wasps did not differ in the levels of OA in the two tests ($G = 3.24$, $df = 1$, $p > 0.05$). It was noted, however, that the wasps rarely investigated (PI) non-active *C. septempunctata*, while they often investigated non-active *H. axyridis*, the difference being significant with both Japanese ($G = 24.85$, $df = 1$, $p < 0.001$), and UK ($G = 6.92$, $df = 1$, $p < 0.001$) ladybirds. The rate of escalation from PI to OA was not significantly different between the two species for Japanese ladybirds ($G = 0.3$, $df = 2$, $p > 0.05$), but was for UK ladybirds ($G = 4.58$, $df = 1$, $p < 0.05$), with greater escalation being observed towards *C. septempunctata*.

Experiment 3

The results (Table 4) show that British wasps have a significant preference for PI towards Japanese *H. axyridis* over *C. septempunctata brucki* ($G = 17.25$, $df = 1$, $p < 0.001$). This is largely a product of a much higher level of activity by the *H. axyridis* in these tests compared to *C. septempunctata brucki*. When the level of PI towards stationary *H. axyridis* and *C. septempunctata brucki* are compared, the level of PI is greater towards the latter, and the difference is not significant ($G = 1.88$, $df = 1$, n.s.). Overall, the rate of escalation from PI through OS to OA was significantly greater for *C. septempunctata brucki* than for *H. axyridis* ($G = 8.95$, $df = 2$, $p < 0.05$). Analysis of G showed this to be due to a higher rate of escalation from OS to OA in *C. septempunctata brucki* than in *H. axyridis*.

In interactions with British ladybirds, the British wasps showed no overall difference in levels of PI, OS or OA towards *C. septempunctata* compared with *H. axyridis* (*G*-test

Table 3 Results of experiment 2 in which Japanese wasps were offered a choice between *C. septempunctata* and *H. axyridis* from either Japan or UK

Ladybird activity	*C septempunctata brucki* (Japan)			*H. axyridis* (Japan)		
	PI	OS	OA	PI	OS	OA
Moving	55	28	18	66	23	15
Not moving	6	4	2	37	27	13
Total	61	32	20	103	50	28
	C. septempunctata (UK)			*H. axyridis* (UK)		
	PI	OS	OA	PI	OS	OA
Moving	72	35	31	70	25	14
Not moving	10	6	3	34	16	7
Total	82	35	31	104	42	21

Whether ladybirds were moving or stationary when first contacted by a wasp is recorded

Table 4 Results of experiment 3 in which British wasps were offered a choice between *C. septempunctata* and *H. axyridis* from either Japan or UK

Ladybird activity	*C septempunctata brucki* (Japan)			*H. axyridis* (Japan)		
(a)	PI	OS	OA	PI	OS	OA
Moving	11	7	3	83	46	9
Not moving	59	30	24	45	19	10
Total	70	37	27	128	65	19
	C. septempunctata (UK)			*H. axyridis* (UK)		
(b)	PI	OS	OA	PI	OS	OA
Moving	17	11	6	38	23	6
Not moving	34	32	4	18	10	3
Total	51	43	10	56	33	9

Whether ladybirds were moving or stationary when first contacted by a wasp is recorded

goodness-of-fit $p > 0.05$ for all comparisons). However, this lack of difference may be the result of differences in the activities of the two species of ladybird. *Harmonia axyridis* was more active than *C. septempunctata*, which is reflected in the higher level of wasp interactions with active *H. axyridis* and stationary *C. septempunctata.*

Successful development of *D. coccinellae* in different hosts

Table 5 gives the numbers of those ladybirds of the different species from different origins that were attacked by Japanese or British wasps and were then retained under suitable conditions that subsequently produced adult wasps. The origin of wasps did not affect the probability of successful development of their progeny in either species of ladybird (*G*-test $p > 0.05$ for all comparisons) (however, note the low successful development in UK *H. axyridis*). When considering wasps from both Japan and Britain together, significantly more wasps developed to adulthood in *C. septempunctata* than in *H. axyridis* ($G = 36.92$, $df = 1$, $p < 0.001$). Moreover, successful development was significantly higher in Japanese *H. axyridis* than in British *H. axyridis* ($G = 6.44$, $df = 1$, $p < 0.05$). Conversely, there was no significant difference in the rate of successful development in *C. septempunctata brucki* compared to *C. septempunctata* ($G = 1.04$, $df = 1$, $p > 0.05$).

Table 5 Number of wasps that emerged from Japanese and UK *H. axyridis* and *C. septempunctata* that were attacked (OA) during experiments

Host species	Number attacked (OA)		Number that produced adult wasps	
	Japanese wasps	UK wasps	Japanese wasps	UK wasps
H. axyridis Japan	36	34	9	9
H. axyridis UK	18	23	3	0
C. septempunctata brucki	38	45	21	22
C. septempunctata	30	33	17	21

Results are given separately for wasps of Japanese and UK origin

Discussion

In this paper, we have described initial tests to investigate interactions between the parasitoid wasp *D. coccinellae* and two coccinellids, *H. axyridis* and *C. septempunctata.*

Comparison of UK field collected sub-samples A and B showed that prevalence levels on hosts retained to allow wasp larvae to emerge, and those dissected soon after collection were similar, indicating that retention did not lead to incomplete ascertainment of unsuccessful parasitizations in the retained (A) sub-samples.

Prevalence level of the wasp was more than twice as high in *C. septempunctata* than in *H. axyridis*. Moreover, the proportion of parasitized individuals in which *D. coccinellae* successfully completed development was significantly lower for *H. axyridis* than for *C. septempunctata*. This may be a result of the host specificity of UK *D. coccinellae* if these wasps are poorly adapted to *H. axyridis*, which only established in Britain in 2004 (Majerus et al. 2006a). However, tests in which UK and Japanese wasps were offered a choice of sympatric and allopatric *H. axyridis*, showed that wasps had no preference for sympatric ladybirds. The same result was found when wasps were tested against UK and Japanese sub-species of *C. septempunctata*. This suggests that the low prevalence of *D. coccinellae* in UK *H. axyridis* is not the result of the unfamiliarity of UK wasps with *H. axyridis*. That said, measures of the proportion of wasps that successfully developed in UK *H. axyridis* compared with Japanese *H. axyridis* showed that the former is a significantly poorer host, irrespective of whether ovipositing wasps were from Japan or the UK.

The reason for the low level of developmental success of *D. coccinellae* in UK *H. axyridis* is unknown and warrants further investigation. However, it is feasible that during the establishment of stocks by biocontrol companies, selection has inadvertently been imposed against *H. axyridis* individuals that have low resistance to *D. coccinellae* through guarantee procedures imposed on the stocks of *H. axyridis* obtained from its native range. Those *H. axyridis* that produced wasps, and those that failed to reproduce rapidly [*D. coccinellae* infection suppresses host ovarian maturation (e.g. Ceryngier and Hodek 1996)] would be discarded from stocks.

In tests where Japanese and British wasps were offered a choice between *H. axyridis* and *C. septempunctata*, interactions were affected by the levels of activity shown by the two species of ladybird. In general, *H. axyridis* were more active than *C. septempunctata*, with the result that wasp encounters with *H. axyridis* were more frequent than encounters with *C. septempunctata*. However, importantly, for both Japanese and UK wasps, rates of escalation from PI or OS to OA were higher towards *C. septempunctata* than *H. axyridis*. These higher escalation rates may explain the higher prevalences of *D. coccinellae* in *C. septempunctata* compared to *H. axyridis* that have been reported in both Japan (Maeta 1969; Kawauchi 1984) and Britain (this study).

Two cautionary notes should be borne in mind in further investigations of this type. First, all wasps used were from *C. septempunctata*. It is possible that natural populations of *D. coccinellae* are sub-divided by host, with matrilines having adaptations to, and preferences to oviposit into, the species of host that they developed in. Whether this is the case has not been tested. However, in tests in which *D. coccinellae* from *C. undecimpunctata* (a suitable host) were offered a choice between *C. undecimpunctata* and *Adalia bipunctata* (L.) (an unsuitable host) (Iperti 1964; Majerus 1994; Ceryngier and Hodek 1996), wasps attacked both species equally (S. Koyama and M. E. N. Majerus, unpublished data).

Second, the activity levels of the ladybirds used in choice tests seemed to affect the frequency of some interactions. It is likely that the lower activity of *C. septempunctata* compared to *H. axyridis* was a consequence of differences in the life-history strategies of the two species. In appropriate conditions, *H. axyridis* will mate and oviposit following eclosion without requirement for a dormant period (Majerus and Roy 2005), while *C. septempunctata* usually requires such a period before becoming reproductively active (Dobrzhanskii 1922a, b). While it is difficult to control coccinellid activity levels, we suggest that in future tests, *C. septempunctata* that had undergone a period of dormancy, and so were reproductively active, should be used.

The initial aims of this study were to consider first, whether *D. coccinellae* is likely to have any significant impact on populations of *H. axyridis* in Europe, and second, whether the spread and increase of *H. axyridis* in Europe is likely to affect *D. coccinellae*. The wasps low prevalence in the field, the lower rate of escalation to OA in *H. axyridis* compared to *C. septempunctata* and the very low successful development of the wasp in UK *H. axyridis* all argue that *D. coccinellae* will have little impact on *H. axyridis* populations in Europe.

Conversely, the spread and increase of *H. axyridis* in Europe may have a detrimental impact on *D. coccinellae*. *Dinocampus coccinellae* attacks a wide range of Coccinellinae species (Ceryngier and Hodek 1996), preferring larger species (Richerson and DeLoach 1972). In locations where the commonest coccinellid encountered by *D. coccinellae* becomes *H. axyridis*, a host in which a high proportion of the parasitoid's eggs fail to develop to adulthood, much of the reproductive output of the *D. coccinellae* populations will be misdirected. For example, one might compare two sites, one containing 3,000 *C. septempunctata* and no *H. axyridis*, the other containing 1,000 *C. septempunctata* and 2,000 *H. axyridis*. Using the prevalence levels observed in UK field samples and the rate of successful development in the A sub-samples, the former site would result in 328 wasps (host population size × prevalence level in *C. septempunctata* × parasite success rate in this host), while the latter would result in 112 wasps (109 from *C. septempunctata* and just three from *H. axyridis*). It thus seems possible that populations of *D. coccinellae* will decline as *H. axyridis* increases and displaces native Coccinellinae. If this is the case, we predict that prevalences of *D. coccinellae* will be greater in suitable native host species, such as *C. septempunctata*, in parts of Europe where *H. axyridis* is not yet established, compared to parts of Europe where *H. axyridis* already predominates. Survey work across Europe should allow this prediction to be verified or refuted. An alternative to this scenario is that *D. coccinellae*, in areas where *H. axyridis* is abundant, will respond to selective pressure imposed by this abundance and become better adapted to this host. In this case, we predict that the proportion of infected *H. axyridis* that produce adult wasps will increase over time. This prediction can also be tested by field sampling over time, with assessment of both prevalence of parasitization and success of parasitization being assayed.

Acknowledgements We thank Ian Wright for technical support, Remy Ware, Sherif El Nagdy, Laura Jane Michie and Sami Hassan for help with culture maintenance, and Yoshiaki Obara, Remy Ware and Christina Majerus for help in collecting samples.

References

Anderson JME, Hales DF, van Brunschot KA (1986) Parasitisation of coccinellids in Australia. In: Hodek I (eds) Ecology of aphidophaga. Academia, Prague and Dr. W. Junk, Dordrecht

Balduf WV (1926) The bionomics of *Dinocampus coccinellae* Schrank. Ann Entomol Soc Am 19:465–489

Brown PMJ, Adriaens T, Bathon H, Cuppen J, Goldarazena A, Hägg T, Kenis M, Klausnitzer BEM, Kovář I, Loomans AJM, Majerus MEN, Nedved O, Pedersen J, Rabitsch W, Roy HE, Ternois V, Zakharov IA, Roy DB (2007a) *Harmonia axyridis* in Europe: spread and distribution of a non-native coccinellid. BioControl (this issue). doi:10.1007/s10526-007-9132-y

Brown PMJ, Roy HE, Rothery P, Roy DB, Ware RL, Majerus MEN (2007b) *Harmonia axyridis* in Great Britain: analysis of the spread and distribution of a non-native coccinellid. BioControl (this issue). doi: 10.1007/s10526-007-9124-y

Cartwright B, Eikenbary RD, Angalet GW (1982) Parasitism by *Perilitus coccinellae* (Hym.: Braconidae) of indigenous coccinellid hosts and the introduced *Coccinella septempunctata* (Col.: Coccinellidae), with notes on winter mortality. Entomophaga 27:237–244

Ceryngier P, Hodek I (1996) Enemies of Coccinellidae. In: Hodek I, Honek A (eds) Ecology of Coccinellidae. Kluwer, Dortrecht

Chapin JB, Brou VA (1991) *Harmonia axyridis* (Pallas), the 3rd species of the genus to be found in the United States (Coleoptera, Coccinellidae). Proc Entomol Soc Wash 93:630–635

Davis DS, Stewart SL, Manica A, Majerus MEN (2006) Adaptive preferential selection of female coccinellid hosts by the parasitoid wasp *Dinocampus coccinellae* (Hymenoptera: Braconidae). Eur J Entomol 103:41–45

Dixon AGF (2000) Insect predator-prey dynamics: ladybird beetles and biological control. Cambridge University Press, Cambridge

Dobrzhanskii FG (1922a) Imaginal diapause in Coccinellidae. Izv Otdel Prikl Entomol 2:103–124

Dobrzhanskii FG (1922b) Mass aggregations and migrations in Coccinellidae. Izv Otdel Prikl Entomol 2:229–234

Geoghegan IE, Thomas WP, Majerus MEN (1997) Notes on the coccinelid parasitoid *Dinocampus coccinellae* (Schrank) (Hymnoptera: Braconidae) in Scotland. Entomologist 116:179–184

Geoghegan IE, Majerus TMO, Majerus MEN (1998) Differential parasitisation of adult and pre-imaginal *Coccinella septempunctata* (Coleoptera: Coccinellidae) by *Dinocampus coccinellae* (Hymenoptera: Braconidae). Eur J Entomol 95:571–579

Geoghegan IE, Chudek JA, Mackay RL, Lowe C, Moritz S, McNicol R, Hunter G, Majerus MEN (2000) Study of the anatomical changes in *Coccinella septempunctata* (Coleoptera: Coccinellidae) induced by diet and by infection with the larvae of *Dinocampus coccinellae* (Hymenoptera: Braconidae) using magnetic resonance microimaging. Eur J Entomol 97:457–461

Gordon RD (1985) The Coleoptera (Coccinellidae) of America north of Mexico. J N Y Entomol Soc 93: 1–912

Hodek I (1996) Food relationships. In: Hodek I, Honek A (eds) Ecology of Coccinellidae. Kluwer, Dortrecht

Iperti G (1964) Les parasites des Coccinelles aphidiphages dans les Basses-Alpes et les Alpes-Maritimes. Entomophaga 9:153–180

Kawauchi S (1984) Ecological studies on the natural enemies of *Coccinella septempunctata brucki*, *Propylea japonica*, and *Scymnus (Neopullus) hoffmanni* (Coleoptera: Coccinellidae). Kurume Univ J 33:63–67

Koch RL (2003) The multicoloured Asian lady beetle, *Harmonia axyridis*: a review of its biology, uses in biological control and non-target impacts. J Insect Sci 3:32

Koide T (1961) Observations on *Perilitus coccinellae* (Schrank). Gensei 1961(11):1–5

Maeta Y (1969) Biological studies on the natural enemies of some coccinellid beetles. I. On *Perilitus coccinellae* (Schrank). Kontyu 37:147–166

Majerus MEN (1994) Ladybirds (New Naturalist 81). HarperCollins, London

Majerus MEN (1997) Parasitization of British ladybirds by *Dinocampus coccinellae* (Schrank) (Hymenoptera: Braconidae). Br J Entomol Nat Hist 10:15–24

Majerus MEN, Roy HE (2005) Scientific opportunities presented by the arrival of the harlequin ladybird, *Harmonia axyridis*, in Britain. Antenna 29:196–208

Majerus TMO, Majerus MEN, Knowles B, Wheeler J, Bertrand D, Kuznetzov VN, Ueno H, Hurst GDD (1998) Extreme variation in the prevalence of inherited male-killing microorganisms between three populations of the ladybird *Harmonia axyridis* (Coleoptera: Coccinellidae). Heredity 81:683–691

Majerus MEN, Kearns PWE, Forge H, Ireland H (1989) Ladybirds as teaching aids: 1 Collecting and culturing. J Biol Educ 23:85–95

Majerus MEN, Geoghegan IE, Majerus TMO (2000) Adaptive preferential selection of young coccinelid hosts by the parasitoid wasp *Dinocampus coccinellae* (Hymnoptera: Braconidae). Eur J Entomol 97:161–164

Majerus MEN, Mabbott P, Rowland F, Roy HE (2006a) The harlequin ladybird, *Harmonia axyridis*, arrives in Britain. Entomol Mon Mag 142:87–92

Majerus M, Strawson V, Roy H (2006b) The potential impacts of the arrival of the harlequin ladybird, *Harmonia axyridis* (Pallas) (Coleoptera: Coccinellidae), in Britain. Ecol Entomol 31:207–215

Obrycki JJ (1989) Parasitization of native and exotic coccinellids by *Dinocampus coccinellae* (Schrank) (Hymenoptera: Braconidae). J Kans Entomol Soc 62:211–218

Parker BL, Whalon ME, Warshaw M (1977) Respiration and parasitism in *Coleomegilla maculata lengi* (Coleoptera: Coccinellidae). Ann Entomol Soc Am 70:984–987

Pell JK, Baverstock J, Roy HE, Ware RL, Majerus MEN (2007) Intraguild predation involving *Harmonia axyridis*: a review of current knowledge and future perspectives. BioControl (this issue). doi: 10.1007/s10526-007-9125-x

Randall K, Majerus MEN, Forge H (1992) Characteristics for sex determination in British ladybirds (Coleoptera: Coccinellidae). Entomologist 111:109–122

Richerson JV, DeLoach CJ (1972) Some aspects of host selection by *Perilitus coccinellae*. Ann Entomol Soc Am 65:834–839

Roy HE, Brown P, Majerus MEN (2006) *Harmonia axyridis*: a successful biocontrol agent or an invasive threat? In: Eilenberg J, Hokkanen H (eds) An ecological and societal approach to biological control. Kluwer, The Netherlands

Snyder WE, Evans EW (2006) Ecological effects of invasive arthropod generalist predators. Ann Rev Ecol Evol Syst 37:95–122

Ware RL, Majerus MEN (2007) Intraguild predation of immature stages of British and Japanese coccinellids by the invasive ladybird *Harmonia axyridis*. BioControl (this issue). doi:10.1007/s10526-007-9135-8

Interactions between the fungal pathogen *Beauveria bassiana* and three species of coccinellid: *Harmonia axyridis*, *Coccinella septempunctata* and *Adalia bipunctata*

Helen Elizabeth Roy · Peter M. J. Brown · Peter Rothery · Remy L. Ware · Michael E. N. Majerus

Abstract *Harmonia axyridis* is a predatory coccinellid, native to central and eastern Asia. It has been available in many countries for use as a biological control agent of pest aphids and scale insects. In many of these countries, including the USA, *H. axyridis* has established. It is now considered an invasive alien species for a number of reasons, including its impact on functional biodiversity. *Beauveria bassiana* is known to be a natural mortality agent of overwintering coccinellids and is a potential candidate for the biological control of *H. axyridis*. In this paper we compare the susceptibility of three species of coccinellid, *H. axyridis* (cultures derived from Japan and UK), *Coccinella septempunctata* and *Adalia bipunctata* to infection by *B. bassiana* (commercial strain GHA) after exposure at three doses (10^5, 10^7, 10^9 conidia ml^{-1}). The two subpopulations of *H. axyridis* (Japan and UK) were more resistant to *B. bassiana* infection than either *A. bipunctata* or *C. septempunctata*. This is exemplified by the median lethal doses at 10 days post-inoculation (LD_{50}) of $10^{6.2}$, $10^{6.0}$, $10^{8.3}$, $10^{9.6}$ conidia ml^{-1} for *A. bipunctata*, *C. septempunctata*, *H. axyridis* (Japan) and *H. axyridis* (UK), respectively. Only doses of 10^9 conidia ml^{-1} resulted in mortality of *H. axyridis*, in contrast, 80% of *C. septempunctata* and 70% of *A. bipunctata* exposed to 10^7 conidia ml^{-1} of *B. bassiana* succumbed to infection. The fecundity (cumulative mean egg production over 10 days) of *A. bipunctata* and *H. axyridis* (UK) was also assessed. The fecundity of *C. septempunctata* could not be assessed because this species requires diapause prior to the onset of reproduction and these studies were on beetles that had recently eclosed (2–8 weeks). *Harmonia axyridis* (Japan) produced no eggs in most treatments including the control and so was excluded from analysis. High dose (10^9 conidia ml^{-1}) inoculation reduced the fecundity of *A. bipunctata* to zero but egg production was similar for individuals inoculated with doses

H. E. Roy (✉) · P. M. J. Brown · P. Rothery
NERC Centre for Ecology and Hydrology Monks Wood, Huntingdon PE28 2LS, UK
e-mail: hele@ceh.ac.uk

H. E. Roy · P. M. J. Brown
Department of Life Sciences, Anglia Ruskin University, East Road, Cambridge CB1 1PT, UK

R. L. Ware · M. E. N. Majerus
Department of Genetics, University of Cambridge, Downing Street, Cambridge CB2 3EH, UK

Biological Control to Invasion: Ladybird Harmonia axyridis. H.E. Roy, E. Wajnberg (eds.),
doi: 10.1007/978-1-4020-6939-0_18, © IOBC 2007

of 10^5, 10^7 conidia ml^{-1} and control individuals. In contrast, all doses of *B. basssiana* reduced *H. axyridis* (UK) egg production dramatically. We discuss these results in relation to the potential for control of *H. axyridis* using *B. bassiana*.

Keywords *Beauveria bassiana* · *Harmonia axyridis* · *Coccinella septempunctata* · *Adalia bipunctata* · Fungal entomopathogen · Coccinellidae · Invasive alien species · Native species · Biological control

Introduction

The fungus *Beauveria bassiana* (Balsamo) Vuillemin (Ascomycota: Hypocreales) is a facultative pathogen infecting species from most insect orders (Keller and Zimmerman 1989). It is ubiquitous in the soil and recently propagules have been reported from the phylloplanes of hedgerow vegetation (Meyling and Eilenberg 2006), elm bark (Doberski and Tribe 1980), conifer plantations (Ormond 2007) and as an endophyte of corn (Lewis et al. 1996). As an insect pathogen it is found naturally infecting many insect species in the field (Riedel and Steenberg 1998; Ormond et al. 2006), particularly those which have a component of their life history associated with the soil, such as some species of over-wintering coccinellid. Indeed, *B. bassiana* is documented as a major mortality factor of seven-spot ladybirds *Coccinella septempunctata* L. (Coleoptera: Coccinellidae) which, in temperate regions, spend the winter months in diapause under leaf litter where they are exposed to this fungal pathogen (Majerus and Kearns 1994; Ormond et al. 2006).

The interactions between *B. bassiana* and coccinellids have been studied from two main perspectives: mortality of overwintering populations exposed to naturally occurring *B. bassiana* (Iperti 1966; Barron and Wilson 1998; Ceryngier and Hodek 1996; Ormond et al. 2006) and mortality of non-target coccinellids exposed to *B. bassiana* based biorational pesticides (James et al. 1995; Pingel and Lewis 1996; Todorova et al. 1996; Roy and Pell 2000; Smith and Krischik 2000; Pell and Vandenberg 2002). There is no doubt that the latter has dominated. Intriguingly the arrival of the harlequin ladybird, *Harmonia axyridis* (Pallas) (Coleoptera: Coccinellidae), as an invasive alien in many countries provides a new perspective on the interactions between coccinellids and *B. bassiana*. Whereas previously infection of coccinellids by *B. bassiana* has been seen as a negative non-target impact it could now be considered as an asset if the target is *H. axyridis*. However, the wide host range of *B. bassiana* should be considered in this context.

Harmonia axyridis is an exotic species in many countries (Brown et al. 2007) and is reported to be the dominant species of coccinellid in some habitats within its non-native range (Brown and Miller 1998; Colunga-Garcia and Gage 1998; Cottrell and Yeargan 1998; Wells and McPherson 1999; Michaud 2002; Cottrell and Shapiro-Ilan 2003). *Harmonia axyridis* is widely considered as an invasive alien species for a number of reasons, including: impacts on native fauna, overwintering aggregations in domestic dwellings and damage to orchard fruits (Majerus et al. 2006). Recent attention has been given to the control of *H. axyridis* (reviewed by Kenis et al. 2007) and *B. bassiana* is considered a potential candidate. However, endemic insect species are known to be susceptible to *B. bassiana* and so the differential susceptibility of these species to *B. bassiana* must be evaluated as a first stage of assessing the risks of this fungus as a possible control agent of *H. axyridis*.

In this paper we assess the susceptibility of *H. axyridis* and two native UK coccinellid species, *Adalia bipunctata* (L.) (Coleoptera: Coccinellidae) and *C. septempunctata*, to

B. bassiana (Mycotrol GHA). In a previous study we demonstrated that adult *H. axyridis* reared in the laboratory from UK field collections differed behaviourally in comparison to adults derived from Japan field collections (Roy et al. 2007a). Therefore, we included UK and Japan subpopulations of *H. axyridis* in our studies. *Adalia bipunctata* has similar life history traits to *H. axyridis* in terms of sites for oviposition, larval development and pupation (diverse but often on deciduous trees). Furthermore both species overwinter in aggregations in buildings (Majerus 1994; Majerus et al. 2006). In contrast, *C. septempunctata* has different habitat preferences for oviposition, larval development and pupation (usually herbaceous layer) and often overwinters under leaf litter or in soil (Majerus 1994; Majerus et al. 2006). *Coccinella septempunctata* is commonly found infected with *B. bassiana* in the field, whereas mycosed *A. bipunctata* are rarely reported (Majerus 1994; Barron and Wilson 1998; Majerus et al. 2006). It is for these reasons that we decided to include these three coccinellid species in our studies.

Most studies that examine the potential of entomopathogens as biological control agents express efficacy as a median lethal dose (LD_{50}) and median lethal time to death (LT_{50}); therefore considering only the mortality of the study species (Inglis et al. 2001; Cottrell and Shapiro-Ilan 2003). More recent attention has focused on the importance of sublethal effects such as alterations to behaviour (Roy et al. 2006a; Roy et al. 2006b; Roy et al. 2007b). A number of studies have shown that fungal-infected hosts produce fewer progeny (Blanford and Thomas 2001; Baverstock et al. 2006). Reduction in fecundity could increase pathogen fitness by diverting host resources for the pathogen although there is, of course, a trade-off to the pathogen because fewer susceptible hosts will consequently be available for infection; although the extent of this is less for a facultative pathogen, such as *B. bassiana*, than a host-specific pathogen.

The aims of our studies were to assess the impact of a commercial isolate of *B. bassiana* (Mycotrol GHA) on the survival of three coccinellid species (*H. axyridis*, *A. bipunctata* and *C. septempunctata*) and the fecundity of two coccinellid species (*H. axyridis* and *A. bipunctata*).

Materials and methods

Insect and fungus cultures

Laboratory reared adult *H. axyridis*, *C. septempunctata* and *A. bipunctata* were used in all experiments. These cultures were obtained from the following field locations: *H. axyridis* from the UK (Battersea Park, London) and Japan (Fuchu, Honshu), *C. septempunctata* and *A. bipunctata* from the UK (Norfolk and Cambridgeshire, respectively). All adult coccinellids had undergone eclosion within 2–8 weeks of the bioassays. The coccinellids were maintained in batches of five individuals within 90 mm diameter single-vented Petri dishes at 22°C (14L:10D). All individuals were fed daily on an excess of adult pea aphids, *Acyrthosiphon pisum* Harris (Hemiptera: Aphididae) which were cultured on broad bean plants *Vicia faba* L. (cultivar: The Sutton).

The GHA strain of *B. bassiana* that is the active ingredient of the product Botanigard® (Laverlam) was used in all bioassays. The culture was stored in 10% glycerol (cryoprotectant) at –86°C until required. After retrieval from –86°C, the fungal plugs were macerated within the glycerol and spread onto Sabouraud Dextrose Agar (SDA). The fungus was then incubated at 20°C (16L:8D) for 2 weeks and subcultured again on to SDA. After a further 2 weeks, *B. bassiana* conidia could be harvested from the plates. The fungal conidia from eight plates were dispersed in 20 ml of 0.03 % Tween 80 and held on ice. Spore concentration was calculated using a Neubauer haemocytometer and suspensions of 10^9, 10^7 and 10^5 conidia ml^{-1} were achieved through dilution in the surfactant Tween 80 (0.03%). All spore suspensions were again assessed with the Neubauer haemocytometer to ensure accuracy of concentration values. All bioassays used spore suspensions that were less than four hours old and had been stored on ice.

Bioassays

Batches of either 1 ml of fungal spore suspension or 1 ml of 0.03 % Tween 80 (control) were decanted into 1.5 ml Eppendorf tubes and stored on ice. Each individual coccinellid was dipped in new spore or control suspension (inverted five times) and transferred to 50 mm single-vented Petri dishes. Each dish contained a filter disc moistened with sterile distilled water and an excess of adult pea aphids. All dishes were placed in trays lined and covered with moistened paper to maintain high humidity for a period of 24 h. After 24 h the moistened filter paper discs were removed from each dish and the moistened paper was also removed from the tray. All individuals were fed daily on an excess of adult pea aphids. Replication of each treatment, and the number of male and female coccinellids within a bioassay varied depending on availability of laboratory cultured coccinellids (Table 1).

Mortality was recorded daily for each individual until 10 days post inoculation (days p.i.). Individual dead coccinellids were transferred to a 50 mm single-vented Petri dish containing a filter disc moistened with sterile distilled water. All dishes were placed in trays lined and covered with moistened paper to maintain high humidity. Dead coccinellids were assessed daily for external fungal growth (sporulation), which could be confirmed as *B. bassiana* on the basis of morphology.

Eggs were counted and removed on a daily basis for 10 days p.i.

Statistical analyses

Mortality

Mortality data were analysed using binary logistic regression in Minitab 14 (Minitab 2003). Control data were excluded from the analysis because of the exceptionally low mortality (only one individual (*A. bipunctata*) across all replicates and this individual showed no evidence of mycosis). Goodness-of-fit was used to compare the logit and the

Table 1 Number of replicates per treatment (Control or *B. bassiana* treatment: 10^5, 10^7 or 10^9 conidia ml^{-1}) for each species (or biotype) of coccinellid: *A. bipunctata*, *C. septempunctata*, *H. axyridis* (Japan) and *H. axyridis* (UK)

Coccinellid species	Control	10^5 *B. bassiana*	10^7 *B. bassiana*	10^9 *B. bassiana*
A. bipunctata	10	10 (6♀:4♂)	10	10
C. septempunctata	20	20	20	20
H. axyridis (Japan)	10 (6♀:4♂)	10 (6♀:4♂)	10 (7♀:3♂)	10 (4♀:6♂)
H. axyridis (UK)	20	20	20 (9♀:11♂)	20 (8♀:12♂)

An equal number of male and females were used for each treatment unless stated

Gompit (complementary log–log) model, and the latter was subsequently used for all analyses. Therefore, the relationship between mortality and $\log_{10}$ dose was described by a binary regression model with Gompit link function. If m_{ij} denotes the mortality rate for the *i*th species with $\log_{10}$ dose x_{ij} the model is:

$$\log_e[-\log_e(1 - m_{ij})] = a_i + bx_{ij}$$

where a_i denotes an intercept term for the *i*th species. The model assumes the same slope (b) across species. This assumption was tested but no differences in slope were detected. Significance tests were adjusted for over-dispersion using the Pearson method, by division of *t*-ratios by $\sqrt{(\chi^2/\text{df})}$, equivalent to 1.21 for 10 day p.i.

The LD_{50} corresponds to 50% mortality and is obtained by setting $m_{ij} = 0.5$, i.e.

$$\log_{10} \text{LD50}_i = -(0.367 + a_i)/b$$

A 95% confidence interval for the $\log_{10}$ LD_{50} is obtained by application of Fieller's theorem (Finney 1971). It should be noted that where an LD_{50} exceeds 10^9 conidia ml^{-1} the value is an extrapolation of the regression curve.

Fecundity

Fecundity (cumulative number of eggs laid per individual) was analysed using repeated measures ANOVA for *A. bipunctata* (Fig. 3) and *H. axyridis* UK (Fig. 4) in Genstat version 10 (Genstat 2007). Counts were transformed using log(count + 1) to approximate the assumptions of constant variance and normality. Significance tests for time effects and the interaction with treatment were adjusted to allow for a non-constant correlation between repeat observations on the same units using the Greenhouse–Geisser correction factor.

It should be remembered that at day 10 p.i. for some species-treatment combinations the replication was low because of high mortality, particularly for *A. bipunctata* inoculated with 10^9 conidia ml^{-1}, where mortality by day 10 was 100%. Therefore, analyses on fecundity of individuals at late stages of *B. bassiana* infection are presented tentatively.

Coccinella septempunctata fecundity was not analysed because this species does not reproduce prior to diapause and all experiments were conducted with 2–8 week old individuals (Majerus 1994). Data for *H. axyridis* (Japan) were also excluded because of a high number of zero values for all individuals (control and treatments).

Results

Mortality

By 5 days p.i., the proportion of coccinellids that had succumbed to *B. bassiana* infection was similar for *C. septempunctata* and *H. axyridis* (Japan) ($t = -1.93$, df = 7, $P = 0.095$; Fig. 1). There was a trend suggesting that a lower proportion of *H. axyridis* (UK) had died in comparison to *H. axyridis* (Japan) ($t = 1.66$, df = 7, $P = 0.07$; Fig. 1). The mortality of *A. bipunctata* was significantly higher compared to *H. axyridis* (Japan) ($t = -4.90$, df = 7, $P < 0.001$; Fig. 1).

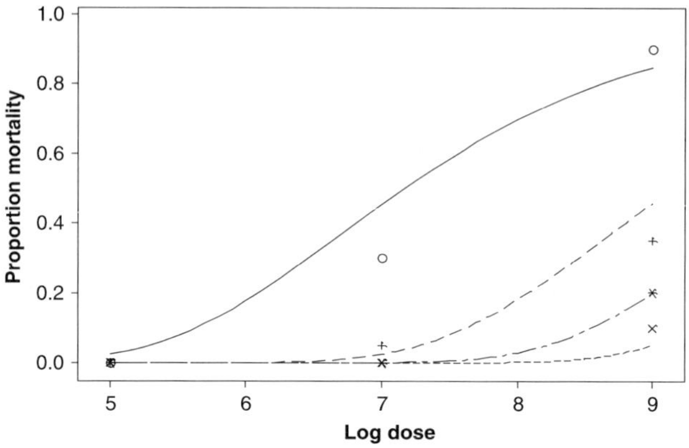

Fig. 1 Proportion mortality at 5 days for four species of coccinellid (*Coccinella septempunctata* (large hash line/+), *Adalia bipunctata* (solid black line/o), *Harmonia axyridis* (Japan) (broken hash line/*) and *Harmonia axyridis* (UK) (small hash line/x)) exposed to three doses of *B. bassiana* (10^5, 10^7 and 10^9 conidia ml^{-1}). Symbols represent experimental data and connected lines represent data as predicted by the Gompit (complementary log–log) model

By 10 days p.i., the proportion of *H. axyridis* (Japan) and *H. axyridis* (UK) that had died from *B. bassiana* infection was statistically similar ($t = 2.05$, df = 7, $P = 0.078$, Fig. 2), although there was a trend suggesting that *H. axyridis* (UK) were more resistant to infection than *H. axyridis* (Japan). Mortality of *C. septempunctata* ($t = -3.55$, df = 7, $P = 0.009$; Fig. 2) and *A. bipunctata* ($t = -2.85$, df = 7, $P = 0.024$; Fig. 2) was significantly greater than observed for *H. axyridis* (Japan and UK). So, in conclusion, the order of susceptibility from highest (greatest proportion dead) to lowest (smallest proportion dead) was *C. septempunctata*, *A. bipunctata*, *H. axyridis* (Japan) and *H. axyridis* (UK). This is apparent from the calculated LD_{50} values (Table 2, Figs. 1 and 2).

Fecundity

There was a significant difference between treatments ($F_{3,8} = 6.6$, $P = 0.015$, Fig. 3), time ($F_{4,32} = 18.84$, $P = 0.001$; Fig. 3) and the treatment × time interaction ($F_{12,32} = 3.64$,

Fig. 2 Proportion mortality at 10 days for four species of coccinellid (*Coccinella septempunctata* (large hash line/+), *Adalia bipunctata* (solid black line/o), *Harmonia axyridis* (Japan) (broken hash line/*) and *Harmonia axyridis* (UK) (small hash line/x)) exposed to three doses of *B. bassiana* (10^5, 10^7 and 10^9 conidia ml^{-1}). Symbols represent experimental data and connected lines represent data as predicted by the Gompit (complementary log–log) model

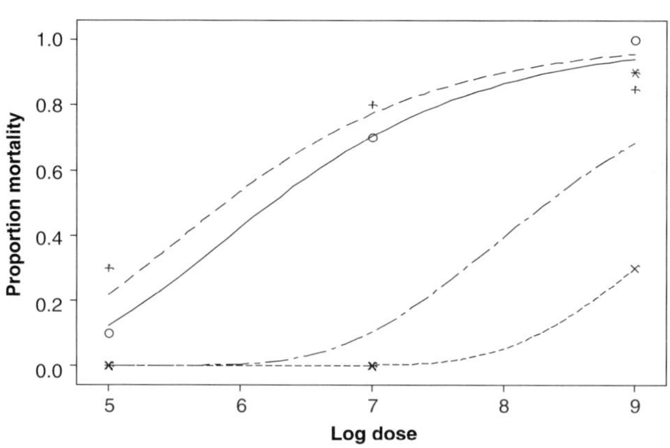

Table 2 Estimated LD_{50} ($\log_{10}$ conidia ml^{-1}) at 10 days with 95% confidence interval for *A. bipunctata*, *C. septempunctata*, *H. axyridis* (Japan) and *H. axyridis* (UK)

Species	LD_{50} (95% confidence interval)
A. bipunctata	$10^{6.2}$ ($10^{5.4}$, $10^{7.3}$)
C. septempunctata	$10^{6.0}$ ($10^{5.0}$, $10^{6.7}$)
H. axyridis (Japan)	$10^{8.3}$ ($10^{7.4}$, $10^{9.5}$)
H. axyridis (UK)	$10^{9.6}$ ($10^{8.9}$, $10^{10.7}$)

$P = 0.031$; Fig. 3) for *A. bipunctata*. The cumulative egg count increased over time for all treatments except the 10^9 spores ml^{-1} (high dose) treatment. Indeed when the zero egg counts observed for the 10^9 spores ml^{-1} (high dose) treatment were excluded from the analysis there was no evidence of either a treatment × time interaction effect ($F_{8,24} = 1.74$, $P = 0.23$), or treatment effect ($F_{2,6} = 0.08$, $P = 0.93$).

There was no significant difference between treatments for *H. axyridis* ($F_{3,7} = 2.73$, $P = 0.124$; Fig. 4) but there was a significant difference between treatments in comparison to the control ($F_{1,7} = 7.46$, $P = 0.023$; Fig. 4). Cumulative egg count increased significantly over time for all treatments including the control ($F_{4,28} = 24.12$, $P = 0.001$; Fig. 4) but there was no significant treatment × time interaction ($F_{12,28} = 0.6$, $P = 0.75$; Fig. 4).

In summary *A. bipunctata* exposed to 10^9 conidia ml^{-1} of *B. bassiana* did not produce any eggs while the fecundity of individuals exposed to 10^5 and 10^7 conidia ml^{-1} of *B. bassiana* was similar to that of control individuals. In contrast the fecundity of *H. axyridis* inoculated with *B. bassiana* was lower, irrespective of dose, than control individuals (Figs. 3 and 4).

Discussion

Numerous studies have demonstrated the physiological susceptibility of coccinellids to *B. bassiana* (Iperti 1966; Todorova et al. 1996; Cottrell and Shapiro-Ilan 2003). In a recent paper, Cottrell and Shapiro-Ilan (2003) highlighted the differential susceptibility of a native coccinellid and an exotic coccinellid to an isolate of *B. bassiana* that had been derived from the native coccinellid. The exotic coccinellid, *H. axyridis*, was less susceptible to *B. bassiana* than the native coccinellid, *Olla v-nigrum* Mulsant (Coleoptera:

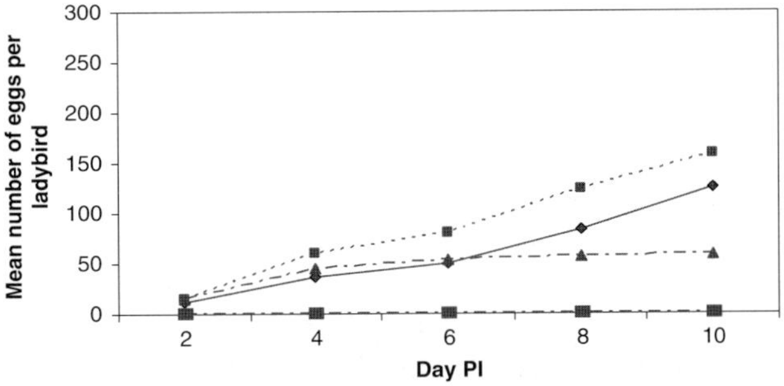

Fig. 3 Cumulative number (mean) of eggs laid per *Adalia bipunctata* over time (day p.i.) after inoculation at day 0 with *B. bassiana* at either a low dose (10^5 conidia ml^{-1}; small squares, dashed line), mid dose (10^7 conidia ml^{-1}; triangle, dashed line), high dose (10^9 conidia ml^{-1}; large square, dashed line) or control (0.03% tween 80; diamond, solid line)

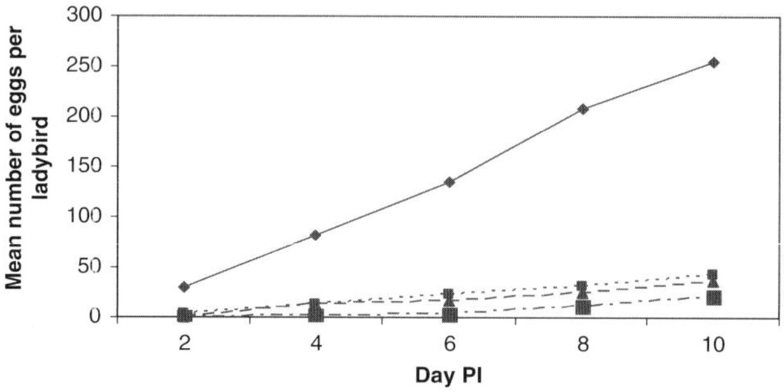

Fig. 4 Cumulative number (mean) of eggs laid per *Harmonia axyridis* (Japan and UK combined) over time (day p.i.) after inoculation at day 0 with *B. bassiana* at either a low dose (10^5 conidia ml^{-1}; small squares, dashed line), mid dose (10^7 conidia ml^{-1}; triangle, dashed line), high dose (10^9 conidia ml^{-1}; large square, dashed line) or control (0.03% tween 80; diamond, solid line)

Coccinellidae). In our study we compared the susceptibility of three species of coccinellid to a commercial isolate (geographic origin unknown) of *B. bassiana* (Mycotrol GHA) and conclude that *H. axyridis* is more resistant to infection than either *A. bipunctata* or *C. septempunctata* (which are both highly susceptible). Mortality of *H. axyridis* did not occur at doses less than 10^9 conidia ml^{-1}, whereas 70 and 80% mortality of *A. bipunctata* and *C. septempunctata* (respectively) occurred after inoculation at 10^7 conidia ml^{-1}. Inoculation at the highest dose (10^9 conidia ml^{-1}) resulted in 100% mortality of *A. bipunctata* and *C. septempunctata*. Interestingly 90% mortality of *H. axyridis* (Japan) was achieved at this high dose but only 30% mortality of *H. axyridis* (UK). Although the difference in mortality of the two *H. axyridis* subpopulations (derived from cultures initiated from UK and Japan collected adults) was not significant it is certainly worth further exploration. In previous work, Roy et al. (2007a) demonstrated a difference in the feeding behaviour of these two subpopulations and concluded that *H. axyridis* (UK) was a stronger intraguild predator than *H. axyridis* (Japan). The observed anomaly between the UK and Japan derived *H. axyridis* suggests that the individuals that have invaded the UK could have a different genetic profile to those in its native range.

We included *A. bipunctata* in our study for two reasons: first, there are no published studies reporting infection of *A. bipunctata* by *B. bassiana* and second, it has similar life history traits to *H. axyridis* and occupies similar habitats for both feeding and overwintering (Majerus 1994; Lanzoni et al. 2004; Majerus et al. 2006). Both *A. bipunctata* and *H. axyridis* are found in a range of habitats but favour deciduous trees (Majerus 1994; Majerus et al. 2006). *Beauveria bassiana* is generally considered to be a soil-borne pathogen (Keller and Zimmerman 1989) with limited distribution on foliage (Meyling and Eilenberg 2006). Therefore, it is widely assumed that coccinellids overwintering in leaf litter, such as *C. septempunctata*, are more likely to be exposed to *B. bassiana* than coccinellids, such as *H. axyridis* and *A. bipunctata*, which spend winter above ground. The lack of reports of interactions between *A. bipunctata* and *B. bassiana* could, perhaps, be linked to the spatial disparity of their habitats. From our studies it can be concluded that *A. bipunctata* is as physiologically susceptible as *C. septempunctata* to infection by *B. bassiana*. This supports the hypothesis that the lack of field observations of mycosed *A. bipunctata* is because they either do not contact sufficiently high doses of inoculum, or are not in abiotic environments suitable for infection. Furthermore, our study contributes a further example of the resilience of *H. axyridis*, compared to other coccinellids, to

B. bassiana. It should be a priority to extend these laboratory studies to the field through a combination of experimental manipulations and observational surveys of natural enemies, including insect pathogens. Indeed the UK on-line Harlequin Ladybird Survey (www.harlequin-survey.org) is now providing the opportunity for contributors to include this information.

Throughout our studies we used only one isolate of *B. bassiana*. A number of recent publications have shown that *B. bassiana* isolates are very genetically variable (Aquino de Muro et al. 2003, 2005; Estrada et al. 2007; Ormond 2007), and concomittantly highly phenotypically variable in parameters such as virulence (Bidochka et al. 2002; Devi and Rao 2006; Talaei-Hassanloui et al. 2006). The natural enemy escape hypothesis is often used as an explanation for the rapid establishment and proliferation of an invasive alien species; the geographical displacement of the species results in escape from predators and parasites and so gives the invading species an advantage over native species that are attacked by a suite of indigenous natural enemies (e.g. Colautti et al. 2004; Carroll et al. 2005). It is possible that an isolate derived from the native range of *H. axyridis* would be more virulent to *H. axyridis* than the isolates studied so far (Bazzocchi et al. 2004). However, it is striking that, despite different approaches to isolate choice in studies on *H. axyridis*, the conclusion that *H. axyridis* is resistant to infection is unanimous (Cottrell and Shapiro-Ilan 2003). It would be interesting to explore the mechanisms underlying the resistance of *H. axyridis* to *B. bassiana*.

Initial studies should focus on the infection process and mechanisms of defence by the host. At high humidity, *B. bassiana* conidia will germinate on the surface of a suitable host and differentiate into an appressorium, which penetrates directly through the cuticle. Insect hosts have a limited range of defences but can, as an initial response, produce melanin at the appressorium penetration site (James et al. 2003). Therefore, it could be hypothesised that melanic individuals, which are common in *H. axyridis*, would be more resistant to infection than non-melanic individuals. However, in preliminary bioassays comparing melanic and non-melanic individuals there was no difference in susceptibility (H. Roy personal observation). It is possible that the defensive chemistry of *H. axyridis* could be adversely affecting fungal development within the haemocoel and this could be investigated in vitro. *Beauveria bassiana* produces a number of secondary metabolites, such as bassianin, bassiacridin, beauvericin, bassianolide, beauverolides, tenellin and oosporein (Strasser et al. 2000; Vey et al. 2001; Quesada-Moraga and Vey 2004). Some of these have toxic effects (beauvericin, bassianin, bassianolide, beauverolides, tenellin) while others have an antibiotic or antifungal role (oosporein) (Strasser et al. 2000; Scholte et al. 2004). Perhaps the toxic alkaloids circulating in the haemocoel of *H. axyridis* are interacting with these secondary metabolites and negating their effects. Considerable further work is required to elucidate this.

Harmonia axyridis inoculated with *B. bassiana* produced fewer eggs than control individuals and this suggests that the fungus is invading the host. This impact on fecundity was not observed for *A. bipunctata* (and not tested for with *C. septempunctata* because a period of diapause is required prior to reproduction and our studies used recently eclosed adults). Most studies on insect pathogens consider only host mortality but recent attention has focussed on sublethal effects, such as reduced fecundity (e.g. Roy et al. 2006a; Roy et al. 2006b). Baverstock et al. (2006) demonstrated a reduction in fecundity of *B. bassiana*-infected *A. pisum* (pea aphid) within 24 h of infection. *Acyrthosiphon pisum* were highly susceptible to the isolate of *B. bassiana* used in these studies (Mycotrol GHA) and it is possible that resources were directed away from host embryogenesis to benefit development of the pathogen. An alternative strategy involves a host maintaining or increasing reproductive output in response to early stages of infection, to ensure that part of their reproductive potential is realised. Blanford and Thomas (2001) observed such a phenomenon in desert locusts, *Schistocerca gregaria* Forskal (Orthoptera: Acrididae), infected with *Metarhizium anisopliae* (Metsch.) Sorokin var *acridum* Milner & Driver (Ascomycota: Hypocreales). In both these studies the host was highly susceptible to the pathogen under investigation. Our studies suggest that the susceptible coccinellid, *A. bipunctata*, is employing the strategy of maintaining egg production throughout the infection process, whereas *H. axyridis*, which suffers much lower mortality as a consequence of *B. bassiana* inoculation, produces fewer eggs after exposure even to low doses (10^5 conidia ml^{-1}). We tentatively propose that *H. axyridis* may be diverting resources from egg production to invest in defence against the pathogen. This intriguing result requires further investigation to see, for example, whether this sublethal effect is consistent across a range of *B. bassiana* isolates and, indeed, for other pathogens. It would also be worth assessing whether the reduction in egg laying persists in the longer term.

The importance of assessing sublethal effects is highlighted through our study. It is clear that *H. axyridis* is more resistant, in terms of mortality, to *B. bassiana* than the other coccinellids included in these experiments and, therefore, it could be concluded that the potential for *B. bassiana* to control *H. axyridis* is negligible. However, including an assessment of fecundity within these experiments has provided further insight into the interactions between coccinellids and *B. bassiana*.

References

Aquino de Muro M, Mehta S, Moore D (2003) The use of amplified fragment length polymorphism for molecular analysis of *Beauveria bassiana* isolates from Kenya and other countries, and their correlation with host and geographical origin. FEMS Microbiol Lett 229:249–257

Aquino De Muro M, Elliott S, Moore D, Parker BL, Skinner M, Reid W, El Bouhssini M (2005) Molecular characterisation of *Beauveria bassiana* isolates obtained from overwintering sites of Sunn Pests (*Eurygaster* and *Aelia* species). Mycol Res 109:294–306

Barron A, Wilson K (1998) Overwintering survival in the seven spot ladybird, *Coccinella septempunctata* (Coleoptera: Coccinellidae). Eur J Entomol 95:639–642

Baverstock J, Roy HE, Clark SJ, Alderson PG, Pell JK (2006) Effect of fungal infection on the reproductive potential of aphids and their progeny. J Invertebr Pathol 91:136–139

Bazzocchi GG, Lanzoni A, Accinelli G, Burgio A (2004) Overwintering, phenology and fecundity of *Harmonia axyridis* in comparison with native coccinellid species in Italy. BioControl 49:245–260

Bidochka MJ, Menzies FV, Kamp AM (2002) Genetic groups of the insect-pathogenic fungus *Beauveria bassiana* are associated with habitat and thermal growth preferences. Arch Microbiol 178:531–537

Blanford S, Thomas MB (2001) Adult survival, maturation and reproduction of the desert locust *Schistocerca gregaria* infected with the fungus *Metarhizium anisopliae* var. *acridum*. J Invertebr Pathol 78:1–8

Brown MW, Miller SS (1998) Coccinellidae (Coleoptera) in apple orchards of eastern West Virginia and the impact of invasion by *Harmonia axyridis*. Entomol News 109:143–151

Brown PMJ, Adriaens T, Bathon H, Cuppen J, Goldarazena A, Hagg T, Kenis M, Klausnitzer BE, Kovar I, Loomans AJ, Majerus MEN, Nedved O, Pedersen J, Rabitsch W, Roy HE, Ternois V, Zakharov I, Roy DB, Ware RL, Majerus MEN (2007) *Harmonia axyridis* in Europe: spread and distribution of a non-native coccinellid. BioControl (this issue). doi:10.1007/s10526-007-9132-y

Carroll SP, Loye JE, Dingle H, Mathieson M, Famula TR, Zalucki MP (2005) And the beak shall inherit—evolution in response to invasion. Ecol Lett 8:944–951

Ceryngier P, Hodek I (1996) Enemies of Coccinellidae. In: Hodek I, Honěk A (eds) Ecology of Coccinellidae, Kluwer Academic Publishers, Dordrecht

Colautti RI, Ricciardi A, Grigorovich I, MacIsaac HJ (2004) Is invasion success explained by the enemy release hypothesis? Ecol Lett 7:721–733

Colunga-Garcia M, Gage SM (1998) Arrival, establishment, and habitat use of the multicolored Asian ladybeetle (Coleoptera: Coccinellidae) in a Michigan land-scape. Environ Entomol 27:1574–1580

Cottrell TE, Yeargan KV (1998) Intraguild predation between an introduced lady beetle, *Harmonia axyridis* (Coleoptera: Coccinellidae) and a native lady beetle, *Coleomegilla maculata* (Coleoptera: Coccinellidae). J Kans Entomol Soc 71:159–163

Cottrell TE, Shapiro-Ilan DI (2003) Susceptibility of a native and an exotic lady beetle (Coleoptera: Coccinellidae) to *Beauveria bassiana*. J Invertebr Pathol 84:137–144

Devi KU, Rao CUM (2006) Alee effect in the infection dynamics of the entomopathogenic fungus *Beauveria bassiana* (Bals) Vuill. on the beetle, *Mylabris pustulata*. Mycopathologia 161:385–394

Doberski JW, Tribe HT (1980) Isolation of entomogenous fungi from elm bark and soil with reference to ecology of *Beauveria bassiana* and *Metarhizium anisopliae*. Trans Br Mycol Soc 74:95–100

Estrada ME, Camacho MV, Benito C (2007) The molecular diversity of different isolates of *Beauveria bassiana* (Bals.) Vuill. as assessed using intermicrosatellites (ISSRs). Cell Mol Biol Lett 12:240–252

Finney DJ (1971) Probit analysis, 3rd edn. Cambridge University Press, Cambridge

Genstat (2007) Genstat for Windows 10. Adept Scientific, Hertfordshire

Inglis GD, Goettel MS, Butt TM, Strasser H (2001) Use of hyphomycetous fungi for managing insect pests. In: Butt TM, Jackson C, Magan N (eds) Fungi as Biocontrol agents: progress, problems and potential. CABI Publishing, Wallingford

Iperti G (1966) Protection of coccinellids against mycosis. In: Hodek I (ed) Ecology of aphidophagous insects. Academia Prague Dr W. Junk, Dordrecht

James RR, Shaffer BT, Croft B, Lighthart B (1995) Field evaluation of *Beauveria bassiana*: its persistence and effects on the pea aphid and a non-target coccinellid in alfalfa. Biocontrol Sci Technol 5:425–438

James RR, Buckner JS, Freeman TP (2003) Cuticular lipids and silverleaf whitefly stage affect conidial germination of *Beauveria bassiana* and *Paecilomyces fumosoroseus*. J Invertebr Pathol 84:67–74

Keller S, Zimmerman G (1989) Mycopathogens of soil insects. In: Wilding N, Collins NM, Hammond PM, Webber JF (eds) Insect-fungus interactions. Academic, London, pp 240–270

Kenis M, Roy HE, Majerus MEN (2007) Current and potential management strategies against *Harmonia axyridis*. BioControl (this issue). doi:10.1007/s10526-007-9136-7

Lanzoni A, Accinelli G, Bazzochi GG (2004) Biological traits and life table of the exotic *Harmonia axyridis* compared with *Hippodamia variegate* and *Adalia bipunctata* (Col., Coccinellidae). J Appl Entomol 128:298–306

Lewis LC, Berry EC, Obrycki JJ, Bing LA (1996) Aptness of insecticides (*Bacillus thuringiensis* and carbofuran) with endophytic *Beauveria bassiana* in suppressing larval populations of the European corn borer. Agric Ecosyst Environ 57:27–34

Majerus MEN (1994) Ladybirds. Harper Collins Publishers, UK

Majerus MEN, Kearns P (1994) Ladybirds. No. 81, New Naturalist Series. HarperCollins, London

Majerus MEN, Strawson V, Roy HE (2006) The potential impacts of the arrival of the Harlequin ladybird, *Harmonia axyridis* (Pallas) (Coleoptera: Coccinellidae), in Britain. Ecol Entomol 31:207–215

Meyling NV, Eilenberg J (2006) Isolation and characterisation of *Beauveria bassiana* isolates from phylloplanes of hedgerow vegetation. Mycol Res 110:188–195

Michaud JP (2002) Invasion of the Florida citrus ecosystem by *Harmonia axyridis* (Coleoptera: Coccinellidae) and asymmetric competition with a native species, *Cycloneda sanguinea*. Environ Entomol 31:827–835

Minitab (2003) MINITAB User's guide 2: data analysis and quality tools. Minitab Inc., State College, PA

Ormond EL (2007) The overwintering interactions of the seven spot ladybird (*Coccinella septempunctata*) and the entomopathogenic fungus *Beauveria bassiana*. Ph.D. Thesis, Anglia Ruskin University, UK

Ormond EL, Thomas APM, Pell JK, Roy HE (2006) Overwintering ecology of *Coccinella septempunctata*, *Beauveria bassiana* and *Dinocampus coccinellae*. In: Walter AH, Rossing WAH, Eggenschwiler L, Poehling H-M (eds) Working Group "Landscape management for functional biodiversity" at Zürich-Reckenholz (Switzerland), 16–19 May 2006 IOBC/wprs Bulletin 29(6), 85–88

Pell JK, Vandenberg JD (2002) Interactions among the aphid *Diuraphis noxia*, the entomopathogenic fungus *Paecilomyces fumosoroseus* and the coccinellid *Hippodamia convergens*. Biocontrol Sci Technol 12:217–224

Pingel RL, Lewis LC (1996) The fungus *Beauveria bassiana* (Balsamo) Vuillemin in a corn ecosystem: its effect on the insect predator *Coleomegilla maculata* De Geer. Biol Control 6:137–141

Quesada-moraga E, Vey A (2004) Bassiacridin, a protein toxic for locusts secreted by the entomopathogenic fungus *Beauveria bassiana*. Mycol Res 108:441–452

Riedel W, Steenberg T (1998) Adult polyphagous coleopterans overwintering in cereal boundaries: winter mortality and susceptibility to the entomopathogenic fungus *Beauveria bassiana*. BioControl 43:175–188

Roy HE, JK Pell (2000) Interactions between entomopathogenic fungi and other natural enemies: implications for biological control. Biocontrol Sci Technol 10:737–752

Roy HE, Brown P, Majerus MEN (2006a) *Harmonia axyridis*: A successful biocontrol agent or an invasive threat? In: Eilenberg J, Hokkanen H (eds) An ecological and societal approach to biological control, Kluwer Academic Publishers, Netherlands

Roy HE, Steinkraus D, Eilenberg E, Hajek A, Pell JK (2006b) Bizarre interactions and endgames: entomopathogenic fungi and their arthropod hosts. Ann Rev Entomol 51:331–357

Roy HE, Baverstock J, Ware RL, Clark SJ, Majerus MEN, Baverstock KE, Pell JK (2007a) Intraguild predation of the aphid pathogenic fungus *Pandora neoaphidis* by the invasive coccinellid *Harmonia axyridis*. Ecol Entomol (in press)

Roy HE, Baverstock J, Pell JK (2007b) Manipulating fungal induced host-altered behaviour: a strategy for pest control? In: Ekesi S, Maniania NK (eds) Use of Entomopathogenic Fungi in Biological Pest Management. Research Signpost, Trivandrum (in press)

Scholte EJ, Knols BGJ, Samson RA, Takken W (2004) Entomopathogenic fungi for mosquito control: a review. J Insect Sci 4:1–24

Smith SF, Krischik VA (2000) Effects of biorational pesticides on four coccinellid species (Coleoptera: Coccinellidae) having potential as biological control agents in interiorscapes. J Econ Entomol 93: 732–736

Strasser H, Abendstein D, Stuppner H, Butt TM (2000) Monitoring the distribution of secondary metabolites produced by the entomogenous fungus *Beauveria brongniartii* with particular reference to oosporein. Mycol Res 104:1227–1233

Talaei-Hassanloui R, Kharazi-Pakdel A, Goettel M, Mozaffari J (2006) Variation in the virulence of *Beauveria bassiana* isolates and its relatedness to some morphological characters. Biocontrol Sci Technol 16:525–534

Todorova SI, Cote JC, Coderre D (1996) Evaluation of the effects of two *Beauveria bassiana* (Balsamo) Vuillemin strains on the development of *Coleomegilla maculata lengi* Timberlake (Col, Coccinellidae). J Appl Entomol 120:159–163

Vey A, Hoagland RE, Butt TM (2001) Toxic metabolites of fungal biocontrol agents. In: Butt TM, Jackson C, Magan N (eds) Fungi as biocontrol agents: progress, problems and potential, CABI Publishing, Wallingford

Wells ML, McPherson RM (1999) Population dynamics of three coccinellids in flue-cured tobacco and functional response of *Hippodamia convergens* (Coleoptera: Coccinellidae) feeding on tobacco aphids (Homoptera: Aphididae). Environ Entomol 28:768–773

Estimation of mortality by entomophages on exotic *Harmonia axyridis* versus native *Adalia bipunctata* in semi-field conditions in northern Italy

Giovanni Burgio · Alberto Lanzoni · Gianumberto Accinelli · Stefano Maini

Abstract A semi-field experiment was carried out in two peach orchards in northern Italy to assess mortality due to predators and parasitoids on the exotic coccinellid *Harmonia axyridis* (Pallas) (Coleoptera: Coccinellidae) in comparison with the native coccinellid *Adalia bipunctata* L. (Coleoptera: Coccinellidae). The experiments were conducted in cages to avoid the possible escape of the exotic ladybird (not yet established in Italy). Two kinds of cage experiments were included: 'exclusion cages' (access by walking predators impeded) and 'free cages' (walking predators free to enter). The cages, containing all the stages of the two ladybird species, were placed in two localities and left for 24 h. All ladybird stages used for the semi-field experiments came from a laboratory rearing. The eggs of *H. axyridis* experienced less mortality than those of *A. bipunctata.* The ant workers were the most frequent predators in 'free cages' but *A. bipunctata* cannibalism on eggs was also detected. Larvae of both coccinellid species were predated equally but larval predation of L1 and L2 was higher in comparison to predation of L3 and L4. Pupae and adults of both exotic and native ladybirds were never attacked by predators. Predation on younger larval stages was higher in the 'free cages' in comparison with 'exclusion cages'. No ladybird parasitisation was observed. The 'free cage' technique seems to provide a standardised and realistic estimation of predation impact but more studies are needed to evaluate ladybird parasitisation in semi-field conditions.

G. Burgio (✉) · A. Lanzoni · G. Accinelli · S. Maini
Dipartimento di Scienze e Tecnologie Agroambientali – Entomologia, Alma Mater Studiorum, Università di Bologna, viale G. Fanin, 42-40127 Bologna, Italy
e-mail: giovanni.burgio@unibo.it

A. Lanzoni
e-mail: alberto.lanzoni2@unibo.it

G. Accinelli
e-mail: gaccinelli@entom.agrsci.unibo.it

S. Maini
e-mail: stefano.maini@unibo.it

Biological Control to Invasion: Ladybird Harmonia axyridis. H.E. Roy, E. Wajnberg (eds.),
doi: 10.1007/978-1-4020-6939-0_19,

Keywords *Adalia bipunctata* · Field cage experiments · *Harmonia axyridis* · Intraguild predation · Mortality factors · Parasitoids · Predators

Introduction

Biotic factors, including intraguild predation (IGP), can play an important role in the establishment of exotic coccinellids, like *Harmonia axyridis* (Pallas) (Coleoptera: Coccinellidae). For this reason quantifying the effects of natural enemies in such a system could contribute to explaining and understanding the relationships between introduced species and native ones. Increased understanding of the dynamics of natural enemies of coccinellids could also contribute to an evaluation of the potential efficacy of biological control agents against an invasive alien coccinellid that has established in a new area.

Although the knowledge of the natural enemies of coccinellids has progressed in recent years, the real impact of the biotic factors on the population changes of coccinellids has been poorly investigated (Hodek and Honek 1996; Lucas 2005). This scenario seems particularly true for the predators of coccinellids, including vertebrates and invertebrates, while parasitoids have received a little more attention (Iperti 1964; Cartwright et al. 1982; Obrycki et al. 1985; Obrycki 1989) and a quantification of their impact was assessed in various part of the world (Iperti 1964, Cartwright et al. 1982; Obrycki et al. 1985; Obrycki 1989; Hodek and Honek 1996; Geoghegan et al. 1998; Okuda and Ceryngier 2000). Furthermore, the role of microorganisms as mortality factors of coccinellids has been poorly investigated with limited information on entomopathogenic protozoans and fungi (Hodek and Honek 1996). Fungal disease is considered to be one of the main biotic factors causing winter mortality of coccinellids (see papers cited in Hodek and Honek 1996) and recently the susceptibility of some species of coccinellids, including *H. axyridis* to infection by *Beauveria bassiana* (Balsamo) Vuillemin has been studied by Roy et al. (2007). Further data on the natural enemies of *H. axyridis* are included in the paper by Kenis et al. (2007).

There is considerable interest in the role and prevalence of IGP among aphidophagous predators, partly because such interactions among predators are thought to influence the efficacy of biological control (Rosenheim et al. 1995). Declines in populations of native aphidophages in North America have been linked to IGP by the exotic *H. axyridis* (Brown and Miller 1998; Colunga-Garcia and Gage 1998). In a study measuring the impact of IGP by *H. axyridis* on the control of the newly established soybean aphid, *Aphis glycines* Matsumura in Michigan, Gardiner and Landis (2007) showed that while the exotic *H. axyridis* does act as an intraguild predator and may contribute to local declines in *Aphidoletes aphidimyza* (Rondani) (Diptera: Cecyiidomidae) and *Chrysoperla carnea* (Stephens) (Neuroptera: Chrysopidae). In European countries, the impact of *H. axyridis* on native coccinellids is uncertain but IGP interactions are thought to occur and to have an influence on aphidophagous guilds. In general, in Europe *H. axyridis* is considered to be an invasive alien species (Brown et al. 2007) and in the UK the IGP of immature stages of British and Japanese coccinellids by the invasive ladybird *H. axyridis* has been demonstrated (Ware and Majerus 2007).

IGP in coccinellid guilds was recently reviewed by Dixon (2000) and a review by Pervez and Omkar (2006) focuses more specifically on the predator–predator interaction involving *H. axyridis*. The problem of IGP in aphidophagous systems was critically commented on by Lucas (2005); by an analysis of the literature this author states that IGP, like cannibalism, seems difficult to observe and quantify in the field and this explains the scarcity of data within this topic. For example Cottrell and Yeargan (1998a, b) report the amount of egg cannibalism by a coccinellid and direct observation of other predators feeding on that coccinellid's eggs.

In conclusion, although some data are available on the impact of natural enemies of coccinellids, the degree of their influence has been poorly investigated. This is particularly true for *H. axyridis*, an exotic species for which no field data on the natural enemies are available in Italy.

The objective of this study was to assess the impact of natural enemies on exotic *H. axyridis*, an East Palearctic species, compared with the native West Palearctic species *Adalia bipunctata* L., using field cages studies. The native species was selected because it is common in hedgerows and orchard trees (Hodek and Honek 1996) and is abundant in northern Italy agroecosystems (Burgio et al. 2004, 2006). Furthermore *A. bipunctata*, as *H. axyridis*, is a cosmopolitan species with strong colonisation potential and is expected to establish in Japan (Kajita et al. 2000; Toda and Sakuratani 2006). A specific objective of the research was to quantify the incidence of the entomophages on all life stages of the two selected coccinellid species, in order to understand if the exotic species display a differential degree of protection against predation or parasitism in comparison with the native one.

Materials and methods

Insect rearing

Coccinellids were reared in the entomological laboratories of Dipartimento di Scienze e Tecnologie Agroambientali (DiSTA), Alma Mater Studiorum, University of Bologna. A culture of *A. bipunctata* was initiated from field-collected specimens; *H. axyridis* was supplied by the commercial insectary Koppert (The Netherlands). The biology and overwintering capacity of this biofactory-reared strain of *H. axyridis* were previously studied in northern Italy (Bazzocchi et al. 2004; Lanzoni et al. 2004). Larvae of the two coccinellid species were fed with frozen eggs of *Ephestia kuehniella* (Zeller). Adults were fed with *Aphis gossypii* Glover that had been reared on potted summer squash (*Cucurbita pepo* L.) plants. Adult coccinellids were maintained in Plexiglas cages (40 × 30 × 45 cm) and larvae in plastic cylinders (Kartell©, Milano, Italy) (diameter = 18 cm, h = 18 cm). Both adults and larvae were kept at 20–25°C with a relative humidity of 60–80% and photoperiod 16L: 8D.

Field cage experiments

Field cage experiments that assessed the impact of natural enemies on the exotic species *H. axyridis* in comparison with the native *A. bipunctata*, were conducted in the summer of 2000 in two localities in Emilia-Romagna region (northern Italy). This period was chosen because the occurrence of aphids and their antagonists are at the maximum population level (Burgio et al. 2004, 2006) in this area. Field cage experiments were chosen to avoid the possible escape of the exotic ladybird, not yet established in Italy (only recently this species was recorded in Turin, Brown et al. (2007), but not yet in Emilia-Romagna region). Two treatments (*H. axyridis* versus *A. bipunctata*) were compared and each composed of all the ontogenic stages (eggs, L1, L2, L3, and L4 larvae, pupae and adults, each stage isolated in a separate cage). In addition to split predation by walking predators from predation by all the predators, two kinds of field exposure were set up for each treatment: 'exclusion cages', where access by walking predators was impeded, and 'free cages', that allowed access to all the predators.

All cages used consisted of a PVC container that varied in size according to the ontogenic stage added: 25 × 18 × 11 cm for each of the preimaginal stages and 33 × 22 × 12.5 cm for the adult stage. Twenty-five eggs in clusters, 10 L1, L2, L3 and L4 larvae, pupae and adults were placed separately into the cages. Some summer squash leaves infested with *A. gossypii* were added in each of the larval and adult cages. To avoid specimens escaping, the inner surface of the cages was covered with PTFE (polytetrafluoroethylene, Algoflon®), an inert substance preventing the insects climbing up but not falling down into the cage. This attribute was previously demonstrated in laboratory tests (unpublished data). Moreover, the adult's hind wings were distally cut. All the cages were stored on open metallic shelves covered by a roof (100 × 40 cm, 100 cm above ground level), in two organic peach orchards in the Bologna province (Imola, 41°18′57″ N, 11°39′36″ E and Mercatale, 44°23′7″ N, 11°25′37″ E). Each shelf was wrapped in garden netting of 1 cm^2 mesh to prevent bird and mammalian predation. The exclusion of walking predators was achieved by covering the base of the shelves containing the 'exclusion cages' with insect glue (The Tanglefoot Company, MI, USA) and placing each cage in a small basin full of water plus a surfactant (liquid soap). In each site, all the cages were randomly arranged on four shelves (two for the exclusion system and two for the free system) every 3 weeks, and a total of seven replicates, from June to September, in the site of Mercatale, and five replicates, from July to September, in the site of Imola were completed. For each replicate cages were left in the fields for 24 h, and then returned to the laboratory. Temperature and relative humidity were continuously monitored in the site of Mercatale.

The number of eggs, larvae of each instar, pupae and adults predated or missing was recorded. Missing specimens were considered as totally preyed, due to the impossibility for predators trapped in the cages to escape. In order to avoid cannibalism events, aphid colonies were introduced in the cages. The alive specimens were reared individually in cylindrical cages (8 cm diameter, 8 cm height) in a climatic chamber at 25 ± 1°C, 75 ± 10% RH and a photoperiod of 16L: 8D to assess parasitization of the coccinellid. Adults were reared on *A. gossypii* for 30 days, and larvae were reared on *E. kuehniella* frozen eggs until adult emergence. Egg-batches were kept at the same conditions until eclosion.

Field sampling of coccinellids

In each site, four field samplings by means of mechanical knock down were carried out in order to collect coccinellids within peach orchards, hedgerows and shrubs. This sampling was employed in order to estimate the coccinellid assemblages and to assess the natural parasitization in the study areas. Four field collections were carried out from July to October. Mechanical knock down consisted of beating branches with a stick and collecting the dislodged insects in a 90 cm diameter funnel for examination in the laboratory. Plants were beaten 100 times on each of five branches of a tree or shrub. Plants and branches were randomly selected at each sampling date. The collected coccinellids were identified to species and reared at the same conditions as described for the field cage experiments.

A χ^2 test in 2 × 2 contingency tables was carried out in order to compare the predation ratio between the treatments (Zar 1984).

Results

Different arthropods entered the field cages and the insects found in both treatments are listed in Table 1. The predators included: ant workers, *C. carnea* larvae and adults of the coccinellids *Hippodamia variegata* (Goeze) and *A. bipunctata*. The wild specimens of *A. bipunctata* were easily detected since they were characterised by uncut hind wings.

The predation on *A. bipunctata* eggs by predators was significantly higher in comparison with the predation on *H. axyridis* eggs, in both sampling sites (Fig. 1). No predation on pupae and adults occurred for either the exotic or native ladybirds. It is likely that the larvae of the native and exotic species were preyed on by generalist predators. For both coccinellid species, L1 and L2 instars were more likely to be eaten in comparison with L3 and L4 (Fig. 1). The percentage predation for all the instars of *A. bipunctata* was not different in comparison with predation on *H. axyridis* ($P > 0.05$, χ^2 test).

In general, predation was higher in the 'free cages' in comparison with the 'exclusion cages', for both coccinellid species and localities (Figs. 2, 3). In both sites, ants were the most abundant generalist predator found in the 'free cages' (Table 1) and we suppose that they play an important role in the predation of coccinellids, as reported also by Hodek and Honek (1996).

In the site of Mercatale, one third instar of *C. carnea* was directly observed in the 'free cage' system and this could have been responsible for the predation of all of the *A. bipunctata* eggs. Also, in two collections in the site of Imola, *A. bipunctata* adults were trapped in the 'free cages'; in these two samplings the predation of eggs was again 100% and we suggest that cannibalism strongly contributed to the egg mortality.

No parasitoids emerged from our exposed coccinellids in spite of the regular presence of parasitoid wasps (Hymenoptera, Apocrita, Parasitica) in the cages, which indicates that these wasps was not impeded from entering the cages. These parasitoids were not identified and they were probably just attracted by aphid honeydew.

Among the naturally occurred species sampled in the field by beating, *H. variegata* was the most abundant coccinellid in both localities (relative abundances were 52.3% (n = 57) in Mercatale and 70.9% (n = 66) in Imola), followed by *A. bipunctata* (32.1% (n = 35) and 1.1% (n = 1), respectively) and *Chilocorus* spp. (11.0% (n = 12) and 24.7% (n = 23), respectively). From the samples of native species collected in field and reared in laboratory, only one parasitoid emerged from one pupa of *A. bipunctata*. The species was identified as *Aprostocetus neglectus* (Domenichini) (Hymenoptera: Eulophidae). The rate of parasitism calculated on native species obtained by field samplings was very low (0.5%) within the period of the experiment.

Discussion

Our cage system for ladybird seems to provide a standardised and realistic estimation of mortality by predation, including IGP predation. These events are difficult to observe and quantify in the field for aphidophagous predators (Lucas 2005) and probably for this reason most of the observations on ladybirds are confined to the laboratory while only a few experiments are conducted in semi-field or field conditions (Obrycki et al. 1998a, b;

Table 1 Number of insect adults trapped in the field cages (*A.b.* = *Adalia bipunctata*, *H.v.* = *Hippodamia variegata*, *C.c.* = *Chrysoperla carnea*)

	Mercatale					Imola				
	Ant workers	Hym. parasitoids	*A.b.*	*H.v.*	*C.c.*	Ant workers	Hym. parasitoids	*A.b.*	*H.v.*	*C.c.*
Harmonia axyridis										
Free egg cages										
Free larval cages (L1–L2)	2	1				2				
Free larval cages (L3–L4)	3					3	1			
Free pupal cages	2									
Free adult cages										
Exclusion egg cages		1								
Exclusion larval cages (L1–L2)				1			1			
Exclusion larval cages (L3–L4)										
Exclusion pupal cages	3									
Exlcusion adult cages			1	1						
Adalia bipunctata										
Free egg cages			1					2		1 larvae
Free larval cages (L1–L2)	2	1				2				
Free larval cages (L3–L4)	2			1		2				
Free pupal cages						2	1			
Free adult cages	2		1	1						
Exclusion egg cages			1	1						
Exclusion larval cages (L1–L2)										
Exclusion larval cages (L3–L4)		1								
Exclusion pupal cages										
Exclusion adult cages	1	1								

Gardiner and Landis 2007). In particular, life tables for two predatory ladybirds, *C. septempunctata* and *H. axyridis*, were constructed through a field experiment (Hironori and Katsuhino 1997) and provided quantification of cannibalism and IGP in relation to these coccinellids. However, many factors concerning larval mortality were not known and so it is difficult to be assured that the most important factors structuring the ladybird guild were cannibalism and IGP (Hironori and Katsuhino 1997). Furthermore, in areas where an

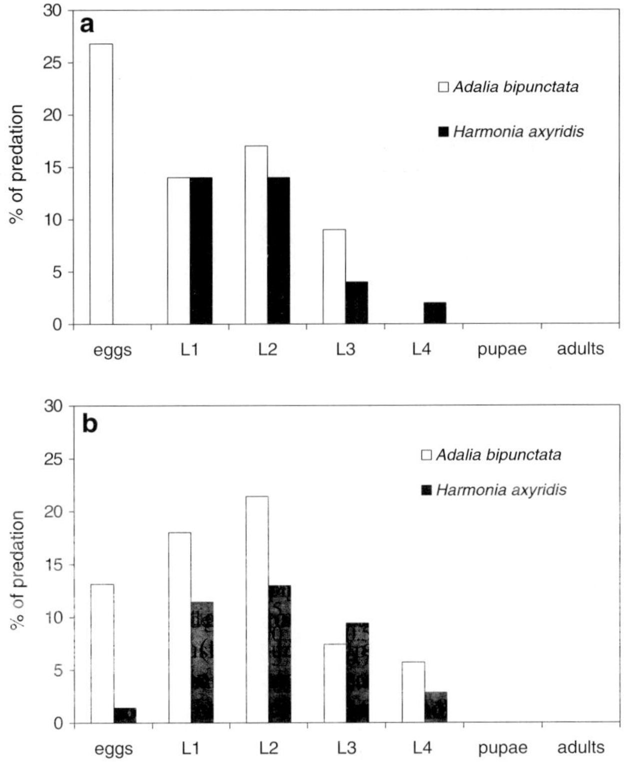

Fig. 1 Total percentage of predation on different life stages (eggs, L1–L4, pupae and adults) of native (*Adalia bipunctata*, *A.b.*) and exotic ladybird (*Harmonia axyridis*, *H.a.*) in field cages at the Imola (**a**) (eggs $\chi^2 = 77.37$; df = 1; $P = 0.00001$; *A.b.* $N = 250$; *H.a.* $N = 250$; L1 $\chi^2 = 0$; df = 1; P = 0; *A.b.* $N = 100$; *H.a.* $N = 100$; L2 $\chi^2 = 0.34$; df = 1; $P = 0.55$; *A.b.* $N = 100$; *H.a.* $N = 100$; L3 $\chi^2 = 2.76$; df = 1; $P = 0.096$; *A.b.* $N = 100$; *H.a.* $N = 100$; L4 $\chi^2 = 2.02$; df = 1; $P = 0.155$; *A.b.* $N = 100$; *H.a.* $N = 100$; pupae *A.b.* $N = 94$; *H.a.* $N = 97$; adults *A.b.* $N = 100$; *H.a.* $N = 100$) and Mercatale (**b**) (eggs $\chi^2 = 35.5$; df = 1; $P = 0.00001$; *A.b.* $N = 350$; *H.a.* $N = 350$; L1 $\chi^2 = 2.39$; df = 1; $P = 0.12$; *A.b.* $N = 139$; *H.a.* $N = 140$; L2 $\chi^2 = 3.37$; df = 1; $P = 0.0663$; *A.b.* $N = 140$; *H.a.* $N = 131$; L3 $\chi^2 = 0.36$; df = 1; $P = 0.54$; *A.b.* $N = 135$; *H.a.* $N = 138$; L4 $\chi^2 = 1.39$; df = 1; $P = 0.23$; *A.b.* $N = 140$; *H.a.* $N = 140$; pupae *A.b.* $N = 138$; *H.a.* $N = 134$; adults *A.b.* $N = 120$; *H.a.* $N = 140$) field sites. Data pooled for collection times

exotic is not yet recorded, only cage studies are recommended to avoid the establishment of the species.

Our 'exclusion cages' did not completely prevent access by walking predators but in general it was discriminated between the predation by walking predators like ants from groups of flying predators. The predators collected, including ant workers, could have been also attracted by aphids within the cages, thus resulting in an overestimation of their importance as coccinellid predators. Among polyphagous predators, ant workers were the most abundant insects found in the cages in our experiment and we speculate that this group plays a fundamental role in the predation of eggs. Ants can affect IGP in aphidophagous systems and this complex topic was reviewed by Lucas (2005). For example ants can prey on aphids, or ants may attend aphids and protect them against predators and parasitoids. Ants most often associated with aphids and coccinellids include *Lasius* spp. and *Formica* spp. (Hodek and Honek 1996) and many observations confirm that ants are

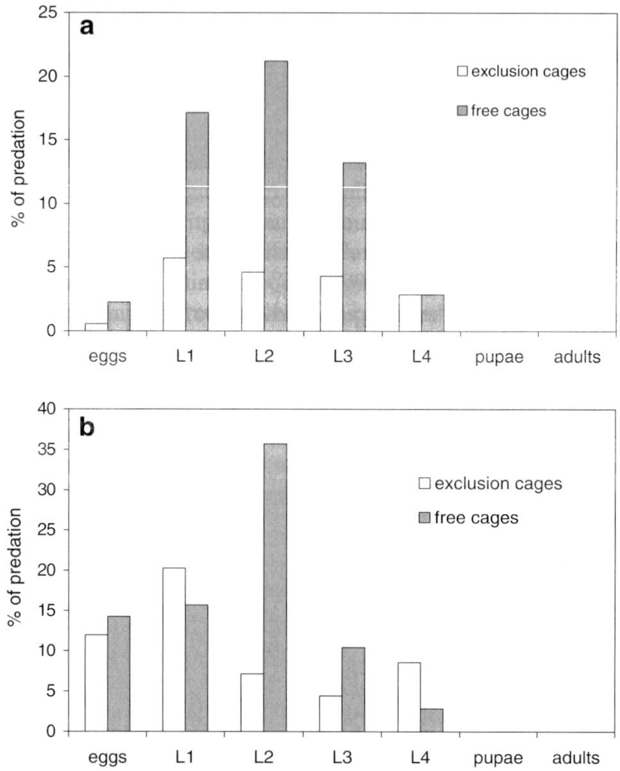

Fig. 2 Percentage of predation on *Harmonia axyridis* (*H.a.*) (eggs, L1–L4, pupae and adults) in exclusion (e.c.) versus free cages (f.c.) in the site of Mercatale (**a**) (eggs $\chi^2 = 1.83$; df = 1; $P = 0.17$; e.c. *H.a. N* = 175; f.c. *H.a. N* = 175; L1 $\chi^2 = 26.04$; df = 1; $P = 0.00001$; e.c. *H.a. N* = 70; f.c. *H.a. N* = 70; L2 $\chi^2 = 26.01$; df = 1; $P = 0.00001$; e.c. *H.a. N* = 65; f.c. *H.a. N* = 66; L3 $\chi^2 = 2.29$; df = 1; $P = 0.13$; e.c. *H.a. N* = 70; f.c. *H.a. N* = 68; L4 $\chi^2 = 0$; df = 1; $P = 0$; e.c. *H.a. N* = 70; f.c. *H.a. N* = 70; pupae e.c. *H.a. N* = 67; f.c. *H.a. N* = 67; adults e.c. *H.a. N* = 175; f.c. *H.a. N* = 175) and on *Adalia bipunctata* (*A.b.*) in exclusion (e.c.) versus free cages (f.c.) in the site of Mercatale (**b**) (eggs $\chi^2 = 0.40$; df = 1; $P = 0.52$; e.c. *A.b. N* = 175; f.c. *A.b. N* = 175; L1 $\chi^2 = 0.49$; df = 1; $P = 0.48$; e.c. *A.b. N* = 69; f.c. *A.b. N* = 70; L2 $\chi^2 = 15.48$; df = 1; $P = 0.0001$; e.c. *A.b. N* = 70; f.c. *A.b. N* = 70; L3 $\chi^2 = 1.79$; df = 1; $P = 0.18$; e.c. *A.b. N* = 68; f.c. *A.b. N* = 67; L4 $\chi^2 = 2.12$; df = 1; $P = 0.14$; e.c. *A.b. N* = 70; f.c. *A.b. N* = 70; pupae e.c. *A.b. N* = 69; f.c. *A.b. N* = 69; adults e.c. *A.b. N* = 70; f.c. *A.b. N* = 50)

hostile to enemies of attended aphids. Predatory ants may attack any insects within their foraging territories (Hodek and Honek 1996) and our data seems to confirm these observations.

In our semi-field study, *C. carnea* was likely responsible for predation on *A. bipunctata* eggs. In fact, *C. carnea* completed preimaginal development when fed *A. bipunctata* eggs but it did not complete preimaginal development on *H. axyridis* eggs (Santi and Maini 2006).

Our data demonstrate that eggs of *H. axyridis* are protected against intraguild predation more than eggs of *A. bipunctata*, because the egg-predation was significantly lower than that of the native species. This evidence confirms the results of laboratory experiments on IGP and cannibalism on *H. axyridis* and *A. bipunctata* (Burgio et al. 2002). In a laboratory study Cottrell (2005) reported that the *H. axyridis* eggs were predated on to a lesser extend

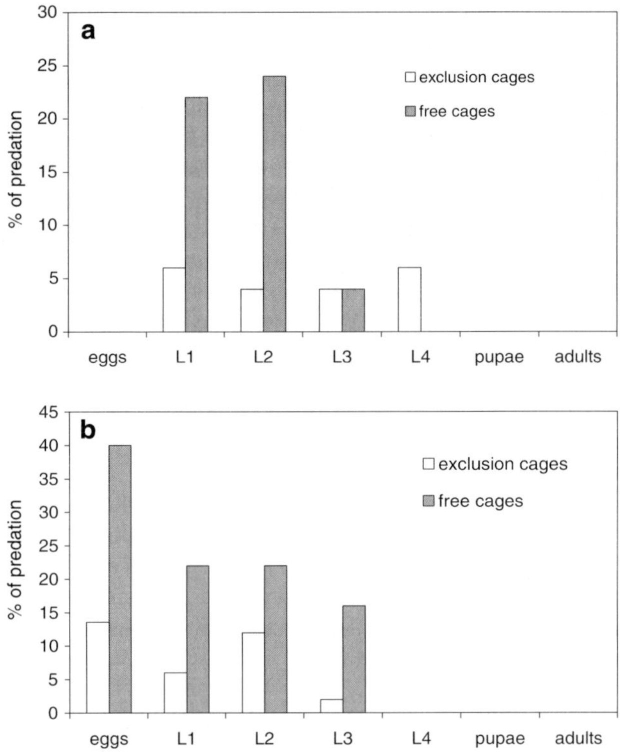

Fig. 3 Percentage of predation on *Harmonia axyridis* (*H.a.*) (eggs, L1–L4, pupae and adults) in exclusion (e.c) versus free cages (f.c.) in the site of Imola (**a**) (eggs e.c. *H.a.* $N = 125$; eggs f.c. *H.a.* $N = 125$; L1 $\chi^2 = 5.32$; df = 1; $P = 0.021$; e.c. *H.a.* $N = 50$; f.c. *H.a.* $N = 50$; L2 $\chi^2 = 8.31$; df = 1; $P = 0.004$; e.c. *H.a.* $N = 50$; f.c. *H.a.* $N = 50$; L3 $\chi^2 = 0$; df = 1; $P = 0$; L4 $\chi^2 = 2.04$; df = 1; $P = 0.15$; e.c. *H.a.* $N = 50$; f.c. *H.a.* $N = 50$; pupae e.c. *H.a.* $N = 47$; f.c. *H.a.* $N = 50$; adults e.c. *H.a.* $N = 50$; f.c. *H.a.* $N = 50$) and on *Adalia bipunctata* (*A.b.*) in exclusion (e.c.) versus free cages (f.c.) in the site of Imola (**b**) (eggs $\chi^2 = 35.2$; df = 1; $P = 0.00001$; e.c. *A.b.* $N = 125$; f.c. *A.b.* $N = 125$; L1 $\chi^2 = 5.32$; df = 1; $P = 0.021$; e.c. *A.b.* $N = 50$; f.c. *A.b.* $N = 50$; L2 $\chi^2 = 1.77$; df = 1; $P = 0.183$; e.c. *A.b.* $N = 50$; f.c. *A.b.* $N = 50$; L3 $\chi^2 = 4.0$; df = 1; $P = 0.045$; e.c. *A.b.* $N = 50$; f.c. *A.b.* $N = 50$; L4 e.c. *A.b.* $N = 50$; f.c. *A.b.* $N = 50$; pupae e.c. *A.b.* $N = 47$; f.c. *A.b.* $N = 47$; adults e.c. *A.b.* $N = 50$; f.c. *A.b.* $N = 50$)

than those of the native species *Coleomegilla maculata* DeGeer and *Olla v-nigrum* (Mulsant). For these reasons *H. axyridis* seems to display a competitive advantage in colonizing new areas, that could be partially responsible for high competitiveness of this species compared to native species in many countries (Elliott et al. 1996; Brown and Miller 1998; Colunga-Garcia and Gage 1998; Gardiner and Landis 2007). The results of the laboratory studies of Cottrell (2005) agree with those of our semi-field trials. Ladybirds defend themselves with bitter-tasting alkaloids, volatile pyrazine and quinolenes (Pasteels et al. 1973; Hodek and Honek 1996; Ware et al. (2007), and surface alkanes (Hemptinne et al. 2000). Agarwala and Dixon (1992) demonstrated that larvae and adults of *Coccinella septempunctata* L. and *A. bipunctata* were reluctant to eat conspecific eggs treated with a water extract of the other species' eggs.

The most important result that emerged from our experiments is that the eggs of the exotic *H. axyridis* are more protected against predation in comparison with those of the native ones. This evidence could explain the success of *H. axyridis* in niches already exploited by other coccinellids; however our findings showed that predation by generalist predators on *H. axyridis* larvae was similar to that observed on *A. bipunctata* larvae.

Other experiments and sampling techniques that allow us to study IGP and cannibalism in the field need to be developed particularly to estimate the impact of parasitoids and general parasitism in the field. We suppose that the exposure time of 24 h in our experiment might be insufficient to estimate the rate of parasitism in field. The parasitism of coccinellids is influenced by the period of the year and by many factors and variables (climatic and biotic, the latter including the relative densities of ladybirds) (Hodek and Honek 1996) and it is probably a local-scale phenomenon. Collection of coccinellid specimens, by manual or beating samplings, should be employed to integrate with cage systems and to avoid an under-estimation of the parasitism due to the consumption of parasitized coccinellids.

Acknowledgements We are grateful to Michele Ruzza for technical assistance during the experiments. This research was funded by ERBIC (Evaluating Risks of Biological Control Introductions into Europe) FAIR5-CT97-3489. A special thank to Helen Roy for the English revision of the manuscript and for useful suggestions.

References

Agarwala BK, Dixon AFG (1992) Laboratory study of cannibalism and interspecific predation in ladybirds. Ecol Entomol 17:303–309

Bazzocchi GG, Lanzoni A, Accinelli G, Burgio G (2004) Overwintering, phenology and fecundity of *Harmonia axyridis* in comparison with native coccinellid species in Italy. BioControl 49:245–260

Brown MW, Miller SS (1998) Coccinellidae (Coleoptera) in apple orchards of eastern West Virginia and their impact of invasion by *Harmonia axyridis*. Ent News 109(2):136–142

Brown PMJ, Adriaens T, Bathon H, Cuppen J, Goldarazena A, Hägg T, Kenis M, Klausnitzer BEM, Kovář I, Loomans AJM, Majerus MEN, Nedved O, Pedersen J, Rabitsch W, Roy HE, Ternois V, Zakharov IA, Roy DB (2007) *Harmonia axyridis* in Europe: spread and distribution of a non-native coccinellid. BioControl (this issue). doi:10.1007/s10526-007-9132-y

Burgio G, Santi F, Maini S (2002) On intraguild predation and cannibalism in *Harmonia axyridis* Pallas and *Adalia bipunctata* L. (Coleoptera: Coccinellidae). Biol Control 24:110–116

Burgio G, Ferrari R, Pozzati M, Boriani L (2004) The role of ecological compensation areas on predator populations: an analysis on biodiversity and phenology of Coccinellidae (Coleoptera) on non-crop plants within hedgerows in Northern Italy. Bull Insectology 57:1–10

Burgio G, Ferrari R, Boriani L, Pozzati M, van Lenteren J (2006) The role of ecological infrastructures on Coccinellidae (Coleoptera) and other predators in weedy field margins within northern Italy agroecosystems. Bull Insectology 59:5–67

Cartwright B, Eikenbary RD, Angalet GW (1982) Parasitism by *Perilitus coccinellae* (Hym.: Braconidae) of indigenous coccinellids hosts and the introduced *Coccinella septempunctata* (Col.: Coccinellidae), with notes on winter mortality. Entomophaga 27:237–244

Colunga-Garcia M, Gage SH (1998) Arrival, establishment, and habitat use of the multicolor Asian lady beetle (Coleoptera: Coccinellidae) in a Michigan landscape. Environ Entomol 27:1574–1580

Cottrell TE (2005) Predation and cannibalism of lady beetle eggs by adult lady beetles. Biol Control 34: 159–164

Cottrell TE, Yeargan KV (1998a) Influence of a native weed, *Acalypha ostryaefolia* (Euphorbiaceae), on *Coleomegilla maculata* (Coleoptera: Coccinellidae) population density, predation, and cannibalism in sweet corn. Environ Entomol 27:1375–1385

Cottrell TE, Yeargan KV (1998b) Effect of pollen on *Coleomegilla maculata* (Coleoptera: Coccinellidae) population density, predation, and cannibalism in sweet corn. Environ Entomol 27:1402–1410

Dixon AFG (2000) Insect predator–prey dynamic: ladybird, beetles and biological control. Cambridge University Press, Cambridge, UK

Elliott N, Kieckhefer R, Kauffman W (1996) Effects on an invading coccinellid on native coccinellids in an agricultural landscape. Oecologia 105:537–544

Gardiner MM, Landis DA (2007) Impact of intraguild predation by adult *Harmonia axyridis* (Coleoptera: Coccinellidae) on *Aphis glycines* (Hemiptea: Aphididae) biological control in cage studies. Biol Control 40:386–395

Geoghegan IE, Majerus TMO, Majerus MEN (1998) Differential parasitization of adult and pre-imaginal *Coccinella septempunctata* (Coleoptera: Coccinellidae) by *Dinocampus coccinellae* (Hymenoptera: Braconidae). Eur J Entomol 95:571–579

Hemptinne JL, Lognay G, Gauthier C, Dixon FG (2000) Role of surface chemical signals in egg cannibalism and intraguild predation in ladybirds (Coleoptera: Coccinellidae). Chemoecology 10:123–128

Hironori Y, Katsuhino S (1997) Cannibalism and interspecific predation in two predatory ladybirds in relation to prey abundance in the field. Entomophaga 42:153–163

Hodek I, Honěk A (1996) Ecology of coccinellidae. Kluwer Academic Publishers, Dordrecht, The Netherlands

Iperti G (1964) Les parasites des coccinelles aphidiphages dans les Alpes-Maritimes et les Basses-Alpes. Entomophaga 9:153–180

Kajita H, Takano F, Yasuda H, Agarwala BK (2000) Effects of indigenous ladybird species (Coleoptera: Coccinellidae) on survival of an exotic species in relation to prey abundance. Appl Entomol Zool 35(4):473–479

Kenis M, Roy HE, Zindel R, Majerus MEN (2007) Current and potential management strategies against *Harmonia axyridis*. BioControl (this issuse). doi:10.1007/s10526-007-9136-7

Lanzoni A, Accinelli G, Bazzocchi GG, Burgio G (2004) Biological traits and life table of the exotic *Harmonia axyridis* compared with *Hippodamia variegata*, and *Adalia bipunctata* (Col., Coccinellidae). J Appl Entomol 128:298–306

Lucas E (2005) Intraguild predation among aphidophagous predators. Eur J Entomol 102:351–364

Obrycki JJ (1989) Parasitization of native and exotic Coccinellids by *Dinocampus coccinellae* (Schrank) (Hymenoptera: Braconidae). J Kans Entomol Soc 32:211–218

Obrycki JJ, Tauber MJ, Tauber CA (1985) *Perilitus coccinellae* (Hymenoptera: Braconidae): parasitization and development in relation to host-stage attacked. Ann Entomol Soc Am 78:852–854

Obrycki JJ, Giles KL, Ormord AM (1998a) Experimental assessment of interactions between larval *Coleomegilla maculata* and *Coccinella septempunctata* (Coleoptera: Coccinellidae) in field cages. Environ Entomol 27:1280–1288

Obrycki JJ, Giles KL, Ormord AM (1998b) Interactions between an introduced and indigenous coccinellid species at different prey densities. Oecologia 117:279–285

Okuda T, Ceryngier P (2000) Host discrimination in *Dinocampus coccinellae* (Hymenoptera: Braconidae). Appl Entomol Zool 35:595–539

Pasteels JM, Deroe C, Tursch B, Braekman JC, Daloze D, Hootele C (1973) Distribution et activites des alcaloides defensifs des Coccinellidae. J Insect Physiol 19:1771–1784

Pervez A, Omkar GM (2006) Ecology and biological control application of multicoloured Asian ladybird, *Harmonia axyridis*: a review. Biocontrol Sci Technol 16:111–128

Rosenheim JA, Kaya HK, Ehler LE, Marois JJ, Jaffee BAJ (1995) Intraguild predation among biological-control agents: theory and evidence. Biol Control 5:303–335

Roy HE, Brown PMJ, Rothery P, Ware RL, Majerus MEN (2007) Interactions between the fungal pathogen *Beauveria bassiana* and three species of coccinellid: *Harmonia axyridis*, *Coccinella septempunctata* and *Adalia bipunctata*. BioControl (this issuse). doi:10.1007/s10526-007-9122-0

Santi F, Maini S (2006) Predation upon *Adalia bipunctata* and *Harmonia axyridis* eggs by *Chrysoperla carnea* larvae and *Orius laevigatus* adults. Bull Insectology 59:53–58

Toda Y, Sakuratani Y (2006) Expansion of the geographical distribution of an exotic ladybird beetle, *Adalia bipunctata* (Coleoptera: Coccinellidae), and its interspecific relationships with native ladybird beetles in Japan. Ecol Res 21:292–300

Ware RL, Majerus MEN (2007) Intraguild predation of immature stages of British and Japanese coccinellids by the invasive ladybird *Harmonia axyridis*. BioControl (this issuse). doi:10.1007/s10526-007-9135-8

Ware RL, Ramon-Portugal F, Magro A, Ducamp C, Hemptinne J-L, Majerus MEN (2007) Chemical protection of *Calvia quatuordecimguttata* eggs against intraguild predation by the invasive ladybird *Harmonia axyridis*. BioControl (this issue). doi:10.1007/s10526-007-9129-6

Zar JH (1984) Biostatistical analysis, 2nd edn. Prentice Hall, Englewood Cliffs, NJ